Springer Proceedings in Complexity

Springer Proceedings in Complexity publishes proceedings from scholarly meetings on all topics relating to the interdisciplinary studies of complex systems science. Springer welcomes book ideas from authors. The series is indexed in Scopus.

Proposals must include the following:

- name, place and date of the scientific meeting
- a link to the committees (local organization, international advisors etc.)
- scientific description of the meeting
- list of invited/plenary speakers
- an estimate of the planned proceedings book parameters (number of pages/articles, requested number of bulk copies, submission deadline)

Submit your proposals to: christoph.baumann@springer.com

More information about this series at http://www.springer.com/series/11637

Naoki Masuda • Kwang-Il Goh • Tao Jia
Junichi Yamanoi • Hiroki Sayama
Editors

Proceedings of NetSci-X 2020: Sixth International Winter School and Conference on Network Science

Springer

Editors
Naoki Masuda
Department of Mathematics
University at Buffalo, State University of New York
Buffalo, NY, USA

Tao Jia
College of Computer and Information Science
Southwest University
Chongqing, China

Hiroki Sayama
Department of Systems Science and Industrial Engineering
Binghamton University, State University of New York
Binghamton, NY, USA

Kwang-Il Goh
Department of Physics
Korea University
Seoul, Korea (Republic of)

Junichi Yamanoi
School of Commerce
Waseda University
Tokyo, Tokyo, Japan

ISSN 2213-8684 ISSN 2213-8692 (electronic)
Springer Proceedings in Complexity
ISBN 978-3-030-38964-2 ISBN 978-3-030-38965-9 (eBook)
https://doi.org/10.1007/978-3-030-38965-9

This Springer imprint is published by the registered company Springer Nature Switzerland AG.
The registered company address is: Gewerbestrasse 11, 6330 Cham, Switzerland

Preface

NetSci-X, the Network Science Society's signature winter conference, extends the popular NetSci conference series to provide an additional forum for a growing community of academics and practitioners working on networks. The series started in 2015. Its sixth edition, NetSci-X 2020, held in Tokyo from January 20 to January 23, 2020, is the first NetSci conference held in Japan, and the third one in Asia after Seoul (NetSci 2016) and Hangzhou (NetSci-X 2018). The conference was hosted by the Waseda Innovation Lab at Waseda University, and took place in its Waseda Campus located in the heart of Tokyo. The conference fostered interdisciplinary communication and collaboration in network science research across computer and information sciences, physics, mathematics, statistics, life sciences, neuroscience, engineering, social sciences, finance, business, and others.

We received 279 submissions and the authors submitted their contributions to either the Proceedings Track or the Abstract Track. Each paper in both tracks was peer-reviewed by at least two (typically three) independent reviewers from an international program committee. The present volume is the collection of Proceedings Track papers that were accepted for publication in this proceedings book and were presented in an oral or poster session.

The great success of NetSci-X 2020 owes to these and other authors, who have come to Tokyo and presented their work. Needless to say, the success also owes to the keynote and invited speakers, who have delivered impressive talks. Our keynote speakers are: Albert-László Barabási, Alain Barrat, Katy Börner, and Meeyoung Cha; our invited speakers are Manlio De Domenico, Shlomo Havlin, Petter Holme, Byungnam Kahng, Joe Labianca, Linyuan Lü, and Misako Takayasu.

In addition to the presenters, NetSci-X 2020 would have been impossible without devoted support by many parties. We first thank Waseda University and the Waseda Innovation Lab, which have fully supported the conference by providing the venue in the center of Tokyo and necessary staff support. We also greatly thank our generous sponsors, including Tateisi Science and Technology Foundation, Sansan, Habitech, Think Lab, Dropbox Paper, Journal of Physics: Complexity, CyberAgent, ZUVA, Springer Nature, and Bureau Van Dijk. We also thank the Network Science Society for their encouragement and support, and Springer Nature for editorial support for this Proceedings volume. Last but not the least, our sincere thanks go to the Program Committee members, each of who has provided quality review reports to roughly seven submissions despite a short deadline, and also to all the Organizing Committee members, who have dedicated themselves to their respective role to realize this conference. We hope that readers of this volume enjoy the papers as well as the participants have enjoyed the main conference.

Program Committee

Laura Maria Alessandretti
Marco Tulio Angulo
Alberto Antonioni
Tomaso Aste
Rodolfo Baggio
James Bagrow
Paolo Bajardi
Stefano Balietti
Carmelo J. A. Bastos Filho
Federico Battiston
Rosa M. Benito
Ginestra Bianconi
Federico Botta
Dan Braha
Colin Campbell
Alessio Cardillo
Claudio Castellano
Matteo Chinazzi
Young Sul Cho
Giulio Cimini
Paolo Cintia
Sean Cornelius
Michael Danziger
Nima Dehmamy
Jean-Charles Delvenne
Fabrizio De Vico Fallani
Riccardo Di Clemente
Wenbo Du
Jordi Duch
Gregorio D'Agostino
Alexandre Evsukoff
Paul Expert
Ying Fan
Daniel Figueiredo
Alexandre P. Francisco
Feng Fu
José Manuel Galán
Edoardo Galimberti
Jianxi Gao
Zhong-Ke Gao
Alexander Gates
James P. Gleeson
Kwang-Il Goh
Jesus Gomez-Gardeñes
Emre Guney
Aleksandar Haber
Aric Hagberg
Philipp Hövel
Agnes Horvat
Yanqing Hu
Junming Huang
Marco Alberto Javarone
Peng Ji
Tao Jia
Hang-Hyun Jo
Kenneth Joseph
Fariba Karimi
Márton Karsai
Tatsuro Kawamoto
Beom Jun Kim
Teruyoshi Kobayashi
Lucas Lacasa
Daniel Larremore
Deok-Sun Lee

Sang Hoon Lee
Balázs Lengyel
Xiang Li
Yang-Yu Liu
Xin Lu
Yifang Ma
Enrico Maiorino
Giuseppe Mangioni
Naoki Masuda
Carolina Mattsson
Matúš Medo
Sandro Meloni
Ronaldo Menezes
Radosław Michalski
Byungjoon Min
Dina Mistry
Osnat Mokryn
Esteban Moro
Satyam Mukherjee
Yohsuke Murase
Zachary Nicolaou
Luca Pappalardo
Juyong Park
Alice Patania
Leto Peel
Tiago P. Peixoto
Giovanni Petri
Carlo Piccardi
Alessandro Pluchino
Peter Pollner
Márton Pósfai
Filippo Radicchi
Luis E. C. Rocha
Zhihai Rong
Giulio Rossetti
Francisco C. Santos
Joaquín Sanz
Piotr Sapiezynski
Hiroki Sayama
Rossano Schifanella
Vedran Sekara
M. Ángeles Serrano
Roberta Sinatra
Per Sebastian Skardal
Tiziano Squartini
Kaiyuan Sun
Michael Szell
Attila Szolnoki
Taro Takaguchi
Marcella Tambuscio
Michele Tizzoni
Janos Török
Emma Towlson
Piet Van Mieghem
Onur Varol
Christian Lyngby Vestergaard
Johannes Wachs
Huijuan Wang
Lin Wang
Pu Wang
Zhi-Xi Wu
Feng Xia
Haoxiang Xia
Wenwu Yu
Jorge G. T. Zanudo
Qian Zhang
Yi Zhao
Anna Zygmunt

Organizing Committee

General Chairs
Hiroki Sayama (Binghamton University, SUNY, USA/Waseda University, Japan)
Junichi Yamanoi (Waseda University, Japan)

Program Chairs
Naoki Masuda (University at Buffalo, SUNY, USA)
Kwang-Il Goh (Korea University, Korea)
Tao Jia (Southwest University, China)

Logistics Chair
Tomomi Kito (Waseda University, Japan)

School Chair
Tatsuro Kawamoto (AIST, Japan)

Local Chapter Chairs
Toshihiro Tanizawa (Kochi National College of Technology, Japan)
Yuichi Ikeda (Kyoto University, Japan)

Publicity/Social Media Chairs
Yukie Sano (University of Tsukuba, Japan)
Sang Hoon Lee (Gyeongnam National University of Science and Technology, Korea)
Toshio Murase (Waseda University, Japan)

Sponsorship Chair
Susumu Nagayama (Hosei University, Japan)

Poster Session Chairs
Teruyoshi Kobayashi (Kobe University, Japan)
Takaaki Aoki (Kagawa University, Japan)

Web Chairs
Sadamori Kojaku (Kobe University, Japan)
Tomokatsu Onaga (Tohoku University, Japan)

Student Support Chair
Taro Takaguchi (Data Labs, LINE Corporation, Japan)

Other Committee Members
Marko Jusup (Tokyo Institute of Technology, Japan)
Yohsuke Murase (RIKEN, Japan)

Buffalo, NY	Naoki Masuda
Seoul, Korea	Kwang-Il Goh
Chongqing, China	Tao Jia
Tokyo, Japan	Junichi Yamanoi
Binghamton, NY	Hiroki Sayama
January 2020	

The original version of this book was revised: Second author's name in chapter 4 has been updated. A correction to this book is available at https://doi.org/10.1007/978-3-030-38965-9_19

Contents

Part I
Structure

Latent Space Generative Model for Bipartite Networks

Demival Vasques Filho and Dion R. J. O'Neale

Abstract Generative network models are useful for understanding the mechanisms that operate in network formation and are used across several areas of knowledge. However, when it comes to bipartite networks—a class of network frequently encountered in social systems, among others—generative models are practically non-existent. Here, we propose a latent space generative model for bipartite networks growing in a hyperbolic plane. It is an extension of a model previously proposed for one-mode networks, based on a maximum entropy approach. We show that, by reproducing bipartite structural properties, such as degree distributions and small cycles, bipartite networks can be better modelled and properties of one-mode projected network can be naturally assessed.

1 Introduction

Generative models are a powerful approach to describe and understand the processes at work during network formation and the mechanisms producing specific network features. They provide the opportunity to simulate real, growing networks, subject to various assumptions about the importance of controlled parameters [1, 2]. Properties like heterogeneous degree distributions, clustering and community formation in real-world systems can be assessed using such models.

D. Vasques Filho (✉)
Leibniz-Institut für Europäische Geschichte, Mainz, Germany

Te Pūnaha Matatini, Department of Physics, University of Auckland, Auckland, New Zealand
e-mail: vasquesfilho@ieg-mainz.de; https://www.ieg-mainz.de/en/institute/people/vasques_filho

D. R. J. O'Neale
Physics Department, University of Auckland, Auckland, New Zealand

Te Pūnaha Matatini, Centre of Research Excellence, Auckland, New Zealand
e-mail: d.oneale@auckland.ac.nz

N. Masuda et al. (eds.), *Proceedings of NetSci-X 2020: Sixth International Winter School and Conference on Network Science*, Springer Proceedings in Complexity,
https://doi.org/10.1007/978-3-030-38965-9_1

Generative models have been developed in many flavours, associated with the different research communities using them [3, 4]. However, the same is not true for generative models for bipartite networks. Studies on statistical models for networks with bipartite structure are rare and even scarcer on generative models. Furthermore, the few studies addressing models of bipartite networks mostly focus on mimicking the properties of their projections only. Structural properties of bipartite networks are relatively neglected, in comparison to the existing body of work on generative models for one-mode networks. For a brief review of network models, see [5].

We propose a latent space model in a hyperbolic plane, based on a maximum entropy approach, as an extension of work done for one-mode networks [6–8]. We focus on recreating structural properties of bipartite networks, namely degree distribution and small cycles. The latter, especially cycles of length four and six—that we will refer to as four-cycles and six-cycles, have a significant effect on the resulting structure of the projected network. Four-cycles are indicative of recurring interactions, affecting the link weight distribution. Six-cycles, in turn, represent triadic closure and have an impact on the projected clustering [9, 10]. We show that, by reproducing such properties, the generative model produces bipartite networks whose one-mode projections naturally display the structures of interest.

The remainder of this paper is organised as follows. In Sect. 2, we examine the adaptation of null models for one-mode networks to generative models of bipartite networks. We discuss the characteristics of such models and how they fail to reproduce the main structural properties we are looking for in bipartite graphs. In Sect. 3, we discuss the popularity vs. similarity model [8] for one-mode networks growing in a hyperbolic plane, based on a maximum entropy approach. In Sect. 4, we introduce our bipartite model and show how it recreates the features of real-world bipartite networks. Finally, we present the main results and the conclusion of the paper in Sect. 5.

2 Null Models

2.1 *Erdős–Rényi*

The original Erdős–Rényi model (ER) [11, 12] considers an ensemble of graphs $\mathcal{G}$, in which every graph $G \in \mathcal{G}$ has a set of nodes U, and $|L|$ links that connect pairs of nodes at random in the network. In a dynamic version of the model, we add a node to the network at every time step t, until $|U|$ nodes are present. The number of links $|L|$, in turn, is controlled by adding m new links to the network for every t, i.e. $|L| = tm$. Each graph has m nodes at $t = 0$ and at each time step a new node with m links is added to the network, being randomly connected to m existing nodes, until $t = |U| - m$.

Based on this reasoning, we create a dynamic bipartite version of the ER model, $\mathrm{B_{ER}}(|U|, |V|, |E|)$, where $|U|$ and $|V|$ are the size of the bottom set of nodes (U) and the top set of nodes (V), respectively, and $|E|$ is the number of bipartite links, as follows:

1. At time $t = 0$, the network has m bottom nodes and m top nodes, without links connecting them.
2. At each time step, a new bottom node and a new top node are added to the network. The new top node chooses, at random, m existing bottom nodes and connects to them. Then, for simplicity, this same process applies to the new bottom node which, in turn, connects to m existing top nodes.
3. Step 2 is repeated until the network has $|U|$ bottom node and $|V|$ top nodes.

We implemented the above algorithm and used it to generate synthetic bipartite networks built with 200,000 time steps and $m = 2$, such that the average degree $\langle d \rangle$ of V and the average degree $\langle k \rangle$ of U are the same. That is $\langle d \rangle = \langle k \rangle = 4$. Due to the generative mechanism, the degree distribution of bottom and top nodes are similar (Fig. 1), and should be the same in the thermodynamic limit. Moreover, the shape of the degree distribution of the projected network follows that of the bottom distribution, just shifted to the right, as shown in [13].

The evolution of the number of small cycles (Fig. 2a) is roughly constant and at low levels, if compared to real-world networks [5, 9]. The same is true for the link weight distribution (Fig. 2b) and the distribution of the clustering coefficient (Fig. 2c) of the projected network. For the former, the absence of heavily weighted links is due to the low number of four-cycles, while for the latter, the low level of clustering is explained by the small number of six-cycles and by the absence of high-degree top nodes in the bipartite network. As expected, the generative version of the Erdős–Rényi model still does not reproduce structural properties of real networks. Let us explore next a preferential attachment bipartite generative model.

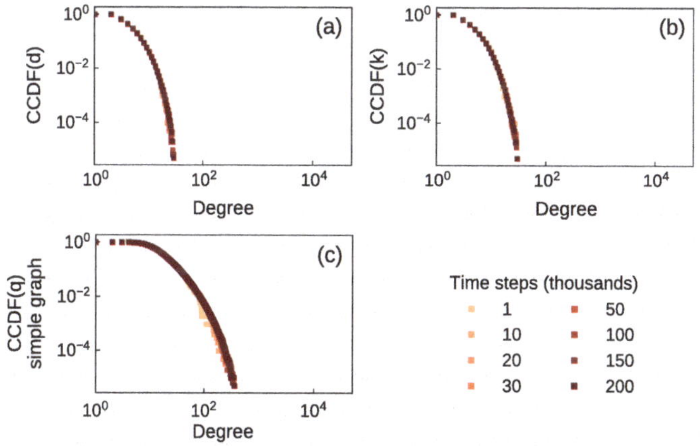

Fig. 1 Degree distributions for the $B_{ER}(|U|, |V|, |E|)$ model. Degree distribution of (**a**) top and (**b**) bottom nodes are peaked and similar, due to the mechanism of link attachment for both sets of nodes. (**c**) Projected networks follow the same degree distribution shape as the bottom node degree distribution, shifted to the right, as shown in [13].

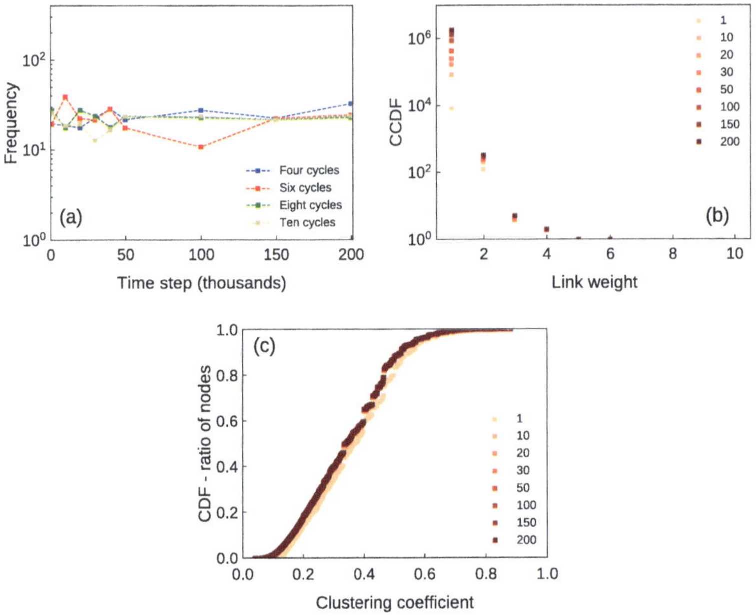

Fig. 2 Evolution of (**a**) cycles in the bipartite network. (**b**) Link weight distribution and (**c**) distribution of the clustering coefficient of the projected network. All for the $B_{ER}(|U|, |V|, |E|)$ model. As the network grows larger, the number of small cycles stays roughly constant. It shows that random bipartite graphs tend to create uniform distributions of cycles size, as shown in [9]. This results in few weighted links in the projected networks, and clustering created mostly by top node degrees instead of six-cycles. CCDF stands for complementary cumulative distribution function. CDF—ratio of nodes is the cumulative distribution for the ratio of nodes with clustering coefficient values lower that the corresponding coefficient in the horizontal axis

2.2 *Preferential Attachment*

Growing networks with preferential attachment has been extensively studied for both one-mode [14–16] and bipartite networks [17–21]. For the latter, to the best of our knowledge, none of these models have addressed bipartite structural properties other than degree distributions. Furthermore, they have not investigated the effects of degree distributions on the one-mode projections.

Our generative model with a preferential attachment mechanism for bipartite networks is a bipartite version of the Barábasi–Albert (BA) model [16]. It follows the same reasoning as the B_{ER} generative model. The only difference is that new nodes now choose to connect to existing nodes from the opposite set with a weighted probability, where the weights are proportional to the degrees of the node in the target set. That is,

$$p_u = \frac{k_u}{\sum_{u'} k_{u'}}, \qquad p_v = \frac{d_v}{\sum_{v'} d_{v'}}, \tag{1}$$

where k_u is the degree of bottom node u and d_v is the degree of top node v.

The model goes as follows:

1. At time $t = 0$, the network has no links with only m bottom and m top nodes.
2. At each time step, a new bottom node and a new top node enter the network. Now, the new top node chooses and connects to m existing bottom nodes, with weighted probability according to Eq. (1). Then, the new bottom node connects to m existing top nodes, using the same formula for calculating the connection probability.
3. Step 2 is repeated until the network reaches $|U|$ bottom and $|V|$ top nodes.

Again, we generate synthetic networks with $|U| = |V| = 200{,}000$, with $m = 2$, and $\langle d \rangle = \langle k \rangle = 4$. In Fig. 3 we can see that, because of the simple preferential attachment mechanism of our model for both sets of nodes, the degree distributions for the top set of nodes, $P_t(d)$, and for the bottom nodes, $P_b(k)$, are the same. The degree distribution of the projected network, $P(q)$, is also heavy-tailed, but is shifted to the right, and shows a flattening similar to that shown in [13], due to the formation of large cliques, a consequence of the high-degree top nodes in B.

The presence of high-degree nodes in the bipartite network increases, albeit only a little, the number of small cycles in the network (Fig. 4a). This is a result of a higher probability of high-degree top and bottom nodes being connected more frequently [9]. However, the observed level of four-cycles is still relatively low compared to that seen in empirical networks [5, 9] and does not create a significant number of weighted links in the projected network G_w, as shown in Fig. 4b. Another consequence of the presence of high-degree top nodes in B can be seen

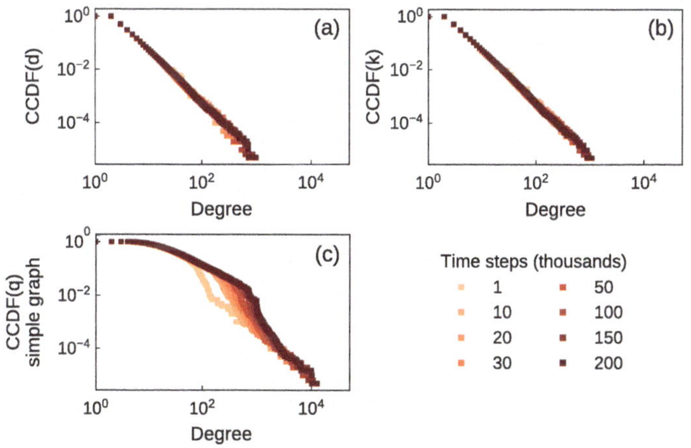

Fig. 3 Degree distributions for the bipartite generative model of the Erdős–Rényi model. Again, the degree distribution of (**a**) top and (**b**) bottom nodes are similar due to the mechanism of network growth. However the BA model creates heavy-tail degree distributions in this case. The behaviour of the (**c**) projected distribution deviates, especially for high-degree nodes, due to the cliques created in the projection

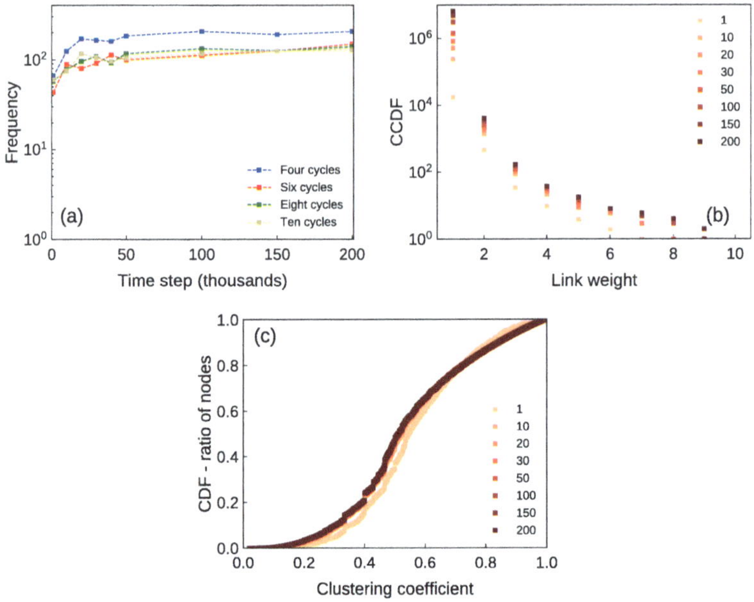

Fig. 4 Evolution of (**a**) cycles in the bipartite network. (**b**) Link weight distribution and (**c**) distribution of the clustering coefficient of the projected network. All for the bipartite BA model. Although the latter creates more small cycles than the B_{ER} model, the presence of such cycles in the network is still very low compared to real-world networks [5, 9]. Still, it produces a few links with some weight and low levels of clustering in the projected networks

in Fig. 4c, where the level of clustering of the projection has increased relatively to the generative $B_{\mathrm{ER}}(|U|, |V|, |E|)$ model.

We have seen in [9] that traditional null models as the B_{ER} model and the configuration model—both static—cannot capture the structural properties of bipartite networks of our interest. Here, we have created synthetic networks with generative bipartite versions of the ER model and of the BA model. Although these dynamic models function well as null models, they do not reproduce the structure of real-world networks either. Hence, more sophisticated models are needed and we move in that direction in the next sections.

3 Hyperbolic Geometry

In a series of three papers [6–8], it was demonstrated that some structural properties found in real-world networks, namely degree heterogeneity (heavy-tail degree distributions) and clustering, can emerge naturally when the network grows in a hyperbolic plane. The authors of [8] used node coordinates in the hyperbolic plane as hidden variables [22–25], characterizing their proposed popularity vs. similarity model as a latent space model.

However, that is not the only important characteristic of this model. The edge probability function chosen by the authors in [7] is the Fermi–Dirac distribution. The reason for that is threefold: firstly, the model incorporates the concepts of the exponential random graph models, through the maximum entropy approach [26]. Secondly, the model is designed for simple graph one-mode networks where links are fermions—for multigraphs, or even weighted networks, the family of connection probabilities chosen would be the Bose–Einstein distribution, where more than one particle (link) could occupy the same energy state (pair of nodes) [26, 27]. Thirdly, using the F-D distribution allows one to make use of existing methods from statistical mechanics in order to understand the hyperbolic plane model, as we shall see next.

In this model, the probability of two nodes being connected is given by Krioukov et al. [7]

$$P(u, u') = \frac{1}{e^{\beta\left(\frac{\zeta}{2}\right)(x_{u,u'} - R)} + 1}, \tag{2}$$

where we have

$$\omega = \frac{E - \mu}{kT} = \beta\left(\frac{\zeta}{2}\right)(x - R). \tag{3}$$

We can now interpret the set of auxiliary fields ω. The hyperbolic distance x between a pair of nodes in the network is the energy level occupied by the fermionic network links; ζ represents the curvature of the hyperbolic plane and plays the role of the Boltzmann constant; and the hyperbolic radius R is the chemical potential. The inverse of temperature, β, acts as a input parameter, which can be used to control node coordinates and influence the strength of preferential attachment in the network, as we will see shortly.

The authors of [8] proposed the one-mode generative model using the hyperbolic space with curvature $K = -\zeta^2 = -4$ (so $\zeta = 2$). In the simplest (micro-canonical) version of the model, each new node connects to the m closest existing nodes, without the use of any connection probability function. However, we are interested in the grand canonical version of the model, where we have an expected number of links, instead of the exact number $|L| = mt$.

The model generates networks according to the following steps [8]:

1. At time $t = 0$, the network is empty.
2. For every time step $t \geq 1$, a new node enters the network with radial coordinate $r_u = \ln t_u$ and angular coordinate θ_u picked from a uniform random distribution on $(0, 2\pi]$.
3. Existing nodes u', with $t_{u'} < t_u$, have their radial coordinates updated as

$$r_{u'}(t) = \alpha r_{u'} + (1 - \alpha)\ln t. \tag{4}$$

The parameter α tunes the tail of the degree distribution. More specifically, the power-law exponent of the degree distribution is given by

$$\gamma = 1 + \frac{1}{\alpha}. \tag{5}$$

That is, when $\alpha = 1$ the radial coordinates are not being updated at all and we have a strong preferential attachment. On the other hand, when $\alpha \to 0$, all nodes move outwards from the center at the same speed, hence, we create a random network.

4. The new node tries to connect to every existing node with probability given by Eq. (2). The hyperbolic distance between a pair of nodes u, u' is given by

$$x_{u,u'} = \frac{1}{2}\,\text{arccosh}(\cosh 2r_{u'} \cosh 2r_u - \sinh 2r_{u'} \sinh 2r_u \cos\theta_{u,u'}), \tag{6}$$

where $\theta_{u,u'} = \pi - |\pi - |\theta_{u'} - \theta_u||$.

Let us take a closer look at the parameters of the model m, T. While T appears in Eq. (2), that is not the case for m. However, just like in the other models, m is a parameter that controls the number of links in the network; it controls the hyperbolic radius R, of Eq. (2), at time t, according to [8]

$$R_t = \ln t - \ln\left[\frac{2T}{\sin T\pi}\,\frac{(1 - e^{-(1-\alpha)\ln t})}{m(1-\alpha)}\right], \tag{7}$$

in such a way that the average degree of the one-mode network still follows $\langle q \rangle = 2m$.

Finally, as T increases, higher energy levels can be occupied by our particles, and more disorder is observed in the system. For our model, this translates to having the probability of connection between distant nodes increasing with T (Eq. (2)). Thus, temperature controls the level of clustering of the network. As $T \to 0$, we reach the strongest levels of clustering, as only nodes positioned closest to each other have high connection probabilities, creating triadic closure in the one-mode network and consequently creating densely connected clusters of nodes. T takes values in the interval (0, 1], which is called the cold regime [7]. At values $T \geq 1$ (hot regime) clustering levels are close to 0, similar to those for the BA model.

In summary, the radial coordinate r and the parameter α determine the amount of preferential attachment in the network, while the angular coordinate θ and the parameter T determine the strength of the clustering. Papadopoulos et al. provide elegant analytical solutions for the model, in [8], along with empirical validation for fitting the model to the Internet, the *E. coli* metabolic network, and the PGP web of trust. However, they note that the model does not reproduce well the actor-movie network because of the over-inflation of connections—the complete subgraphs—created by the co-occurrence network they are considering. In other words, their proposed model fails to replicate a one-node projection of a bipartite social network. That is why we propose a bipartite version of the model in the next section.

4 Bipartite Generative Model

We consider a bipartite generative model with two sets of nodes, U and V, growing in the same plane, with the constraint that nodes of the same set cannot be connected in the bipartite network. We take the artifacts (top nodes) to be the nodes creating new links in the network, while the bottom nodes attract such links. In this way, artifacts only connect to agents in the time step when they enter the network. This process that we choose mimics, for instance, the processes of the scientific network, where papers do not gain links to additional authors after appearing in the network, but authors can continue to produce new publications (with potential co-authors) throughout their careers.

The model goes as follows:

1. At time $t = 0$, the network in empty.
2. For every time step $t \geq 1$, a new top node v and a new bottom node u enter the network with radial coordinates $r_v = \ln t_v$ and $r_u = \ln t_u$; and angular coordinates θ_v and θ_u, drawn at random from a uniform distribution on $(0, 2\pi]$.
3. Existing bottom nodes update their radial coordinates according to Eq. (4) (top nodes have fixed radial coordinates).
4. The new top node v connects to bottom nodes with a probability given by

$$p_{(u,v)} = \frac{1}{e^{\frac{(x_{u,v} - R_u)}{T}} + 1}. \tag{8}$$

As before, the parameters of the model are m, α and T. Because top nodes are not attracting links, their degree distribution will always be roughly the same for every α and T. Moreover, the degree distribution of top nodes will always keep the same shape, but will be right-shifted as m increases.

There is, however, a way to relax the constraint of top nodes always having a similar degree distribution. This can be done simply by drawing a value for m, in each time step, from a probability distribution. As m changes, it affects the chemical potential of the system, given by Eq. (7). Bigger values of m result in higher R which, in turn, increases the connection probability in the network. Hence, the shape of the tail of the degree distribution of top nodes can be easily tuned, while keeping the same $\langle d \rangle$.

The degree distribution of bottom nodes does not need any additional mechanism as the tail of the distribution is controlled by the parameter α (Fig. 5a, b). Preferential attachment is guaranteed based on the time when nodes appear in the network. This is explained by the radial coordinate, since early nodes are positioned closer to the origin of the hyperbolic plane, and therefore have a higher probability of being

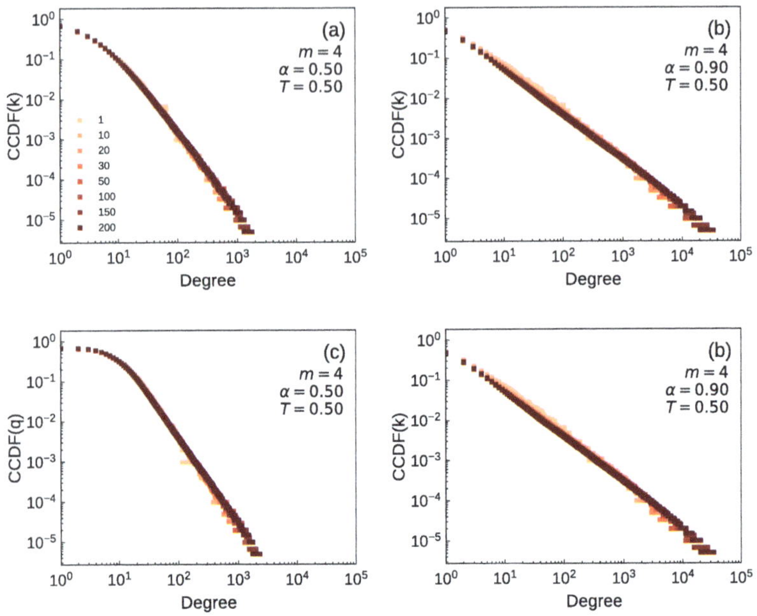

Fig. 5 Bottom and projected degree distributions for synthetic networks built with variations of parameter α. (**a**) $m = 4$, $\alpha = 0.50$ and $T = 0.50$ for bottom and projected distributions, respectively; (**b**) change of parameter $\alpha = 0.90$. (**c**) and (**d**): The same as (**a**) and (**b**), however for projected networks, respectively. We can clearly see the effect of the parameter α controlling the radial coordinates of the bottom nodes and, therefore, the level of preferential attachment in the network. Best fit for the degree distribution of bottom nodes gives us (**a**) $\gamma = 2.87$ and (**b**) $\gamma = 2.09$, compared to the predicted values $\gamma = 3$ and $\gamma = 2.10$ according to the analytical solution given by Eq. (5). Moreover, the degree distributions of the projected networks are similar to the degree distribution of bottom nodes. This is due to the fact that bottom distributions are more right-skewed than top distributions [13]

closer to more nodes in the network, according to Eq. (6). From Eq. (4), we see that when $\alpha \to 1$ we have strong preferential attachment. As α decreases, so does the weight in the tail of the probability distribution, following Eq. (5). On the other hand, for $\alpha \to 0$, the positions of every bottom node are updated, at each time step, moving to the edge of the hyperbolic disc (Figure 1c of [8]). Hence, all nodes have the same connection probability as given by Eq. (8), which characterises a random network regime for the bottom nodes. The analytical solution for the degree distributions of bottom nodes follows that of [8].

The shape of the projected degree distributions (Fig. 5c, d) is in agreement with results shown in [13], following the degree distribution of bottom nodes. In [13], we present analytical solutions for the expected projected degree distributions, according to the degree distribution of top and bottom nodes. Also, we discuss in [13] the difficulties of finding closed-form expressions for projections of bipartite networks with heterogeneous degree distributions.

In order to control the number of small cycles present in the network, we use the last parameter of the model, T, the temperature of the system. Similarly to the case of one-mode networks, where T tunes clustering (the number of triangles), for the bipartite version, T primarily controls the presence of four-cycles, but also the presence of six-, eight- and 10-cycles. At lower temperatures, nodes that are closer in the plane have higher probabilities of being connected, favouring the presence of small cycles. We can see a substantial increase in the number of four-cycles in the network, widening the gap between them and the other small cycles, as shown in Fig. 6a, b.

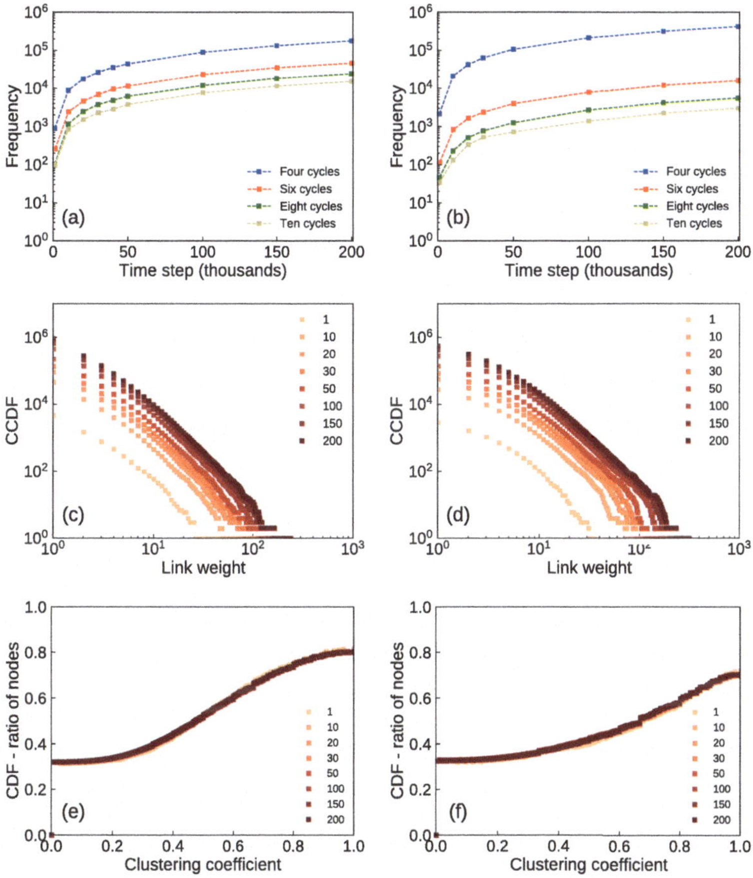

Fig. 6 (**a**) and (**b**) Evolution of small fundamental cycles; (**c**) and (**d**) Link weight distributions; (**e**) and (**f**) Distributions of the clustering coefficient. Left column for network with temperature $T = 0.50$ and right column with $T = 0.10$. At higher temperatures the presence of four-cycles is smaller, and the gap between their frequency and the frequency of other cycles reduces. We can clearly see how more four-cycles in the network shift the link weight distribution. Note that the number of fundamental six-cycles decreases, yet the clustering coefficient increases. This is the effect of the substantial rise in the number of four-cycles, at low temperature, which hides larger cycles (Fig. 7), hence widening the gap between four-cycles and larger cycles

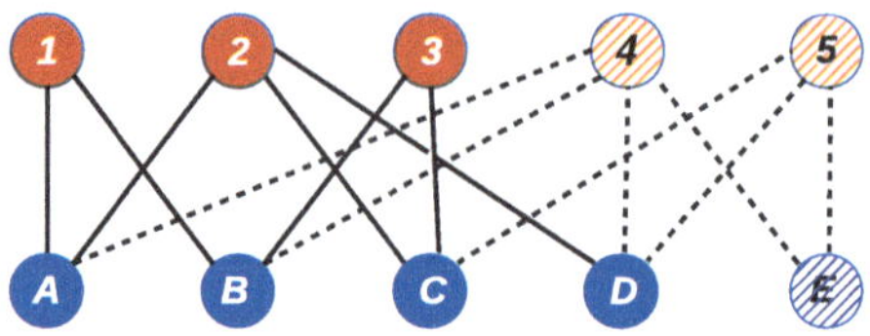

Fig. 7 Schematic of how connections creating new four-cycles change the cycle basis of the graph and hide cycles of higher order. The initial cycle basis of the toy graph has one six-cycle only, $C = \{(1B3C2A)\}$. With node 4 connected to A, B, D and E; and node 5 connected to C, D and E, the new cycle basis is the set $C = \{(2D4A), (4D5E), (2C5D), (1B4A), (3C5D4B)\}$. Although we have two six-cycles in the network, (1B3C2A) and (3C5D4B), just the latter appears in the cycle basis

The effect of the increase of four-cycles in the network is twofold. First, a larger number of four-cycles means an increase in recurrent interactions between pairs of nodes, shifting the link weight distribution to the right (Fig. 6c, d). Second, clustering in the projected network is stronger, even though the number of six-cycles in the cycle basis is smaller than in the case with higher temperature. That is, the change in temperature drastically changes the cycle basis of the network and the wider gap between the cycles, as mentioned above, hides the increase in the actual number of six-cycles in the network. The cycle basis is the set of cycles from which combinations can be made to create all other cycles in the graph. An example of how four-cycles can hide six-cycles in the network is shown in Fig. 7. We do not count the exact number of cycles of each size because counting cycles in a graph is a NP-complete problem [28], i.e. it cannot be solved in polynomial time, and it is computationally too expensive to do so in large networks like ours.

Our proposed model does present a limitation that has yet to be overcome: none of the parameters of the model (α, T, and m) seem to control degree-assortativity of the projected networks. Every bipartite network we built resulted in neutral degree-assortative projections. This is true even when we chose a heavy-tailed probability distribution to pick values of m (not shown here), which turns the degree distribution of top nodes more right-skewed. The random characteristic of the model cannot capture the social factors driving assortativity in real-world projected networks.

However, the popularity vs. similarity model still represents well one-mode networks that are projections of a bipartite structure, otherwise as stated by its own creators in Section III C of [8]. The bipartite version of the model can replicate the original bipartite network structures, such as the degree distribution of top and bottom nodes and small cycles, as observed in real-world networks [5, 9]. As a result, the expected structural properties of projected networks (degree, clustering coefficient, and link weight distributions) naturally arise as part of the projection. We also notice that to model only the projection can be a misleading process. Instead, one should take a step back and consider modelling the bipartite network first, and only then, create its projection.

5 Conclusion

In this work, we have introduced a generative model for bipartite networks, in order to better understand their structural properties. It is imperative that projections created using bipartite networks assessed with such a model can display features such as heavy-tailed degree and link weight distributions, and the high level of clustering, that are present in real one-mode networks.

By extending and adapting the popularity vs. similarity model proposed in [8] to bipartite networks, we can control degree distribution of top nodes with a simple choice of a probability distribution. On the other hand, the tail of the degree distribution of bottom nodes is tuned by the parameter α of the model, ranging from peaked Poisson-like distributions to heavy-tailed power-law distributions. The frequency of the presence of small cycles can be tuned by controlling the temperature T of the system. Therefore, we can recover degree distributions and the frequency of small cycles found in empirical bipartite networks.

Then, the structural properties of projected networks are straightforwardly inferred by building the projection out of the modelled bipartite network, except degree-assortativity. With both degree distributions of the bipartite network and the frequency of four-cycles, we naturally assess the resulting degree and link weight distributions of projections as found in real one-mode networks. The same is true for the clustering coefficients. Due to the high frequency of six-cycles in the network, the level of clustering in the projected one-mode network is not only the minimum level due to high-degree top nodes, as we discussed above.

Finally, in contrast to the claim by the creators of the popularity vs. similarity model for one-mode networks [8] that the model does not represent certain types of collaboration networks well, we have shown otherwise. The type of networks they referred to are actually one-mode projections of bipartite networks. We strengthened our claim, first presented in [5] and [9] and in agreement with other works [29, 30], that such projections should not be modeled directly, without taking in account the underlying bipartite structure. One should always consider the original bipartite network to assess the properties of networks in such cases.

References

1. Orsini, C., Dankulov, M.M., Colomer-de Simón, P., Jamakovic, A., Mahadevan, P., Vahdat, A., Bassler, K.E., Toroczkai, Z., Boguñá, M., Caldarelli, G. et al.: Quantifying randomness in real networks. Nat. Commun. **6**, 8627 (2015)
2. Denny, M.J.: The importance of generative models for assessing network structure. Soc. Sci. Res. Net. SSRN 2798493 (2016)
3. Goldenberg, A., Zheng, A.X., Fienberg, S.E., Airoldi, E.M.: A survey of statistical network models. Found. Trends Mach. Learn. **2**(2), 129–233 (2010)
4. Jacobs, A.Z., Clauset, A.: A unified view of generative models for networks: models, methods, opportunities, and challenges (2014). arXiv:1411.4070
5. Vasques Filho, D. Structure and dynamics of social bipartite and projected networks. Ph.D. thesis. The University of Auckland, 2018

6. Krioukov, D., Papadopoulos, F., Vahdat, A., Boguñá, M.: Curvature and temperature of complex networks. Phys. Rev. E **80**(3), 035101 (2009)
7. Krioukov, D., Papadopoulos, F., Kitsak, M., Vahdat, A., Boguná, M.: Hyperbolic geometry of complex networks. Phys. Rev. E **82**(3), 036106 (2010)
8. Papadopoulos, F., Kitsak, M., Serrano, M.A., Boguñá, M., Krioukov, D.: Popularity versus similarity in growing networks. Nature **489**(7417), 537–540 (2012)
9. Vasques Filho, D., O'Neale, D.R.: Bipartite networks describe R&D collaboration between institutions (2019). arXiv:1909.10977
10. Opsahl, T.: Triadic closure in two-mode networks: Redefining the global and local clustering coefficients. Soc. Netw. **35**(2), 159–167 (2013)
11. Erdős, P., Rényi, A.: On random graphs I. Publ. Math. Deb. **6**, 290–297 (1959)
12. Erdős, P., Rényi, A.: On the evolution of random graphs. Publ. Math. Inst. Hung. Acad. Sci **5**(1), 17–60 (1960)
13. Vasques Filho, D., O'Neale, D. R.: Degree distributions of bipartite networks and their projections. Phys. Rev. E **98**(2), 022307 (2018)
14. Dorogovtsev, S.N., Mendes, J.F.F., Samukhin, A.N.: Structure of growing networks with preferential linking. Phys. Rev. Lett. **85**(21), 4633 (2000)
15. Newman, M.E.J.: Clustering and preferential attachment in growing networks. Phys. Rev. E **64**(2), 025102 (2001)
16. Barabási, A.-L., Albert, R.: Emergence of scaling in random networks. Science **286**(5439), 509–512 (1999)
17. Peruani, F., Choudhury, M., Mukherjee, A., Ganguly, N.: Emergence of a non-scaling degree distribution in bipartite networks: a numerical and analytical study. Europhys. Lett. **79**(2), 28001 (2007)
18. Dahui, W., Li, Z., Zengru, D.: Bipartite producer–consumer networks and the size distribution of firms. Phys. A **363**(2), 359–366 (2006)
19. Batagelj, V., Brandes, U.: Efficient generation of large random networks. Phys. Rev. E **71**(3), 036113 (2005)
20. Guillaume, J., Latapy, M.: Bipartite graphs as models of complex networks. Phys. A **371**(2), 795–813 (2006)
21. Chojnacki, S., Kłopotek, M.A.: Scale invariant bipartite graph generative model. In: nternational Joint Conferences on Security and Intelligent Information Systems. Lecture Notes in Computer Science, pp. 240–250 Springer, Berlin (2012)
22. Binder, J., Koller, D., Russell, S., Kanazawa, K.: Adaptive probabilistic networks with hidden variables. Mach. Learn. **29**(2–3), 213–244 (1997)
23. Boguná, M., Pastor-Satorras, R.: Class of correlated random networks with hidden variables. Phys. Rev. E **68**(3), 036112 (2003)
24. Serrano, M.A., Krioukov, D., Boguná, M.: Self-similarity of complex networks and hidden metric spaces. Phys. Rev. Lett. **100**(7), 078701 (2008)
25. Wu, X., Wang, W., Zheng, W.X.: Inferring topologies of complex networks with hidden variables. Phys. Rev. E **86**(4), 046106 (2012)
26. Park, J., Newman, M.E.J.: Statistical mechanics of networks. Phys. Rev. E **70**(6), 066117 (2004)
27. Garlaschelli, D., Di Matteo, T., Aste, T., Caldarelli, G., Loffredo, M.I.: Interplay between topology and dynamics in the world trade web. Eur. Phys. J. B **57**(2), 159–164 (2007)
28. Safar, M., Mahdi, K., Farahat, H., Albehairy, S., Kassem, A., Alenzi, K.: Approximate cycles count in undirected graphs. Int. J. Comput. Int. Sys. **7**(2), 305–311 (2014)
29. Newman, M.E., Park, J.: Why social networks are different from other types of networks. Phys. Rev. E **68**(3), 036122 (2003)
30. Larremore, D.B., Clauset, A., Jacobs, A.Z.: Efficiently inferring community structure in bipartite networks. Phys. Rev. E **90**(1), 012805 (2014)

BiMLPA: Community Detection in Bipartite Networks by Multi-Label Propagation

Hibiki Taguchi, Tsuyoshi Murata, and Xin Liu

Abstract Community detection in networks, namely the identification of groups of densely connected nodes, has received wide attention recently. A bipartite network is a special class of networks, where there are two types of nodes, and edges exist between different types of nodes only. In bipartite networks, there are two ways to define communities, i.e., the one-to-one correspondence communities and the many-to-many correspondence communities. The latter naturally represents the cluster structures in the bipartite networks. However, few methods aim at detecting the many-to-many correspondence communities. In this paper, we propose a multi-label propagation algorithm BiMLPA for this purpose. Our new algorithm overcomes the limitations of previous approaches and has several desired properties, such as speed and stability. Experimental results on both synthetic networks and real-world networks demonstrate that BiMLPA outperforms previous approaches. We provide source code at https://github.com/hbkt/BiMLPA.

1 Introduction

Cluster structures or communities, where the edges within communities are dense and between communities are sparse, are commonly observed in many networks. For example, a collection of papers discussing the same topic arise in paper citation

---

H. Taguchi (✉)
Department of Computer Science, Tokyo Institute of Technology, Tokyo, Japan

National Institute of Advanced Industrial Science and Technology, Tokyo, Japan
e-mail: taguchi@net.c.titech.ac.jp

T. Murata
Department of Computer Science, Tokyo Institute of Technology, Tokyo, Japan
e-mail: murata@c.titech.ac.jp

X. Liu
National Institute of Advanced Industrial Science and Technology, Tokyo, Japan
e-mail: xin.liu@aist.go.jp

N. Masuda et al. (eds.), *Proceedings of NetSci-X 2020: Sixth International Winter School and Conference on Network Science*, Springer Proceedings in Complexity,
https://doi.org/10.1007/978-3-030-38965-9_2

networks; circles of friends appear in social networks. Community detection [8] provides insight into how the networks are internally organized and is a key step for analyzing network data. Therefore, community detection is one of the important and hot topics for network analysis in these days [8].

Raghavan et al. proposed the Label Propagation Algorithm (LPA) [27] for detecting communities in unipartite networks. LPA is simple and fast. The computational complexity is near-linear with the number of edges. However, there are some drawbacks such as instability. To overcome these issues, researchers proposed some variant algorithms. For example, LPAm [1] is a label propagation algorithm for maximizing the modularity [26], which is the most popular function to evaluate community partitions. LPAm+ [18] prevents LPAm from stuck in local maxima by combining LPAm and multi-step greedy agglomerative algorithm (MSG) [28]. LPA is also extended to detect overlapping communities. COPRA [10] is an extension of LPA by allowing nodes to have more than one label. SLPA [32], BMLPA [31], MLPA [5] are some of the variants of COPRA.

It is not uncommon to find real-world networks with two types of nodes and edges connecting only pairs of nodes of different types. An example is a network of authors and papers, where edges connect authors and the papers they have authored. There is no edge between any pair of authors or papers, and such a network is called bipartite.

Community detection in bipartite networks has received wide attention recently. There are two ways to define communities in bipartite networks, one-to-one correspondence communities and many-to-many correspondence communities, respectively [20]. One-to-one correspondence communities are the groups of nodes that are densely connected. This is the same as the definition of community in unipartite networks. Figure 1a shows an example. The two types of nodes are located on the top and bottom, respectively. A community, as surrounded by the dotted line, is composed of both types of nodes.

On the one hand, many-to-many correspondence communities are the groups of nodes that are of the same type and have similar link patterns [24]. As shown in Fig. 1b, there are two communities for the top nodes and three communities for the bottom nodes. The nodes in each community have a similar link pattern. For example, the nodes in the second bottom community (in the middle) all have dense connections with the two top communities.

The many-to-many correspondence community structure is a natural description of the heterogeneity in the real world because humans or objects always have multi-faceted properties. Take the user-book network of an online book store as an example. An edge exists if a user bought a book. Note that the users have multiple reading interests. As a result, the college student community majoring in computer science will correspond to both programming-related and mathematics-related book communities, while the college student community majoring in physics will correspond to both mathematics-related and physics-related book communities. As shown in Fig. 2, this creates a many-to-many correspondence community structure.

There are two main approaches to detecting communities in bipartite networks. The first one is to compress a bipartite network into a one-mode projection network

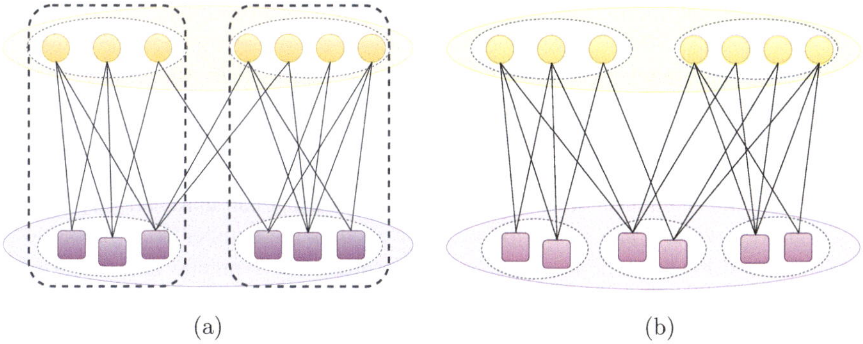

Fig. 1 The examples of the two definitions of community in bipartite networks. (**a**) The one-to-one correspondence community. (**b**) The many-to-many correspondence community

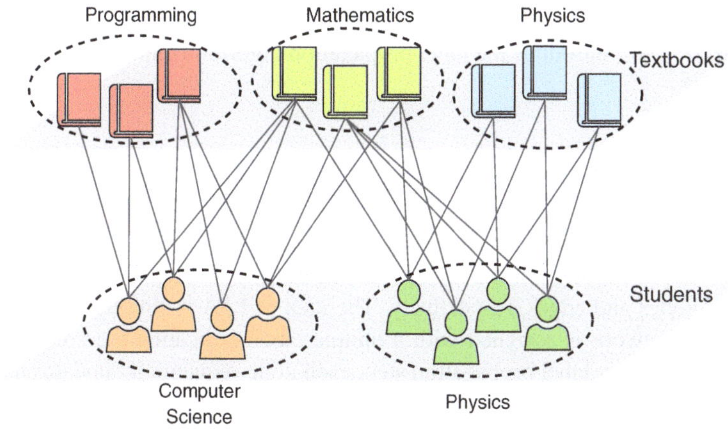

Fig. 2 An example of many-to-many correspondence community structure in the user-book network

and apply a standard community detection algorithm [30, 33]. This approach is usually not as good as handling the original network because the projected network is less informative than the original one. The second one, which overcomes this issue, is to extend unipartite algorithms such as the Louvain method [3], the stochastic block model [11] for the bipartite structure [14, 34].

LPA [27] has also been extended to deal with bipartite networks. Improved LPA [17] updates the labels of the two types of nodes iteratively to avoid the oscillation problem. LP&BRIM [16] combines LPA and BRIM algorithm. LPAb [1] is an algorithm to optimize Barber's bipartite modularity. LPAb+ [19] further extended LPAb to escape stuck in local maxima. Moreover, Li et al. [15] proposed a quantitative function for evaluating partition density in bipartite networks and designed a heuristic algorithm called BiLPA. LPA is also used to optimize quantitative functions other than Barber's modularity [2, 15]. However, none of them aims at detecting the many-to-many correspondence communities.

Inspired by the multi-label propagation for detecting overlapping communities [31], we propose BiMLPA to detect the many-to-many correspondence communities in bipartite networks. Based on the idea of multi-labels, we design a label propagation procedure that is specific for bipartite networks. BiMLPA overcomes the limitations of existing approaches and has the following advantages: (1) BiMLPA can detect communities with many-to-many correspondence, (2) BiMLPA is fast and has near-linear time complexity, (3) BiMLPA can automatically detect communities without any a priori knowledge such as the number of communities, and (4) BiMLPA is a stable algorithm. Experiments in synthetic and real-world networks demonstrate that BiMLPA outperforms the previous algorithms.

2 Related Work

In this section, we introduce the label propagation algorithm and its variants.

2.1 LPA

Raghavan et al. proposed LPA [27] for detecting communities in unipartite networks. A unipartite network can be represented as $G = (V, E)$, where V and E are the set of nodes and edges, respectively. The idea of LPA is simple: initially, each node in the network is assigned with a unique label, indicating the community it belongs to. At every label propagation step, each node sequentially updates its label to a new one which is the most frequent label among its neighbors. Formally, the label updating rule for a node v is

$$l'_v = \underset{l}{\operatorname{argmax}} \sum_{u \in N(v)} \delta(l_u, l) \tag{1}$$

where l_u is the label of node u, l'_v is the new label of node v, $N(v)$ is the set of nodes that are neighbors of v, and δ is the Kronecker's delta. If there are multiple labels satisfy Eq. (1), one is chosen randomly from them. The label updating continues until convergence. Finally, the nodes with the same label are grouped and identified as communities. LPA is fast, effective, and conceptually simple.

2.2 Variants of LPA

Many researchers modified LPA to improve the performance or apply it to other types of networks.

Modularity-Specialized LPA One weakness of LPA is that the algorithm is not stable and is sensitive to the order in which node labels are updated in each step. Sometimes LPA may even end up with a trivial solution where all nodes are identified in the same community. To overcome this problem, Barber et al. [1] and Liu et al. [18] extended LPA by modifying the label updating rule so that modularity [26] can be maximized during the label updating process.

LPA for Bipartite Networks Directly applying LPA to bipartite networks will cause the oscillation problem, where the algorithm cannot converge. Liu et al. proposed improved LPA [17] and LP&BRIM [16] which avoid the oscillation problem. However, these algorithms are for detecting one-to-one communities.

LPA for Overlapping Communities LPA is also extended for detecting overlapping communities. BMLPA [31] is a multi-label propagation algorithm to detect overlapping communities in unipartite networks. In BMLPA, each node can have multiple labels with weights.

3 BiMLPA

In this section, we introduce BiMLPA, community detection in bipartite networks using a multi-label propagation algorithm. A bipartite network can be represented as $G = (U, V, E)$, where U and V are disjoint sets of nodes, and E is the set of edges. Each edge connects $u \in U$ and $v \in V$. $N(v)$ is the set of neighbor nodes of v, i.e., $N(v) = \{u|(u, v) \in E\}$.

3.1 Extension to Multi-Labels

We extend the label propagation algorithm from single-label to multi-labels. In other words, each node v is accompanied by a label set:

$$L_v = \{(l_{v1}, w_{v1}), \cdots, (l_{vt}, w_{vt})\} \tag{2}$$

$$\sum_i w_{vi} = 1 \tag{3}$$

where t is the number of labels that v has, l_{vi} is the i-th label of v, and w_{vi} is the weight of label l_{vi}. For simplicity, labels are sorted in descending order of the weights, i.e., $w_{v1} \geq w_{v2} \geq \cdots \geq w_{vt}$.

3.2 Label Propagation Procedure

We introduce the procedure for propagating the multi-labels. Suppose each node is assigned with a label set. Figure 3 is an example of how we update the labels for a node v. It is composed of three steps.

Step 1 We process the label set L_u for each $u \in N(v)$. Specifically, we first reduce the number of labels in L_u to λ.

$$L'_u = \begin{cases} \{(l_{u1}, w_{u1}), \cdots, (l_{u\lambda}, w_{u\lambda})\} & if \text{ the number of labels} > \lambda \\ L_u & otherwise \end{cases} \tag{4}$$

Note that labels are sorted in descending order of weights, so we preserve the top λ labels that have the largest weights in L_u. Then, we adjust the weight for each label in L_u. To reduce the influence from hub nodes, we discount the weight by the coefficient based on the node degree,

$$L'_u \leftarrow \left\{\left(l, \frac{w}{\sqrt{d_u}}\right) | (l, w) \in L'_u\right\} \tag{5}$$

where d_u is the degree of u.

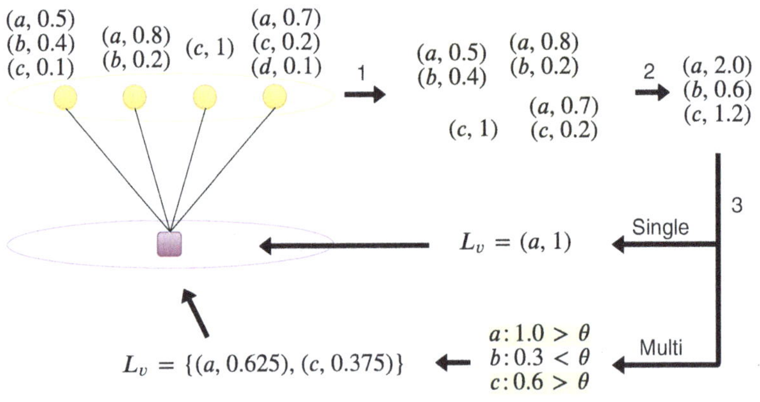

Fig. 3 An example of propagating labels from the top yellow nodes to the bottom purple node. The parameters are $\lambda = 2$ and $\theta = 0.5$. Step 1: we reduce the number of labels in each label set of the top nodes. In this case, the max number of labels is $\lambda = 2$. So we eliminate $(c, 0.1)$. Step 2: we propagate labels from the top nodes to the bottom node. The bottom node's label weight is the sum of the weights of the corresponding top nodes. Step 3: in Single-Label propagation, we choose one label which has the largest weight. In this case, label a is chosen, and the new label set L_v is $(a, 1)$. In Multi-Label propagation, we divide all weights by the maximum weight 2.0 and choose labels that satisfy the condition Eq. (8). In this case, the label a and c are kept, and the weights are normalized

Step 2 We propagate the labels to v from its neighbors, with the weights being the sum of the weights that v's neighbors have. Specifically,

$$L'_v = \{(l'_{v1}, w'_{v1}), \cdots, (l'_{vt'}, w'_{vt'})\} \tag{6}$$

$$w'_{vi} = \sum_{\substack{(l,w)\in L'_u \\ u\in N(v)}} w \cdot \delta(l, l'_{vi}) \tag{7}$$

where t' is the number of labels in L'_v and δ is the Kronecker's delta.

Step 3 We trim the labels in L_v. Note that at the end of step 2, all the labels of $v's$ neighbors are propagated to v, resulting in an excessive number of labels. To trim labels, we introduce the following two strategies.

Multi-Label Propagation We keep labels whose weights are above some threshold. Let $w_{\max}$ be the maximum value of the weight in $L'_v f$, we keep the pair (l, w) if its weight w satisfies the following condition:

$$\frac{w}{w_{\max}} \geq \theta \tag{8}$$

where $\theta \in [0, 1)$ is a threshold parameter. Finally, the weights of remaining labels are normalized and the label set is updated.

$$L_v^{new} = \{(l_{v1}^{new}, w_{v1}^{new}), \cdots, (l_{vt}^{new}, w_{vt}^{new})\} \tag{9}$$

$$w_{vi}^{new} = \frac{w_{vi}}{\sum_j w_{vj}} \tag{10}$$

Single-Label Propagation We only keep one label which has the largest weight.

$$(l_{\max}, w_{\max}) = \underset{(l,w)\in L'_v}{\operatorname{argmax}}\, w \tag{11}$$

Then we update the label set:

$$L_v^{new} = \{(l_{\max}, 1)\} \tag{12}$$

3.3 *Algorithm*

Now we present our algorithm BiMLPA. It contains the following four steps.

1. **Initialization**
   For each node $u \in U$, we assign a unique label with weight set to 1. For each node $v \in V$, we assign an empty label set. In other words,

$$L_u = \{(u, 1)\}, \quad L_v = \{\} \tag{13}$$

2. **Multi-Multi LP**
   We propagate labels in the Multi-Label propagation manner. We first propagate labels from U to V, and then from V to U. This process is repeated until convergence. The convergence is reached when there is no change in terms of labels in any label set (not in terms of weights).
3. **Multi-Single LP**
   We propagate labels from U to V in the Multi-Label propagation manner, then from V to U in the Single-Label propagation manner. As a result, nodes in U have only one label whereas nodes in V have more than one label. This process is repeated until convergence (in the same sense as the Multi-Multi LP).
4. **Post-Processing**
   We identify a community as a group of nodes whose label sets are the same in terms of labels (the label weights do not need to be matched).

3.4 Complexity Analysis

We analyze the time complexity of Multi-Multi LP and Multi-Single LP described in Sect. 3.3. Suppose n is the number of nodes, m is the number of edges, and λ is the parameter. Since the number of neighbor nodes is $O(m/n)$, we assume the maximum number of labels is $O(\lambda m/n)$. Therefore, sorting and choosing λ labels takes $O((\lambda m/n)\log(\lambda m/n))$ (Eq. (4)). Multiplying the coefficient (Eq. (5)) and summing the weight for each label of all neighbors (Eqs. (6) and (7)) take linear time, i.e., $O(\lambda m/n)$.

- **Multi-Label Propagation**
  Choosing the label which has the largest weight and checking whether a label satisfies an inequality (Eq. (8)) takes $O(\lambda m/n)$. Updating the label set takes less than $O(\lambda m/n)$ (Eqs. (9) and (10)).
- **Single-Label Propagation**
  Choosing one label which has the largest weight (Eq. (11)) takes $O(\lambda m/n)$. Updating the label set takes $O(1)$ (Eq. (12)).

Therefore, the time complexity of updating a label set in both Multi-Label and Single-Label Propagation is $O((\lambda m/n)\log(\lambda m/n))$. Since there are n nodes in the network, the time complexity per iteration of both Multi-Multi LP and Multi-Single LP is $O(\lambda m \log(\lambda m/n))$, which is near-linear to the number of edges.

4 Experiments

In this section, we present experiments for testing the performance of BiMLPA. The basic scheme is as follows. (1) We collected four networks including real-world and synthetic networks, with known communities (true communities) for each one. (2) We apply BiMLPA and baseline algorithms to these networks to detect the communities. (3) We evaluate the performance by the normalized mutual information (NMI) [6] that quantifies the similarity between the detected communities and the true ones. Specifically, suppose L1 is the detected community partition and L2 the true partition. If L1 and L2 match completely, we have the maximum NMI value of 1, whereas if L1 and L2 are entirely independent of one another, we have the minimum value of 0. Such kind of testing is widely used by other researchers dealing with the community detection problem.

4.1 Datasets

We use two real-world networks and two synthetic networks in our experiments. The properties of the networks are shown in Table 1. Visualization of the networks with different colors for distinguishing the true communities is shown in Fig. 4. We can find that all of them have many-to-many correspondence communities.

The first real-world network is the Southern Women network [7], which consists of 18 women and 14 social events as nodes. Each edge represents that a woman participated in an event. This network is commonly used as the benchmark for community detection in bipartite networks. The second one is the Malaria network [13], which consists of 297 genes and 806 amino acid substrings. Each edge represents that a gene contains a substring.

Besides, we generated two synthetic networks according to Larremore's procedure [14]. The network can be represented by $\omega = \alpha\omega^{\text{planted}} + (1-\alpha)\omega^{\text{random}}$, where ω is the adjacency matrix, ω^{planted} is the adjacency matrix generated according to the true community structure, ω^{random} is the adjacency matrix of the random bipartite network, and α is the mixing parameter, which adjusts the level of noise in the network. In our experiments, α is set to 0.9.

Table 1 The properties of the networks

	Network	$\|U\|$	$\|V\|$	$\|E\|$	$\|C_U\|$	$\|C_V\|$
Real-World	Southern women	18	14	89	2	3
	Malaria	297	806	2965	3	3
Synthetic	Network 1	70	30	1094	2	3
	Network 2	60	60	1680	3	4

$|U|$ and $|V|$ are the number of nodes in U and V, respectively. $|E|$ is the number of edges in the network. $|C_U|$ and $|C_V|$ are the number of communities in U and V, respectively

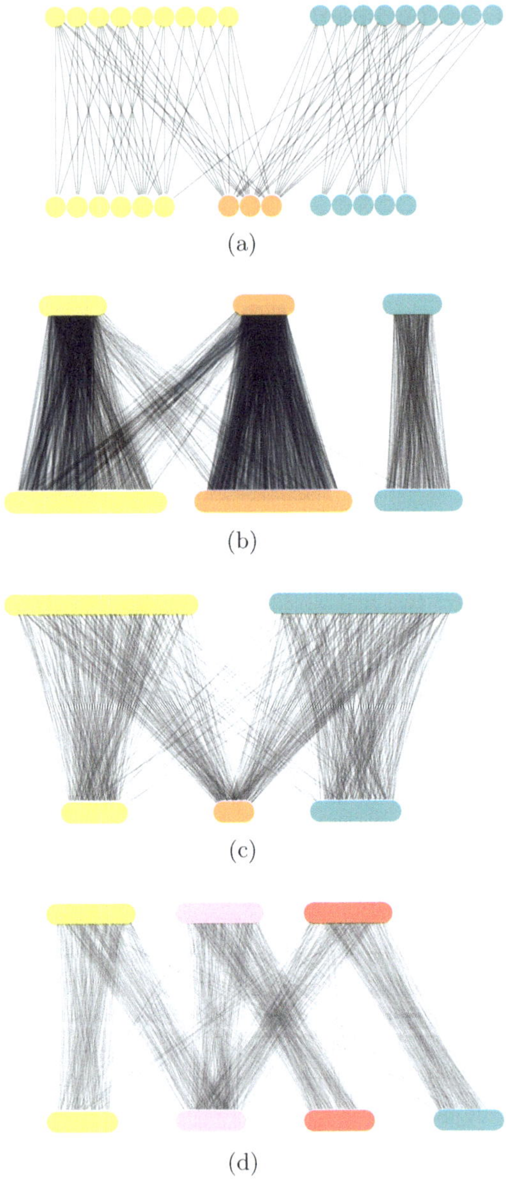

Fig. 4 Visualization of the networks used in our experiments. The nodes in the same community are painted with the same color. (**a**) Southern Women network. (**b**) Malaria network. (**c**) Synthetic network 1. (**d**) Synthetic network 2

4.2 Baselines

We compare BiMLPA with the following three baselines.

- LPAb+ [19] : This method optimizes Barber's modularity using a single-label propagation algorithm.

- biSBM [14] : This method formulates a bipartite stochastic block model and detects communities.
- ComSim [30] : This approach generates projection networks, whose edges are weighted and the weights represent the similarity of the same type of nodes, then applies a standard community detection technique.

LPAb+ is designed for detecting one-to-one correspondence communities. In contrast, biSBM and ComSim are designed for detecting many-to-many correspondence communities. The implementations of biSBM[1] and ComSim[2] written in C++ are available.

4.3 Settings

We set the parameters of each algorithm as follows. LPAb+ is parameter-free. In biSBM, there are two parameters K_a and K_b, which represent the number of communities to be detected in node sets U and V, respectively. We directly set them as the numbers of true communities. In ComSim, we use common neighbors as the similarity for projection. BiMLPA has two parameters, θ and λ. θ is searched from 0.0 to 0.9 with an increment of 0.1 and λ is searched from 3 to 10.

4.4 Results

We apply each algorithm 100 times in each network. Table 2 shows the results of the average and the standard deviation of NMI.

Table 2 The average and the standard deviation of the NMI scores

		BiMLPA	LPAb+	biSBM	ComSim
Southern	U	**1.0** ± 0.0	0.5430 ± 0.0559	**1.0** ± 0.0	0.6794 ± 0.0953
	V	**1.0** ± 0.0	0.7447 ± 0.0681	**1.0** ± 0.0	0.1489 ± 0.3090
Malaria	U	**0.6605** ± 0.0	0.5106 ± 0.0195	0.1491 ± 0.1098	0.4549 ± 0.0064
	V	**0.6330** ± 0.0	0.4508 ± 0.0160	0.1419 ± 0.0989	0.4043 ± 0.0046
SynNet 1	U	**0.9061** ± 0.0	0.6961 ± 0.1199	0.8128 ± 0.0	0.7771 ± 0.1321
	V	**1.0** ± 0.0	0.7562 ± 0.1337	**1.0** ± 0.0	0.0 ± 0.0
SynNet 2	U	**1.0** ± 0.0	0.9394 ± 0.0	**1.0** ± 0.0	0.8915 ± 0.0297
	V	0.9066 ± 0.0	0.6700 ± 0.0	**0.9112** ± 0.0	0.5178 ± 0.0307

U and V represent the types of nodes
Bold values indicates the best accuracy for each dataset

[1]http://danlarremore.com/bipartiteSBM/.
[2]https://github.com/rtackx/ComSim.

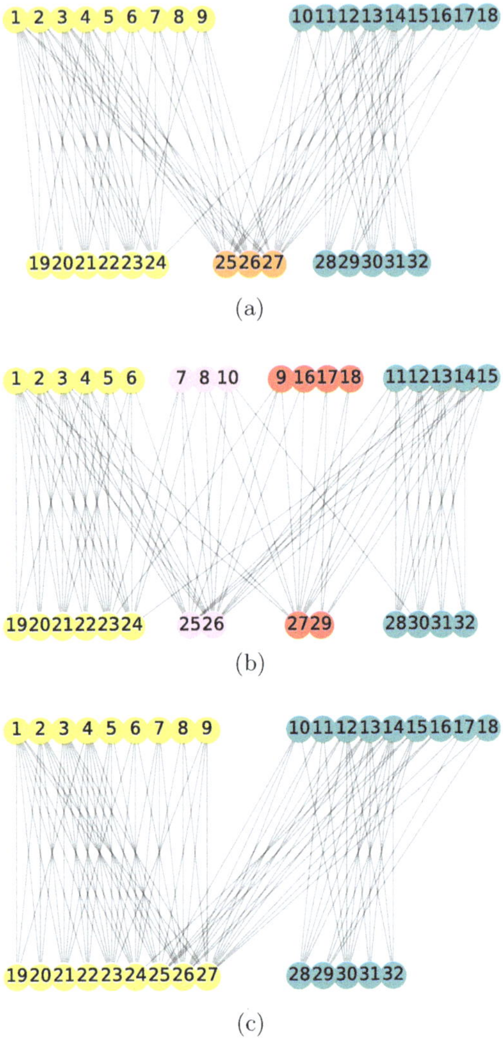

Fig. 5 Visualization of the communities detected by different algorithms in the Southern Women network. Woman nodes are located at the top and event nodes are located at the bottom. (**a**) BiMLPA and biSBM. (**b**) LPAb+. (**c**) ComSim

We can find that BiMLPA outperforms the others in the Southern Women network, the Malaria network, and synthetic Network 1. In synthetic Network 2, BiMLPA is also comparable to biSBM.

As an example, Fig. 5 illustrates the communities detected by different algorithms in the Southern Women network. In the true partition, there are two communities for the woman nodes (woman community 1: nodes 1–9; woman community 2: nodes 10–18) and three communities for the event nodes (event community 1: nodes 19–24; event community 2: nodes 25–27; event community 3: nodes 28–32). Both BiMLPA and biSBM successfully detected the true communities with NMI = 1, as shown in Fig. 5a. We can find that the communities have many-to-many correspondence. LPAb+ detected four woman communities and four event communities. The communities have a one-to-one correspondence, as shown in Fig. 5b. Although ComSim is designed for many-to-many correspondence communities, it fails to detect the true communities. As shown in Fig. 5c, the algorithm mixed the event community 1 and 2 and put nodes 19–27 into one community.

It is worth mentioning that BiMLPA is a stable algorithm. One notorious drawback of the label propagation algorithm is that it introduces some random factors and the result can be quite different in different runs. Indeed, the standard deviation of NMI indicates that the baselines including LPAb+ are unstable. In contrast, the standard deviation of BiMLPA is always 0, indicating a fixed result. This is because that the randomness of the label updating is suppressed due to the extension from single-label to multi-labels.

5 Conclusion

In this paper, we proposed BiMLPA, a novel multi-label propagation algorithm for detecting communities in bipartite networks. BiMLPA can handle communities with many-to-many correspondence. The algorithm consists of two types of label propagation: Multi-Multi LP and Multi-Single LP. BiMLPA is fixed and overcomes the unstable issue of LPA. Moreover, BiMLPA is fast and has near-linear time complexity. Experiments on real-world and synthetic networks demonstrate that BiMLPA outperforms previous algorithms such as LPAb+, biSBM, and ComSim.

As future work, there are two problems. One is the appropriate measure for evaluating communities with many-to-many correspondence in bipartite networks. Although researchers have presented measures such as link-pattern based communities [20–22] and bipartite modularity [25, 29], they have limitations such as not scalable to large networks. Another issue is the relation between multi-label propagation algorithms and algorithms based on graph neural networks [12] for graph embedding [4, 9, 23]. Both of them are based on aggregating information from neighbors. The difference is that a multi-label propagation algorithm uses discrete vectors as labels while graph neural networks use continuous vectors. How to unify them is an interesting and challenging problem. These will be left for our future work.

Acknowledgements We would like to thank JST CREST (Grant Number JPMJCR1687), JSPS Grant-in-Aid for Scientific Research(B) (Grant Number 17H01785) and JSPS Grant-in-Aid for Early-Career Scientists (Grant Number 19K20352).

References

1. Barber, M., Clark, J.: Detecting network communities by propagating labels under constraints. Phys. Rev. E **80**, 026129 (2009)
2. Beckett, S.J.: Improved community detection in weighted bipartite networks. In: Royal Society Open Science (2016)
3. Blondel, V.D., Guillaume, J.L., Lambiotte, R., Lefebvre, E.: Fast unfolding of communities in large networks. J. Stat. Mech: Theory Exp. **2008**(10), P10008 (2008). https://doi.org/10.1088/1742-5468/2008/10/p10008
4. Cai, H., Zheng, V.W., Chang, K.C.C.: A comprehensive survey of graph embedding: problems, techniques and applications. IEEE Trans. Knowl. Data Eng. **30**(9), 1616–1637 (2018)
5. Dai, Q., Guo, M., Liu, Y., Liu, X., Chen, L.: MLPA: Detecting overlapping communities by multi-label propagation approach. In: 2013 IEEE Congress on Evolutionary Computation, pp. 681–688 (2013)
6. Danon, L., Guilera, A.D., Duch, J., Arenas, A.: Comparing community structure identification. J. Stat. Mech: Theory Exp. **2005**(9), P09008 (2005)
7. Davis, A., Gardner, B., Gardner, M.: Deep South; a Social Anthropological Study of Caste and Class. University of Chicago Press, Chicago (1941)
8. Fortunato, S.: Community detection in graphs. Phys. Rep. **486**(3), 75–174 (2010). https://doi.org/10.1016/j.physrep.2009.11.002
9. Goyal, P., Ferrara, E.: Graph embedding techniques, applications, and performance: a survey. Knowl.-Based Syst. **151**, 78–94 (2018)
10. Gregory, S.: Finding overlapping communities in networks by label propagation. New J. Phys. **12**(10), 103018 (2010)
11. Karrer, B., Newman, M.E.J.: Stochastic blockmodels and community structure in networks. Phys. Rev. E **83**, 016107 (2011). https://doi.org/10.1103/PhysRevE.83.016107
12. Kipf, T.N., Welling, M.: Semi-supervised classification with graph convolutional networks. In: Proceedings of ICLR (2017)
13. Larremore, D., Clauset, A., Buckee, C.: A network approach to analyzing highly recombinant malaria parasite genes. PLoS Comput. Biol. **9**(10), 1–12 (2013)
14. Larremore, D., Clauset, A., Jacobs, A.: Efficiently inferring community structure in bipartite networks. Phys. Rev. E Stat. Nonlinear Soft Matter Phys. **90**(1), 012805 (2014)
15. Li, Z., Wang, R., Zhang, S., Zhang, X.: Quantitative function and algorithm for community detection in bipartite networks. Inf. Sci. **367–368**, 874–889 (2016)
16. Liu, X., Murata, T.: Community detection in large-scale bipartite networks. In: 2009 IEEE/WIC/ACM International Joint Conference on Web Intelligence and Intelligent Agent Technology, vol. 1, pp. 50–57 (2009)
17. Liu, X., Murata, T.: How does label propagation algorithm work in bipartite networks? In: 2009 IEEE/WIC/ACM International Joint Conference on Web Intelligence and Intelligent Agent Technology, vol. 3, pp. 5–8 (2009)
18. Liu, X., Murata, T.: Advanced modularity-specialized label propagation algorithm for detecting communities in networks. Phys. A Stat. Mech. Appl. **389**(7), 1493–1500 (2010)
19. Liu, X., Murata, T.: An efficient algorithm for optimizing bipartite modularity in bipartite networks. Comput. Intell. Intell. Inf. **14**(4), 408–415 (2010)
20. Liu, X., Murata, T.: Evaluating community structure in bipartite networks. In: 2010 IEEE Second International Conference on Social Computing, pp. 576–581 (2010)
21. Liu, X., Murata, T.: Detecting communities in k-partite k-uniform (hyper)networks. J. Comput. Sci. Technol. **26**(5), 778–791 (2011)
22. Liu, X., Murata, T.: Extracting the mesoscopic structure from heterogeneous systems. In: Proceedings of the 22nd ACM Conference on Hypertext and Hypermedia (HT), pp. 211–220. ACM, New York (2011)

23. Liu, X., Murata, T., Kim, K.S., Kotarasu, C., Zhuang, C.: A general view for network embedding as matrix factorization. In: Proceedings of the Twelfth ACM International Conference on Web Search and Data Mining, pp. 375–383 (2019)
24. Long, B., Xu, X., Zhang, Z., Yu, P.S.: Community learning by graph approximation. In: Seventh IEEE International Conference on Data Mining (ICDM 2007), pp. 232–241 (2007)
25. Murata, T.: Detecting communities from bipartite networks based on bipartite modularities. In: 2009 International Conference on Computational Science and Engineering, vol. 4, pp. 50–57 (2009)
26. Newman, M., Girvan, M.: Finding and evaluating community structure in networks. Phys. Rev. E **69**, 026113 (2004)
27. Raghavan, U., Albert, R., Kumara, S.: Near linear time algorithm to detect community structures in large-scale networks. Phys. Rev. E Stat. Nonlinear Soft Matter Phys. **76**, 036106 (2007)
28. Schuetz, P., Caflisch, A.: Efficient modularity optimization by multistep greedy algorithm and vertex mover refinement. Phys. Rev. E **77**, 046112 (2008). https://doi.org/10.1103/PhysRevE.77.046112
29. Suzuki, K., Wakita, K.: Extracting multi-facet community structure from bipartite networks. In: 2009 International Conference on Computational Science and Engineering, vol. 4, pp. 312–319 (2009)
30. Tackx, R., Tarissan, F., Guillaume, J.: ComSim: a bipartite community detection algorithm using cycle and node's similarity. In: Cherifi, C., Cherifi, H., Karsai, M., Musolesi, M. (eds.) Complex Networks and Their Applications VI, Cham, pp. 278–289 (2018)
31. Wu, Z., Lin, Y., Gregory, S., Wan, H., Tian, S.: Balanced multi-label propagation for overlapping community detection in social networks. J. Comput. Sci. Technol. **27**(3), 468–479 (2012)
32. Xie, J., Szymanski, B., Liu, X.: SLPA: uncovering overlapping communities in social networks via a speaker-listener interaction dynamic process. In: 2011 IEEE 11th International Conference on Data Mining Workshops, pp. 344–349 (2011)
33. Zhou, T., Ren, J., Medo, M., Zhang, Y.C.: Bipartite network projection and personal recommendation. Phys. Rev. E **76**, 046115 (2007). https://doi.org/10.1103/PhysRevE.76.046115
34. Zhou, C., Feng, L., Zhao, Q.: A novel community detection method in bipartite networks. Phys. A Stat. Mech. Appl. **492**, 1679–1693 (2018). https://doi.org/10.1016/j.physa.2017.11.089

Connected Graphs with a Given Degree Sequence: Efficient Sampling, Correlations, Community Detection and Robustness

John H. Ring IV, Jean-Gabriel Young, and Laurent Hébert-Dufresne

Abstract Random graph models can help us assess the significance of the structural properties of real complex systems. Given the value of a graph property and its value in a randomized ensemble, we can determine whether the property is explained by chance by comparing its real value to its value in the ensemble. The conclusions drawn with this approach obviously depend on the choice of randomization. We argue that keeping graphs in one connected piece, or component, is key for many applications where complex graphs are assumed to be connected either by definition (e.g. the Internet) or by construction (e.g. a crawled subset of the World-Wide Web obtained only by following hyperlinks). Using an heuristic to quickly sample the ensemble of small connected simple graphs with a fixed degree sequence, we investigate the significance of the structural patterns found in real connected graphs. We find that, in sparse networks, the connectedness constraint changes degree correlations, the outcome of community detection with modularity, and the predictions of percolation on the ensemble.

Putting measurements into context is a crucial part of any network analysis. Suppose that we have some real network at our disposal and that we know the value of some of its properties—say its level of transitivity or its homophily with respect to some property [1]. It is clear that these values do not make much sense in and of themselves. Knowing that a network G_1 "has transitivity C," tells us far less than knowing that "network G_1 is more transitive than network G_2." In the first instance, we merely have an arbitrary number; it only begins to make sense once compared

J. H. Ring IV · L. Hébert-Dufresne (✉)
Vermont Complex Systems Center, University of Vermont, Burlington, VT, USA

Department of Computer Science, University of Vermont, Burlington, VT, USA
e-mail: Laurent.hebert-dufresne@uvm.edu

J.-G. Young
Center for the Study of Complex Systems, University of Michigan, Ann Arbor, MI, USA

N. Masuda et al. (eds.), *Proceedings of NetSci-X 2020: Sixth International Winter School and Conference on Network Science*, Springer Proceedings in Complexity,
https://doi.org/10.1007/978-3-030-38965-9_3

to other measurements, either implicitly (e.g., via a normalization) or explicitly, as in the second instance.

Null modeling is a technique whereby we create an ensemble of graphs that preserves some of the properties of the original network data, while randomizing the others. Computing the expected value of a property for the graphs of this ensemble gives a baseline against which to compare an original measurement. The baseline can be used to formulate a statistical test telling us which properties of an input G are "surprising" with respect to the model, and which are not. For example, if the transitivity $C(G)$ of a network takes roughly the same value under a null model that preserves the degree sequence $d(G)$ of the input G, then it is, as far as transitivity is concerned, typical of graphs with this degree sequence. Conversely, if the value of $C(G)$ greatly differs in the randomized ensemble, then G is atypical: The degree sequence $d(G)$ does not explain $C(G)$.

Similar inference can be made for any choice of null models and properties. Hence the more models we have, the more we can control for various features of real networks. Fixing only the number of vertices and edges leads to the classical random graph model of Erdős–Rényi [2, 3]. Further constraining the ensemble to graphs with a fixed degree distribution corresponds to generating graphs from the well-known Configuration Model (CM) [4–6]. Notwithstanding sampling problems [7], exponential random graphs [8] can be used to limit the ensemble to graphs with precise patterns and correlation structure. A wealth of other models allow to control for, e.g., arbitrary mesoscale patterns [9], degree correlations [10], or a centrality structure [11, 12].

All the models above are defined in terms of simple *local* connection rules. The main reason for this choice is that simple local rules usually lead to simple sampling algorithms and make the models analytically tractable [7]. A less desirable consequence of mathematical convenience, however, is that it dictates the models we have. Relying only on convenient models can leave important blind spots in our analyses, because unwieldy connection rules can—and do—lead to critically different null models [13].

An important example of unwieldy constraint, which will be the focus of the present paper, is connectedness. A few recent studies have shown that the connected subsets of random graphs can have a significantly different structure than the entire random graph itself [14–16]. Connectedness is, without a doubt, an important aspect to control for in null models, since it is so frequently found in real systems. It can arise for at least three different reasons. One, it may be a simple matter of perspective: If a food web (or a power grid) is split into independent components, then we are likely to consider them as separate food webs (or power grids) since what occurs in one component does not affect the other. Two, global connectivity may stem from the definition of the network: Such is the case of the Internet, which is a unique, global, connected network of computers. Three, some networks can never split into disconnected pieces because of the way they are sampled [17], including, for example: subsets of the World Wide Web obtained by crawlers that do not teleport [18], or social networks sampled by recursive nomination [19, 20]. If we do not use an appropriate random graph ensemble as a null model in our analysis, we are likely to overestimate the significance of some results or miss other important structural features.

The focus of our paper is twofold. First, we propose a simple heuristic to generate samples from the connected Configuration Model (connected CM), the natural null model for connected graphs with a fixed degree sequence. There are already algorithms that solve related problems—see Sect. 1 for an overview—but they are either inefficient or not adapted to the version of the problem we aim to tackle. Our algorithm is described in Sect. 2 and made available online in a reference implementation.[1] Second, we use this algorithm to quantify the impact of connectedness on applications that rely on comparison with random graphs, in Sect. 3. We show that using the connected version of the CM can change degree correlations, the outcome of community detection, and lead to qualitatively different predictions of percolation on graphs.

1 Connected Configuration Model and Related Work

It is somewhat imprecise to speak of *the* connected configuration model (CM) because there are, in fact, many versions of the classic configuration model [21]. In this paper, we will consider the so-called microcanonical variant, that assigns an identical probability to all graphs that contain a single component and whose nodes $1, \ldots, N$ have degrees $d = (d_1, d_2, \ldots, d_N)$, and zero probability to all other graphs. Formally, if we denote by $\Omega_N(d)$ the ensemble of connected graphs with degree sequence d, the probability of observing any given graph G with degree sequence d' is, under this model,

$$P(G) = \begin{cases} \frac{1}{|\Omega_N(d')|} & \text{if } d(G) = d', \\ 0 & \text{otherwise} \end{cases} \tag{1}$$

where $|\cdot|$ denotes the cardinality of the ensemble. This definition of the connected CM is known as microcanonical by opposition to the canonical definition where the degree of nodes are fixed only on average, instead of exactly.

It turns out sampling that from this ensemble is much more challenging than sampling from the equivalent ensemble without the connectedness constraint [21]. There are a few approaches that solve the problem in different ways.

A sampling algorithm for generic connected models was recently proposed by Gray et al. [16]. Their idea is to first generate a (typically unconnected) initial graph from the model without connectedness constraints. They then add arbitrary links until the graph becomes connected, and they finally run a simple Monte-Carlo Markov chain (MCMC) algorithm with a target distribution that is now the

[1] Available at https://gitlab.com/jhring/connected_cm.

connected version of the model. This method unfortunately does not work when the target model is the *microcanonical* connected CM. The input degree sequence is modified when edges are added, and the resulting graphs therefore have zero probability under the connected model, and the MCMC can never leave the initial state, let alone approach a region of high probability.

Another recent algorithm proposed by Tishby et al. also makes use of the unconnected version of the CM as its starting point [15]. This approach differs from the previous one in what it does next. The idea, in this case, is to discard the disconnected components of the initial graph and to keep the largest one, which then constitutes a sample from the connected model. Tishby et al. show that by engineering the parameters of the generating (unconnected) ensemble, one can obtain the desired size and degree distribution for the large component, in expectation. However, this approach does not control the degree sequence exactly, and it is therefore again not applicable to the microcanonical CM.

The only method that specifically addresses the problem of sampling from the microcanonical CM is an algorithm by Viger and Latapy [22, 23] and its predecessor, proposed by Gkantsidis et al. [24]. Starting from an initial connected configuration generated (for example an input graph or constructed with a graphical test [25]), their algorithm proposes local randomizations that preserve connectivity *and* the degree sequence (double edge-swaps [21]). Since certifying connectivity is expensive, they only verify that the graph has remained connected every so often, and backtrack when necessary. They show that a good choice of monitoring interval can speed the computation up by a large constant factor [24].

2 Efficient Heuristic

Our algorithm is most closely related to the methods of Gkantsidist et al. [24] and Viger and Latapy [23]. In fact, it makes use of the same set of basic operations, but we swap their order to obtain a significant speed-up on smaller graphs such that we can quickly sample millions of instances.

To generate a graph with sequence d, we proceed as follows. (see Algorithm 1). We first use the Havel-Hakimi algorithm to create a graph with the appropriate degree sequence [25, 26]. We then shuffle its edges by applying T double edge swaps on random pairs of edges, where T is a tunable parameter. By construction, these swaps preserve the degree sequence and can be shown to generate ergodic and uniform chains over the set of graphs with a fixed degree sequence (with no connectedness constraints) [21]. Note that we disallow self-loops parallel edges—we are interested in the *simple* graph ensemble where these do not occur.

The innovation of our method lies in a final step where we connect our newly-shuffled graph with double edge swaps that connect components. These edge swaps are carried out with edges drawn uniformly at random within two different components. The components are themselves selected at random, with probability proportional to the number of nodes they contain. The algorithm terminates as soon as the graph becomes connected.

Algorithm 1 Connected configuration model

```
1: procedure CONNECTED_CM(DEG_SEQ)
2:     G ← havelhakimi(DEGSEQ)
3:     while swaps < T do
4:         swaps += randomedgeswap(G)
5:     while numberofcomponents(G) > 1 do
6:         c1 ← randomcomponent(G)
7:         c2 ← randomcomponent(G)
8:         e1 ← randomedge(c1)
9:         e2 ← randomedge(c2)
10:        G.swapedges(e1, e2)
```

All the steps of this algorithm are efficient. Constructing the initial graph and ran- domizing it with double edge swaps takes linear time in the number of edges, i.e., the sum of degrees $\sum_i k_i$ [21]. The connection step, step 3, also admits an efficient implementation: We precompute a list of components (say with union-find), and we then update the list every time an edge swap successfully merges two components. Verifying that a swap connects two components is not too costly, because we only need to verify that the swap did not disconnect the nodes from their original components. Altogether, exact sampling of small connected graphs (less than 200 nodes) with method of Viger and Latapy takes at best 50 times longer than with our approximate sampling. On larger graphs (e.g. over a couple of thousand nodes), our sampling heuristic currently requires similar or slightly larger amount of time as exact sampling, because we have yet to optimize our implementation. That said both approaches were found to scale linearly with graph size on perfect trees; we thus expect the reduction in overhead to be worth it in all regimes once optimized for larger graphs.

The price to pay for this speedup is exact uniformity over $\Omega_N(d)$. The proofs of Ref. [23], for example, do not generalize to our method. This is due to the fact that the swaps we make in step 3 are "unidirectional," in the sense that they move us from a space of disconnected graphs towards a space of connected ones, with no possibility of ever backtracking; classical proof techniques, in contrast, demonstrate the ergodicity of *reversible* chains over an ensemble. That said, our heuristic at least guarantees that (a) the degree sequence is preserved and (b) the graph is connected, and it finds these graphs rapidly in most cases. Furthermore, as we now show, a uniformity test and a comparison with the slower (but provably exact) algorithm of Viger and Latapy suggest that our algorithm samples from the target distribution to a close approximation.

As a first verification that our heuristic approximately generates graphs from the correct ensemble, we carry out a simple uniformity test. We consider the short degree sequence $k = (1, 1, 1, 2, 2, 2, 3)$, and enumerate all the labeled connected

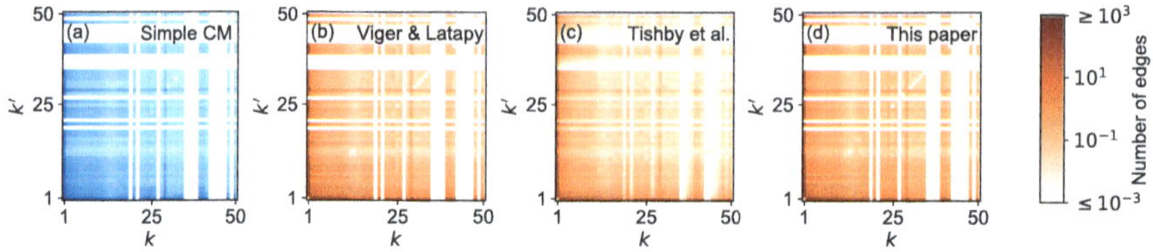

Fig. 1 Expected correlation matrix $\boldsymbol{e}$ for a real network (Norwegian board of directors [27]), as calculated empirically using 1000 draws from: (**a**) the simple CM, (**b**) the exact connected microcanonical CM [22], (**c**) the canonical connected CM [15], and our heuristics. Some features of the exact connected microcanonical sampling are recovered by our heuristics but not by the approach of Ref. [15]. Most relevant to our case studies is the weak diagonal, a signature of the disassortativity (negative degree correlations) of the true connected CM ensemble

Table 1 Frequencies of the isomorphism classes of connected graphs sharing the degree sequence $k = (1, 1, 1, 2, 2, 2, 3)$, compared against their empirical frequencies computed from a 10^6 graph samples

Isomorphism class	Uniform	Viger & Latapy	Our heuristic
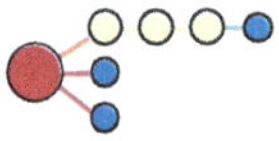	0.3	0.304	0.321
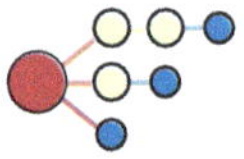	0.6	0.599	0.609
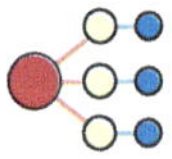	0.1	0.098	0.070

The calculation takes about 90 s with our non-uniform heuristic, versus hours with the exact sampler

graphs that are compatible with it—60 in this case. We then identify all the isomorphism classes up to a relabeling—there are three, see Fig. 1—and count the number of graphs in each class. The frequencies of the isomorphism classes are reported in Table 1. Since the microcanonical connected CM is a uniform distribution over the 60 labeled graphs (and not the isomorphism classes), a useful sampling algorithm should generate graphs with isomorphism classes that closely follow the true frequencies calculated in Table 1. And indeed, we find empirical frequencies that are not exactly equal to the true ones, but the deviations are small.

To further validate our heuristic, we carry out a second test where we randomize the connections of the giant component of a large connected graph. When the graph is large, there are far too many isomorphism classes to calculate their true frequencies, let alone evaluate their empirical frequencies to a reasonable degree of accuracy. Hence, we turn to a different test and analyze the correlations between the neighborhood of nodes, conditioned on their degrees. As we will see later in Sect. 3, this correlation structure intervenes in many of the applications of the connected CM as a null model—it is therefore essential to get it right.

We show the results in Fig. 1, with matrices $\boldsymbol{e}(k, k')$ whose entries $e(k, k')$ are the expected number of edges between nodes of degree k and k'. We use as input the degree sequence of a real network of Norwegian directors sitting on at least one corporate board together [27]. This graph is large, comprising 4475 nodes and 4652 edges, and its degree sequence is non-trivial, with degrees ranging from 1 to 552. We calculate the matrices using four sampling algorithms: The classical double-edge swap algorithm that samples from the simple CM (for the sake of comparison); the algorithm of Viger and Latapy (exact); the algorithm of Tishby et al. (canonical connected CM); and our heuristic. The results confirm that the connected constraints change the correlation structure significantly (compare Fig. 1a, d). They also show that hard constraints on the degree lead to a markedly different correlation matrix (compare Fig. 1b, c). And finally, they demonstrate that our heuristic finds correlations close to the exact ones (compare Fig. 1b, d).

3 On the Impact of Being Connected

3.1 Assortativity in Connected Graphs

Degree correlations aim to capture mixing patterns in a network [28]. Generally, these degree correlations can be captured by "joint-degree measures" such as the probability $\boldsymbol{e}(k, k')$ that a random edge joins nodes of degree k' and k', a measure of correlation we have used in Fig. 1. For any random graph with degree distribution p_k, we can expect the degree of a node at the end of a random edge to be distributed according to $q_k = kp_k/\langle k \rangle$ since a node of degree k participates to k times more edges than a node of degree 1. In a fully random graph with a given degree sequence, i.e., one drawn from the unconnected canonical CM, we could therefore expect $e_{\mathrm{CM}}(k, k') \propto q_j q_{k'}$, but that would assume that the degrees of neighboring nodes are uncorrelated. To correct for possible deviations from the CM one can therefore measure $e(k, k')$ from real datasets.

Edge matrices like $\boldsymbol{e}(k, k')$ can be used to parametrize random graph ensembles, as done in some degree-correlated version of the Configuration Model [10]. In practice, it is more parsimonious to coarse-grain this information and only specify a measure of correlation, the assortativity coefficient [28]. For a graph G, we write

$$r_{\mathrm{CM}}(G) = \frac{1}{\sigma_q^2} \sum_{k,k'} kk' \left[e(k, k') - q_k q_{k'} \right], \tag{2}$$

where $\sigma_q^2 = \sum k^2 q_k - \left(\sum k q_k\right)^2$ is the variance of the distribution q_k. Importantly, the negative term in Eq. (2) corresponds to correlations we would expect in the CM, meaning that we control for correlations that would emerge naturally given only the degree sequence of the graph. In other words, we control for our expectations from the CM.

In real graphs we typically do not expect $r = 0$ which would correspond to a typical graph drawn from the CM. Instead, as a general rule of thumb [28, 29], it is typically accepted that social graphs tend to display assortative mixing, or positive degree correlations $r > 0$, where high degree nodes tend to more connected than you would expect at random Conversely, technological graphs tend to display diassortative mixing, or negative degree correlations $r < 0$, where high degree nodes connect to low degree nodes more often than expected at random.

Technological and social datasets are also typically collected in very different ways. As mentioned in our introduction, some data on technological graphs, such as the structure of the Internet or power grids, tend to be connected per definition. For example, a power grid with two disconnected components would typically be considered as two distinct graphs. In other cases, like subsets of the World-Wide Web, data tends to produce connected graphs because they are often collected by crawling edges.

Given conventional wisdom on assortative mixing in complex graphs and on their different expected connectedness across domains, we ask: How does connectedness affect assortativity? In Fig. 2 we illustrate how connectivity might matter, using two trees as a simple example. Controlling for the wrong expectation means we might find a signal where there is none.

To study the interplay between assortativity and connectedness in real graphs, we calculate a new assortativity coefficient based on the connected CM rather than the typical CM null model. Because of the linearity of Eq. (2) we can write

$$r_{\mathrm{CCM}}(G) = r_{\mathrm{CM}}(G) - \langle r_{\mathrm{CM}}(G_{\mathrm{CCM}})\rangle \tag{3}$$

where we are controlling for our expectations using the connected CM (right-hand side), by measuring the classic assortativity coefficient (first term of left-hand side) and subtracting deviations between the new and old null models (second term of right-hand side). The final term is therefore the classic assortativity coefficients averaged over many graphs drawn from our sampling algorithm for the connected CM.

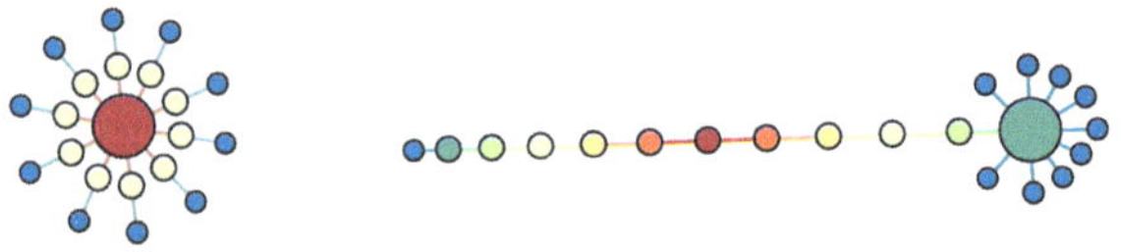

Fig. 2 Two simple examples of connected trees with the same degree sequence. The most assortative this degree sequence be, while remaining connected, is as a stretched star (left graph) with assortativity coefficient of $r = -0.23$. The most disassortative this degree sequence can be is as a star combined with a single chain (right graph), leading to $r = -0.50$. Without looking at the graph or degree sequence in detail, one might naturally conclude that both of these graphs are strongly diassortative. The disassortativity of trees stems from their high number of leaves that cannot be connected without disconnecting the graph

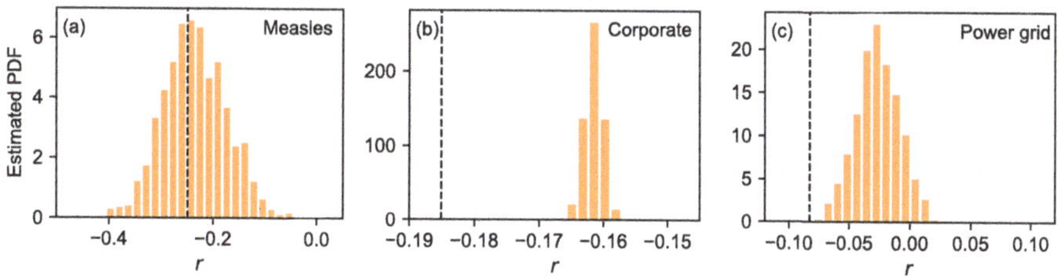

Fig. 3 Distribution of the degree correlation, for random graphs drawn from the connected CM, with degree sequences taken from three real graphs of increasing densities. (**a**) Phylogenetic tree of the Measles [30], of mean degree $\langle k \rangle = 1.99$, with 175 nodes and 174 edges. (**b**) Graphs of Norwegian directors sitting on at least one board together [27], of mean degree $\langle k \rangle = 2.08$, with 4475 nodes and 4652 edges. (**c**) Power grid of Poland during the winter [31], of mean degree $\langle k \rangle = 2.42$, with 2383 nodes and 2886 edges. The classic assortativity coefficients are shown as vertical lines at (**a**) −0.249, (**b**) −0.185, and (**c**) −0.0827; and become, under the connected CM, (**a**) −0.013, (**b**) −0.023 and (**c**) −0.056

We calculate the distribution of this corrected coefficient for a few real connected graphs in Fig. 3. As expected, on real trees such as the phylogenetic tree of the measles virus (Fig. 2a), we find strong disassortativity based on the classic assortativity coefficient $r_CM(G)$ (shown as a vertical line). This disassortativity, however, is not surprising when we compare it to the distribution of assortativity found for graphs generated from the correct, connected, null model (histogram). In sparse social systems with few loops (Fig. 2b), such as a corporate graph of board directors used in our example of Fig. 1, we similarly remove most of the signal when using the connected null model even if that graph can still be rejected as not being a typical instance of the connected CM. Finally, in denser cases such as power grids (Fig. 2c), we would conclude that the graph structure is disassortative by using either the CM or connected CM, because the null distribution is centered around $r = 0$ for the connected CM (and because the classical assortativity effectively compares the assortativity of real graphs to a distribution centered on $r = 0$).

3.2 Community Detection in Connected Graphs

Community detection [32]—and more generally mesoscopic pattern extraction [33]—refers to a wide variety of methods whose goal is to find structurally similar groups of nodes in a network, given only the structure of the network as input. Formally, these methods find assignments of the nodes of G to K communities of similar nodes, assigning precisely one community σ_i to each node $i = 1, \ldots, N$. They are among the most useful methods of network science, because communities can help us understand networks at the exploratory data analysis stage, act as the input of other network analysis methods [34], or even help us identify the fundamental building blocks of networks [35].

Modularity maximization methods were among the first proposed community detec- tion algorithms, and have since gained prominence among practitioners because of their clarity and ease-of-use [32, 36, 37]. These methods use a specific form of objective functions—modularities $Q(\sigma)$—to quantify the quality of partitions. The most general modularity function can be written as [38]

$$Q_P(\sigma) = \frac{1}{m} \sum_{i<j} \left(a_{ij} - \langle a_{ij} \rangle_P \right) \delta_{\sigma_i \sigma_j}, \tag{4}$$

where m is the number of edges, where the sum runs over all pairs of nodes, and where $\delta_{\sigma_i \sigma_j}$ is the Kronecker delta, equal to 1 if $\sigma_i = \sigma_j$ (i.e., node i and j are in the same community) and to zero otherwise. The two terms of Eq. (4) denote, respectively: Whether there is an edge between nodes i and j and the expected number of edges between the node i and j under the null model P, noted as $\langle a_{ij} \rangle_P$. According to this equation, a partition is deemed good if there are many more edges connecting nodes that are inside the same community than what we would expect to observe by chance, *given a null model P* for the graph. A specific modularity maximization *algorithm* consists of a particular choice of null model P and of maximizer[2] [33]—a search strategy for the space of possible partitions that converges to the partition with maximal modularity.

The canonical (unconnected) CM is almost always used as the null model P [36, 37], in part because the expectation $\langle a_{ij} \rangle$ is then given by the simple formula

$$\langle a_{ij} \rangle_{\mathrm{CM}} = \frac{k_i \, k_j}{2m}, \tag{5}$$

where k_i is the degree of node i and $m = \frac{1}{2} \sum_i k_i$ is the total number of edges. But, as we have already argued, this null model is not always the most natural choice [21]. In fact, it is known that different choices of model will tend to resolve different types of communities [33, 41], with no obvious optimal choice on all inputs [42]. Futhermore, there are known applications where switching to a model P tailored to the class of graphs at hand leads more accurate and relevant inference results [21, 43]. So we ask: How do modularity-based algorithms behave when we choose the *connected* ensemble of graphs with fixed degree sequence as our null model?

To answer this question, we need to evaluate the average $\langle a_{ij} \rangle_{\mathrm{CCM}}$ appearing in Eq. (4), with the connected CM as the null model. Just as in Sect. 3.1, we opt for a numerical average, computed with the efficient heuristic introduced in Sect. 2. If there are on average $e(k, k')$ edges connecting nodes of degrees k and k', then the expected number of edges between two nodes i and j of degree k_i and k_j is given by

[2] Many standard optimization methods are capable of fulfilling the role of optimizer: Spectral embedding [37, 39] and greedy maximization [40] are well-known examples.

$$\langle a_{ij}\rangle_{\text{CCM}} = \begin{cases} e(k_i, k_j)/n_{k_i} n_{k_j} & \text{if } i \neq j \text{ and } k_i \neq k_j \\ e(k_i, k_j)/n_{k_i}(n_{k_i} - 1) & \text{if } i \neq j \text{ and } k_i = k_j \\ 0 & \text{otherwise,} \end{cases} \quad (6)$$

where n_k is the number of nodes of degree k in the graph, and where the second expression accounts for the fact that there are no self-loops in the ensemble.

With Eq. (6) for $\langle a_{ij}\rangle_{\text{CCM}}$, we can readily evaluate the connected modularity of the different partitions of a network, and run optimization algorithms to find the best one. The number of communities K is not known a priori, so we run a double optimization[3] where we first find the best partition $\sigma^*(K)$ for many choices of K, and then find the K^* for which this optimal partition has the largest modularity. The results of one such experiment are shown in Fig. 4, where we find the communities of the network of directors (see also Figs. 1 and 3), using the standard modularity Q_{CM} (see Eq. (5)) and the connected modularity Q_{CCM} (see Eq. (6)).

We find a few differences between the analysis ran with the unconstrained and constrained CM as null model. Perhaps most noticeable is the fact that the optimal connected modularity stays close to its maximum for a much larger range of K than the standard modularity (Fig. 4a). This suggests that the number of communities is not as clearly defined once we account for connectivity. Furthermore the optimal partition $\sigma^*(K^*)$ under the connected CM occurs at a slightly larger number of communities K, which allows the algorithm to (correctly) resolve a few more communities (see Fig 4b, c). When we actually inspect these best partitions, however, we find that communities essentially consists of one node of high degree and its degree one neighbors—regardless of the choice of null model. This is a consequence of the fact that placing a node of degree one and its neighbor in the same communities is always good when the networks are near-trees. The two models do not put the exact same penalty on these connections, but the difference is not large enough to alter the optimal clustering choice significantly. And as a result, the communities are qualitatively unchanged when we switch the null model from the CM to the connected CM (the normalized reduced mutual information (nRMI) [46] of the optimal partitions is 0.93, while comparing the best partitions of size K^* yield nRMI $= 0.95$ ($K^* = 80$) and nRMI $= 0.98$ ($K^* = 106$), on a scale of 0–1).

3.3 *Robustness of Connected Graphs*

Percolation is a simple stochastic process in which edges or nodes are randomly removed from an existing network. It is an obvious model for the robustness of

[3]This strategy works thanks to the regularization properties of the modularity: The partition in a single community ($K = 1$) has zero modularity due to the negative contributions of the null model, and so does the partition in $K = n$ communities of 1 node since no two nodes share a community. It follows that there is an optimum K^* somewhere in between these two extremes.

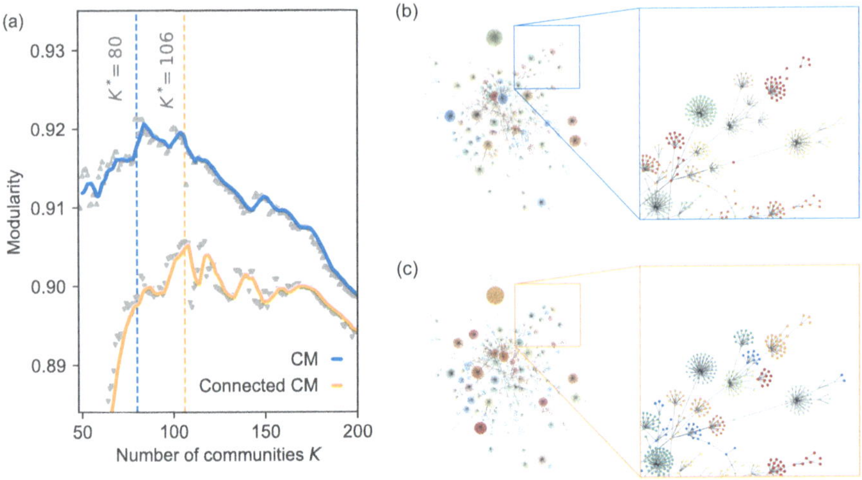

Fig. 4 Outcome of modularity maximization on the network of directors. (**a**) Modularity of the best partition in K communities found by using the CM and the connected CM as null models. We add a moving window average to guide the eye. The optima are identified with vertical lines. (**b**–**c**) Visualization [44] of the optimal partitions when we use as null model (**b**) the CM, and (**c**) the connected CM. The community of a node is indicated by a non-unique combination of node color and shape (but no two adjacent communities are shown with the same combination). These partitions are found as follows. For a fixed K, we first embed the nodes of G in $\mathbb{R}^K$, using the K largest eigenvectors of the modularity matrix $\boldsymbol{B} = \boldsymbol{A} - \langle \boldsymbol{A} \rangle_P$ [39], defined as the difference between the adjacency matrix $\boldsymbol{A}$ and the average adjacency matrix $\langle \boldsymbol{A} \rangle_P$ under null model P. We then create many candidate partitions by running agglomerative clustering on the embedded nodes [45], for various choices of affinity (ℓ_2 norm and cosine similarity) and linkage strategy. We finally select the candidate partition that maximizes the modularity as the optimum for that K. The partitions shown in (**b**–**c**) are the ones that maximize the modularity over all K

graphs to random failures, but also useful to study dynamical processes such as epidemics [47]. While one can certainly simulate percolation on a specific real graph to study its robustness or its ability to sustain an epidemics, it is often instructive to also study the percolation process on a series of random graph ensembles, because this lets us evaluate the impact of different structural properties on the outcome. Hence for example, an epidemic on a real graph could be first compared to the outcome of the percolation on Erdős–Rényi graphs with the same density, to evaluate the impact of density alone; then compared to the outcome of percolation on graphs drawn from the CM, to evaluate the impact of the contact distribution; and so on for higher order models.

Enforcing connectedness can, again, make a big difference on the outcome of percolation. To show this, we consider bond percolation where a fraction $p \in [0, 1]$ of edges are randomly removed from a network instance. We compute the size of the largest connected component (LCC) after this removal has taken place, and investigate its dependency on p. The original connectedness of a real system is

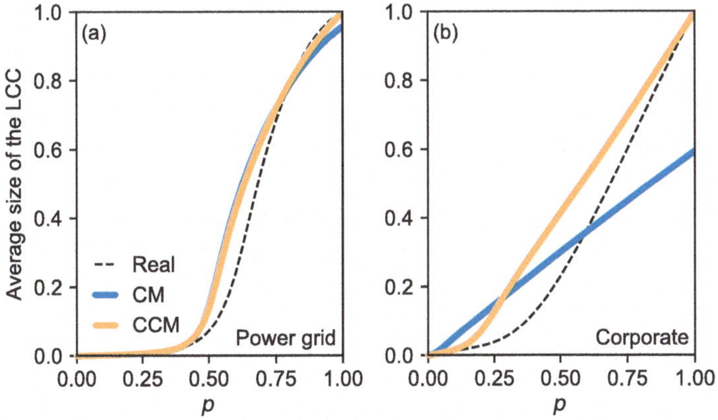

Fig. 5 Percolation on the CM and connected CM ensembles with degree sequences taken from real systems, compared with percolation on these systems. The size of the largest connected component (LLC) is shown as a function of the fraction of removed edges p

critical to account for, since the size of the LCC *after* some edges are removed is obviously bounded by the size of the LCC *before* this removal processes.

In Fig. 5, we show the outcome of bond percolation for two graphs studied in previous case studies: The social network of board directors sitting on common boards of Norwegian public limited companies, and the structure of a Polish power grid. In the first, a percolation model can be used to study pathways for information flows; and in the second, a naive model of robustness to failing power lines. We find that the connected CM delays the onset of the connected phase (i.e. the LCC starts growing at higher occupation probabilities) and forces fixation to full connectivity as the occupation probability goes to one. The delayed onset of the connected phase is most likely a consequence of the disassortativity of connected treelike graphs. The full connectivity at high occupation probability is a trivial consequence of the ensembles containing only connected graphs. Perhaps more interestingly, the connected CM also captures the convex relationship between the size of the LCC and the occupation probability at the onset of connectedness. This convex relationship is often a consequence of core–peri- phery structure, both in dense and sparse graphs, as shown in Refs. [11, 12]. Our results here show that the degree sequence and the connectedness can also explain this feature, and more parsimoniously so.

4 Discussion

In this paper, we have proposed an efficient heuristic that can generate samples from the connected configuration model with hard degree constraints. We have shown that this heuristic is fast on small graphs, and that it always returns a graph as long as the input degree sequence is graphical. By way of three case studies, we have then demonstrated how this algorithm can be used to analyze real systems where connectedness is key. In doing so, we have found, for example, that while graphs with a low density can seem disassortative at first glance, their disassortativity can be explained by the fact that they have to be connected in the first place. In the same manner, we have also found that connectivity can alter the communities found by modularity maximization algorithms or the interpretation of a percolation process.

Throughout the second part of this paper, we have found time and time again that connectedness constraints matter the most when the density of the modeled system is low. And this was, in a sense, expected. After all, two classical results from graph theory show that a high average degree automatically implies connectedness for a large portion of the nodes of a completely random graph [2] as well as for random graphs with fixed degree sequence [5]. Or in other words: Dense random graphs are already connected, whether we ask them to be or not. It follows that imposing connectedness is most important in the regime of sparse graphs. When modeling a sparse system, therefore, we should take special care and ask ourselves: Was this graph expected to be connected? Is there a sampling or construction mechanism that forces us to observe this graph in one piece? These questions help guide our choice of null model. And as we have shown, this choice can certainly change our conclusions—so beware.

Acknowledgements This work is supported by the James S. McDonnell Foundation (JGY) and Grant No. DMS-1622390 from the National Science Foundation (LHD). The authors contributed equally.

References

1. Newman, M.E.J.: Networks: An Introduction. Oxford University Press, Oxford (2010)
2. Erdős, P., Rényi, A.: Publ. Math. **6**, 290 (1959)
3. Gilbert, E.N.: Ann. Math. Stat. **30**(4), 1141 (1959)
4. Connor, E.F., Simberloff, D.: Ecology **60**(6), 1132 (1979)
5. Molloy, M., Reed, B.A., Random Struct. Algoritm. **6**(2/3), 161 (1995). https://doi.org/10.1002/rsa.3240060204
6. Newman, M.E.J., Strogatz, S.H., Watts, D.J.: Phys. Rev. E **64**, 026118 (2001). https://doi.org/10.1103/PhysRevE.64.026118
7. Coolen, T., Annibale, A., Roberts, E.: Generating Random Networks and Graphs. Oxford University Press, Oxford (2017)
8. Frank, O., Strauss, D.: J. Am. Stat. Assoc. **81**(395), 832 (1986)
9. Olhede, S.C., Wolfe, P.J.: Proc. Natl. Acad. Sci. U. S. A. **111**(41), 14722 (2014)
10. Vázquez, A., Moreno, Y.: Phys. Rev. E **67**(1), 015101 (2003)
11. Hébert-Dufresne, L., Allard, A., Young, J.G., Dubé, L.J.: Phys. Rev. E **88**(6), 062820 (2013)
12. Allard, A., Hébert-Dufresne, L.: Phys. Rev. X **9**(1), 011023 (2019)
13. Orsini, C., Dankulov, M.M., Colomer-de Simón, P., Jamakovic, A., Mahadevan, P., Vahdat, A., Bassler, K.E., Toroczkai, Z., Boguná, M., Caldarelli, G., et al.: Nat. Commun. **6**, 8627 (2015)
14. Tishby, I., Biham, O., Katzav, E., Kühn, R.: Phys. Rev. E **97**(4), 042318 (2018)
15. Tishby, I., Biham, O., Katzav, E., Kühn, R.: (2018). arXiv:1810.02198

16. Gray, C., Mitchell, L., Roughan, M.: J. Complex Netw. **7**(6), 896–912 (2019). https://doi.org/10.1093/comnet/cnz011
17. Crane, H.: Probabilistic Foundations of Statistical Network Analysis. Chapman and Hall/CRC, Boca Raton (2018)
18. Barabási, A.L., Albert, R.: Science **286**(5439), 509 (1999)
19. Erickson, B.H.: Sociol. Methodol. **10**, 276 (1979)
20. Lee, S.H., Kim, P.J., Jeong, H.: Phys. Rev. E **73**(1), 016102 (2006)
21. Fosdick, B.K., Larremore, D.B., Nishimura, J., Ugander, J.: SIAM Rev. **60**(2), 315 (2018)
22. Viger, F., Latapy, M.: (2005). arXiv:cs/0502085
23. Viger, F., Latapy, M.: J. Complex Netw. **4**(1), 15 (2015)
24. Gkantsidist, C., Mihail, M., Zegura, E.: In: Proceedings of the Fifth Workshop on Algorithm Engineering and Experiments, vol. 111, p. 16. SIAM, Philadelphia (2003)
25. Havel, V.: Casopis Pest. Mat. **80**, 477 (1955)
26. Hakimi, S.L.: J. Soc. Ind. Appl. Math. **10**(3), 496 (1962). https://doi.org/10.1137/0110037
27. Seierstad, C., Opsahl, T.: Scand. J. Manag. **27**(1), 44 (2011)
28. Newman, M.E.J.: Phys. Rev. Lett. **89**(20), 208701 (2002)
29. Newman, M.E., Park, J.: Phys. Rev. E **68**(3), 036122 (2003)
30. Hadfield, J., Megill, C., Bell, S.M., Huddleston, J., Potter, B., Callender, C., Sagulenko, P., Bedford, T., Neher, R.A.: Bioinformatics **34**(23), 4121 (2018)
31. Zimmerman, R.D., Murillo-Sánchez, C.E., Thomas, R.J.: IEEE Trans. Power Syst. **26**(1), 12 (2010)
32. Fortunato, S., Hric, D., Phys. Rep. **659**, 1 (2016)
33. Young, J.G., St-Onge, G., Desrosiers, P., Dubé, L.J.: Phys. Rev. E **98**(3), 032309 (2018)
34. Hébert-Dufresne, L., Allard, A., Young, J.G., Dubé, L.J.: Sci. Rep. **3**, 2171 (2013)
35. Riolo, M.A., Newman, M.: (2019). arXiv:1908.09867
36. Newman, M.E.J., Girvan, M.: Phys. Rev. E **69**(2), 026113 (2004)
37. Newman, M.E.: Proc. Natl. Acad. Sci. U. S. A. **103**(23), 8577 (2006)
38. Reichardt, J., Bornholdt, S.: Phys. Rev. E **74**(1), 016110 (2006)
39. White, S., Smyth, P.: In: Proceedings of the 2005 SIAM International Conference on Data Mining, pp. 274–285. SIAM, Philadelphia (2005)
40. Blondel, V.D., Guillaume, J.L., Lambiotte, R., Lefebvre, E.: J. Stat. Mech. Theory Exp. **2008**(10), P10008 (2008)
41. Traag, V.A., Van Dooren, P., Nesterov, Y.: Phys. Rev. E **84**(1), 016114 (2011)
42. Peel, L., Larremore, D.B., Clauset, A.: Sci. Adv. **3**(5), e1602548 (2017)
43. Expert, P., Evans, T.S., Blondel, V.D., Lambiotte, R.: Proc. Natl. Acad. Sci. U. S. A. **108**(19), 7663 (2011)
44. Peixoto, T.P.: Figshare (2014)
45. Pedregosa, F., Varoquaux, G., Gramfort, A., Michel, V., Thirion, B., Grisel, O., Blondel, M., Prettenhofer, P., Weiss, R., Dubourg, V., et al.: J. Mach. Learn. Res. **12**, 2825 (2011)
46. Newman, M., Cantwell, G.T., Young, J.G.: (2019). arXiv:1907.12581
47. Newman, M.E.J.: Phys. Rev. E **66**(1), 016128 (2002)

An Allometric Scaling for the Number of Representative Nodes in Social Networks

Liang Zhao and Tianyi Peng

Abstract This paper studies the scale of the size of a representative node set in social networks. First, a simple distance-based representative model is proposed. Then, with two small-world like assumptions which are widely observed in large-scale online social networks, it is shown that the size R of such a representative set satisfies an allometric scaling $R \propto n^{\gamma}$, where n is the size of the network and γ is a constant such that $0 \leq \gamma < 1$. In particular, a theoretical analysis using Dunbar's Number as the average degree of nodes suggests $1/3 \leq \gamma \leq 5/9$ for large-scale real social networks. This is the *first* theoretical model that can explain the phenomenon that the number of congressional representatives scales to the $\frac{2}{5}$-th power (i.e., $\gamma = 2/5$) of the population in real world. It also suggests that, in order to represent (or to influence) a majority in a social network, a surprisingly small (sublinear) number of representatives is enough. For instance, the number is a few thousands for Facebook which has more than two billions users. This demonstrates how easy to spread information in social networks.

1 Introduction

In real world, we often want to know the size of an appropriate representative node set for a network. For example, assuming Facebook wants to have a global users' meeting, it must be first decided how many representatives should be invited.

The original version of this chapter was revised: Second author's name has been updated. A correction to this chapter is available at https://doi.org/10.1007/978-3-030-38965-9_19

L. Zhao (✉)
Graduate School of Advanced Integrated Studies in Human Survivability (Shishu-Kan), Kyoto University, Kyoto, Japan
e-mail: liangzhao@acm.org

T. Peng
Lab for Information and Decision Systems, Massachusetts Institute of Technology, Cambridge, MA, USA

N. Masuda et al. (eds.), *Proceedings of NetSci-X 2020: Sixth International Winter School and Conference on Network Science*, Springer Proceedings in Complexity,
https://doi.org/10.1007/978-3-030-38965-9_4

In general there is a trade-off on this number: more representatives means more democratic but less efficient. The question then asks, assuming there is no physical limitations, *what is an appropriate number of representatives*?

This question arises naturally in social networks. Another, more important example is the number of congressional representatives, which has attracted numerous studies for many centuries. For instance, James Madison, father of the Constitution, stated the question (the trade-off) as the following:

> However small the Republic may be, the Representatives must be raised to a certain number, in order to guard against the cabals of a few; and however large it may be, they must be divided to certain number, in order to guard against the confusion of a multitude.
>
> — The Federalist, No. 10 (1788)

Notice that the social network in the congressional representatives case is *hidden*—we know the existence of a social network connecting us and it plays an important role in election but we do not know its topology, except for some structures like *small-world* and *scale-free*. In fact, the same is true for Facebook and other social networks—usually their topology data is not available or closed.

The missing of exact topology data drives us to *non-algorithmic* studies. Let us first review traditional studies for the congressional case, then consider the general case. Let n denote the population and R be the size of congressional representatives. There exist two types of theories in the literature of politics:

1. **Cube Root theory**: $R = n^{1/3}$. This scale of $1/3$ was statistically derived from real congressional data and population (see [4, 11, 21, 22]).
2. **Square Root theory**: $R \propto n^{1/2}$. This $1/2$ scale was obtained by using models from game theory ([18], well-known as the *Penrose's law*), or from the perspective of both economics and game theory (see [1, 8]).

Unfortunately, these theories do not work well. Cube Root matches real data better than Square Root but it has no guarantee—in fact, latest study shows that $2/5$ is a better scale [1]. On the other hand, Square Root has guarantee in theory, but it does not match real congressional data [1].

The reason why they do not work well is simple: they have no consideration on the structure of the hidden social network. In other words, if we could utilize the structure of the social network, we may find a better theory on the number of congressional representatives, appropriate representatives (or influencers) for Facebook, etc. This was the motivation of this study.

In this paper, we introduce a simple, distance-based representing model and provide a non-algorithmic study[1] to reveal a better scale on the size of representative set from the network aspect. Speaking precisely, we show $R \propto n^{\gamma}$ for some constant γ ($0 \leq \gamma < 1$) independent of n, under some small-world like assumptions which are widely observed in large-scale online social networks, For large-scale

[1]We remark that it is possible to calculate a representative node set with respect to our model for given networks. Detail on the algorithm is omitted due to page limit.

real social networks, we show a better estimation $1/3 \leq \gamma \leq 5/9$ by using the Dunbar's Number [6, 7] as the average degree of nodes. This is the *first* theoretical model that explains the phenomenon $R \propto n^{2/5}$ on the scale of the congressional representatives ([1], see also Sect. 4). It also suggests that, in order to represent (or influence) a majority in a social network, a surprisingly small (sublinear) number of representatives is enough. For instance, it is a few thousands for Facebook which has more than 2 billions users. This demonstrates how easy to spread information in social networks.

The rest of the paper is organized as follows. We first discuss some related works in the rest of this section, then provide a preliminary in Sect. 2. The model including a pyramid structure of social networks and a theoretical analysis on the scale are provided in Sect. 3 with an empirical study in Sect. 4. Finally we conclude and show future works in Sect. 5.

Related Works

As a related work, Kempe et al. [12] considered an *Influence Maximization* (IM) problem which, given an integer K, asks to find a seed set S of K nodes that maximizes the influence of S with respect to some diffusion model. For example, *Independent Cascade* (IC) is a classic diffusion model in which a node v is activated (i.e., influenced) by its incoming neighbors w independently by *given* probabilities $p_{w,v}$. Another *Linear Threshold* (LT) model, on the other hand, activates a node if sufficient number of its incoming neighbors are active. See [12, 14, 23] for more detail on the IM approach.

Unfortunately, this IM approach does not work for our problem because it is algorithmic, hence cannot provide us a scale on the size of the seed set and the size of its influence domain (it provides nothing until the algorithm terminates). Other difficulties in applying the IM approach to representatives include designing of *fixed* probabilities $p_{w,v}$, the time complexity (see [14, Table 1]) and, above all, hidden network topology.

Other related topics include the *dominating set* problem (see [9]) and the so-called hierarchical structure of networks [5, 15, 16, 19]. In particular, our pyramid structure (see Sect. 3) is a growing hierarchical structure. Nevertheless, existing studies could not provide an answer to our question.

2 Preliminary

Let $G = (V, E)$ be a network with a set V of n nodes and a set E of m edges. We use $dist(u, v)$ to denote the distance between two nodes u and v, i.e., the minimum number of edges required to connect u and v (called the *geodesic distance* in [17]). For a set W of nodes and an integer $k \geq 0$, let

$$\mathcal{N}_k(W) = \{v \in V \mid \exists w \in W, dist(w, v) \leq k\} \quad (1)$$

denote the set of nodes that can be reached from some node in W via at most k edges. Thus $\mathcal{N}_0(W) = W$, and $\mathcal{N}_k(W) \setminus \mathcal{N}_{k-1}(W)$ is the neighbor nodes of $\mathcal{N}_{k-1}(W)$ for $k \geq 1$. If W consists of only one node v, we may write $\mathcal{N}_k(v)$ instead of $\mathcal{N}_k(\{v\})$ for simplicity. Let $|\cdot|$ denote the size of a set. We first define a *democratic index* κ for G by the following equation:

$$\kappa = \max\{k \geq 0 \mid \forall v \in V, |\mathcal{N}_k(v)| < n/2\}. \tag{2}$$

Index κ measures the difficulty to represent (or influence) a majority of a network by a *single* node. Clearly it is also the *smallest* k such that $|\mathcal{N}_{k+1}(v)| \geq n/2$ for some v. For instance, $\kappa = 0$ for cliques and stars showing it is easy to represent (influence) a majority in cliques and stars. On the other hand, $\kappa = n/4$ for a path network showing it is difficult for some single node to represent (influence) a majority in a path network. See an illustration in Fig. 1.

Next, let us define the *democratic number* λ—which will be the proposed appropriate number of representative nodes—by

$$\lambda = \min\{|W| : |\mathcal{N}_\kappa(W)| \geq n/2\}, \tag{3}$$

i.e., the minimum size of a node set that can dominate a majority of nodes with distance κ (notice $\lambda \geq 2$ by the definition of κ). The definitions Equations (2) and (3) were first proposed by the first author in [24] who also did empirical studies on a number of real networks in the SNAP datasets [13]. The result showed a surprising phenomenon $\kappa = 2$ for many large-scale social networks (Table 1), implying that in real world, it is much easier to represent (influence) a majority than we thought.[2]

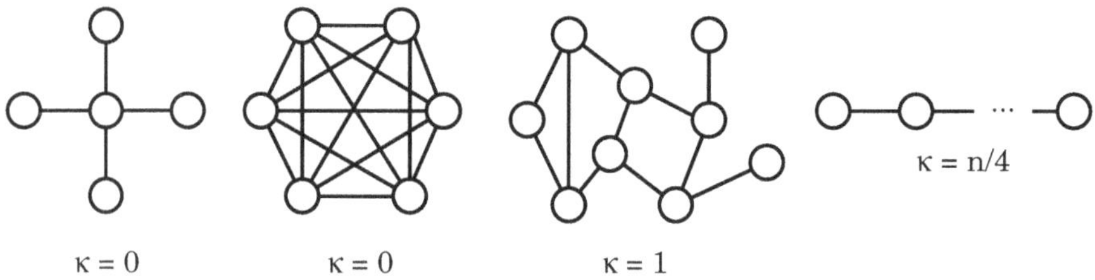

Fig. 1 Illustration of the democratic index κ which measures the difficulty to represent (influence, control, spread information to, etc.) a majority of nodes by a single node

[2]We remark that κ can be found in polynomial time. In fact, running breadth-first searches (see [17], Section 10.3.5) for all nodes can find the answer in $O(mn + n^2)$ time [24]. The study [24] used a faster subroutine in [25]. On the other hand, similar to the IM problem, finding λ for a given network and $\kappa \geq 1$ is NP-hard [25].

3 Model

With the definitions of κ and λ, let us make some assumptions on the structure of social networks then consider the size of a representative set.

3.1 Conjecture 1 (Small-World Phenomenon on κ): $\kappa \propto \log n$

We first conjecture $\kappa \propto \log n$ for social networks (or more generally, for small-world networks). This is reasonable as social networks are usually small-world and κ must not be a constant yet cannot be larger than the diameter (i.e., the average distance between two nodes) $D \propto \log n$.

For exact values for real networks, the first author calculated κ for some social networks in SNAP datasets (Table 1). Notice the estimated diameters D and 90-percentile effective diameter D_{90} are quite different from instances, whereas $\kappa = 2$ holds for all large-scale networks studied. For even more larger networks, we have no data so we cannot provide an exact result but since the average distance of Facebook was estimated to be 4.57 in 2016 ([3], with 1.59 billion active users), it is highly possible that $\kappa = 2$ also holds for Facebook.

3.2 Conjecture 2: Existence of Pyramid Structure of Representing

We next conjecture *that there exists a constant $c > 1$, independent of W and n, such that it is highly possible for a set $W \subseteq V$ of small (sublinear) size, $|\mathcal{N}_k(W)| \propto |W|c^k$ holds for $k = 0, 1, \ldots, \kappa$.*

This assumption is natural for *efficient* representing, since each node in W, on average, contributes to an exponential growing of the influence (estimated by $|\mathcal{N}_k(W)|$) with respect to k, i.e., $|\mathcal{N}_k(W)|/|W| \propto c^k$. It is equivalent to the existence of a pyramid (i.e., a growing hierarchy) structure based on distances from W. See

Table 1 Index κ for some social networks in the SNAP datasets [13], where D is the estimated diameter and D_{90} is the estimated 90-percentile effective diameter

Network	n	m	D	D_{90}	κ
ego-Facebook	4039	88,234	8	4.7	1
loc-Gowalla	196,591	950,327	14	5.7	2
com-Youtube	1,134,890	2,987,624	20	6.5	2
com-Orkut	3,072,441	117,185,083	9	4.8	2
com-Friendster	65,608,366	1,806,067,135	32	5.8	2

Values κ are from [24]. Notice that Index κ is much more stable that D and D_{90}

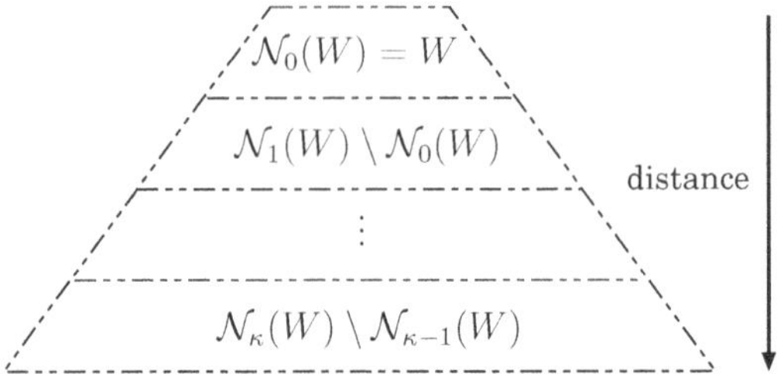

Fig. 2 An illustration of pyramid structure of nodes based on distances from a seed set W. From the upper layer, the i-th layer consists of nodes of distance $i-1$ from W (therefore, while omitted in the illustration, edges are available between two adjacent layers or inside the same layer). It is called pyramid since $|\mathcal{N}_k(W)| \propto |W|c^k$ for some constant $c > 1$, which is equivalent to $|\mathcal{N}_k(W) \setminus \mathcal{N}_{k-1}(W)| \approx c|\mathcal{N}_{k-1}(W) \setminus \mathcal{N}_{k-2}(W)|$

an illustration in Fig. 2. We remark that similar assumption has been widely used so far, e.g., in Barabasi [2, (Section 3.8)] and Newman [17, (Chapter 12.5)] where $|W| = 1$ and c is the average degree. Our assumption is somewhat stronger than them (as $|W| > 1$). For that purpose, we will provide an empirical study in Sect. 4.

3.3 An Allometry on the Size of Representative Nodes

Now we consider the (scale of the) size of representative nodes using the above two conjectures. We claim that λ, which we use as the appropriate size of a representative set, satisfies $\lambda \propto n^\gamma$ for some γ independent of n, $0 \le \gamma < 1$.

This can be roughly observed as follows. Firstly, Conjecture 1 $\kappa \propto \log n$ implies the existence of a constant $\alpha > 0$ such that $\kappa = \alpha \log n$. Then, let $\lambda = |W^*|$. Hence $|\mathcal{N}_\kappa(W^*)| \ge n/2$. By Conjecture 2 $|\mathcal{N}_\kappa(W^*)| \propto |W^*|c^\kappa$, we have

$$\lambda = |W^*| \propto \frac{|\mathcal{N}_\kappa(W^*)|}{c^\kappa} \propto \frac{n}{c^\kappa} = \frac{n}{c^{\alpha \log n}} = \frac{n}{(c^{\log n})^\alpha} = \frac{n}{(n^{\log c})^\alpha} = n^{1-\alpha \log c}.$$

Let $\gamma = \max\{0,\ 1 - \alpha \log c\}$. Since $\alpha > 0$ and $c > 1$, it is clear that $0 \le \gamma < 1$, showing the claimed argument.

The above analysis is not strict. In the following we give a strict analysis. We suppose $\kappa = \Theta(\log n)$, i.e., there exist constants α_1, α_2, $0 < \alpha_1 \le \alpha_2$, such that $\alpha_1 \log n \le \kappa \le \alpha_2 \log n$. This is a strict statement for Conjecture 1. We also suppose $|\mathcal{N}_k(W^*)| = \Theta(|W^*|c^k)$, i.e., there exist constants β_1, β_2, $0 < \beta_1 \le \beta_2$, such that $\beta_1 \le |\mathcal{N}_\kappa(W^*)|/(|W^*|c^\kappa) \le \beta_2$. This is a strict statement for Conjecture 2 (notice that we are only interested in W^* and $k = \kappa$). Then we can have the next theorem.

Theorem 1 $\exists c_1 > 0$, $c_2 > 0$, $0 \le \gamma_1 \le \gamma_2 < 1$, *s.t.* $c_1 n^{\gamma_1} \le \lambda = |W^*| \le c_2 n^{\gamma_2}$.

Proof Let $\alpha = \kappa/\log n$. By the above assumption, $0 < \alpha_1 \le \alpha \le \alpha_2$. Let $\beta = |\mathcal{N}_\kappa(W^*)|/(|W^*|c^\kappa)$. We have $0 < \beta_1 \le \beta \le \beta_2$. By the definition of W^*,

$$\frac{n}{2\beta c^\kappa} \le \lambda = |W^*| \le \frac{cn}{2\beta c^\kappa}.$$

Since $c^\kappa = c^{\alpha \log n} = n^{\alpha \log c}$, we have

$$\frac{1}{2\beta} n^{1-\alpha \log c} \le \lambda \le \frac{c}{2\beta} n^{1-\alpha \log c}.$$

Let $c_1 = 1/(2\beta_2)$, $c_2 = c/(2\beta_1)$, $\gamma_1 = \max\{0,\ 1 - \alpha_2 \log c\}$, $\gamma_2 = \max\{0,\ 1 - \alpha_1 \log c\}$. The theorem is proved. □

Now let us give a rough estimation on γ for real social networks. Firstly, observing $\kappa = 2$ for n from 10^6 to 10^9 (see Table 1), we have $\alpha = \kappa/\log n$ (for 10-based logarithm but any base has the same result) ranges from $\alpha_1 = 2/9$ to $\alpha_2 = 2/6 = 1/3$. Next let us assume $c = 100$ from the Dunbar's number [6, 7][3] Then, since $\log c = \log_{10} 100 = 2$, we see $\gamma = 1 - \alpha \log c$ ranges from $\gamma_1 = 1 - 2/3 = 1/3$ to $\gamma_2 = 1 - 4/9 = 5/9$, i.e.,

$$1/3 \le \gamma \le 5/9. \tag{4}$$

We note that the $\gamma = 2/5$ was first observed from real congressional data by [1].

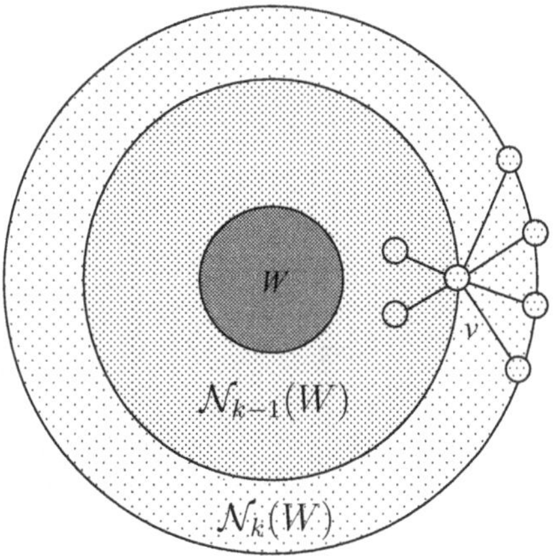

Fig. 3 An illustration of the estimation of c: It should be less than Dunbar's number because some of the edges from nodes v in $\mathcal{N}_{k-1}(W)$ connects to nodes in $\mathcal{N}_{k-1}(W) \setminus \mathcal{N}_k(W)$, thus $c = |\mathcal{N}_k(W)|/|\mathcal{N}_{k-1}(W)|$ should be less than the Dunbar's number (i.e., the total number of edges connect to a node). In this study, we estimated c by 100 comparing to the Dunbar's number 150. An empirical study will be given in Sect. 4

[3] Here we assume $c = 100$ which is less than Dunbar's number 150. This is because not all edges can be used to expand the influence—some of them may connect nodes already counted. See Fig. 3, also Chapter 12.5 of [17] for discussion.

4 Empirical Study

4.1 On the Pyramid Structure of Social Networks

We first study the pyramid structure in Conjecture 2, i.e., values of c and β. For that purpose we used a Facebook network B-anon [20] consisting of 3,097,165 nodes and 23,667,394 edges, which was retrieved from http://networkrepository.com/socfb-B-anon.php. For this network, we found $\kappa = 2$ and an upper-bound 231 on λ using the algorithm in [25]. We then randomly selected $9 \times 5 = 45$ sets W of nodes of sizes 2^p for $p = 0, 1, \ldots, 8$, each 5 samples (notice $2^8 = 256$ is larger than the upper-bound of λ we found). For the representing (influence) distance, we studied $k = 1, 2, 3$ around $\kappa = 2$.

Assuming $\beta = |\mathcal{N}_\kappa(W)|/(|W|c^\kappa)$, we calculated $\mathcal{N}_\kappa(W)$ for all sampled W and k using breadth-first search. The regression result is shown in Fig. 4, in which we have $c = e^{3.7160} = 41.1$ and $\beta = e^{-0.8461} = 0.43$ (the minimum value of $|\mathcal{N}_\kappa(W)|/|W|$ was 2). Therefore we confirmed the existence of the pyramid structure in the studied social network.

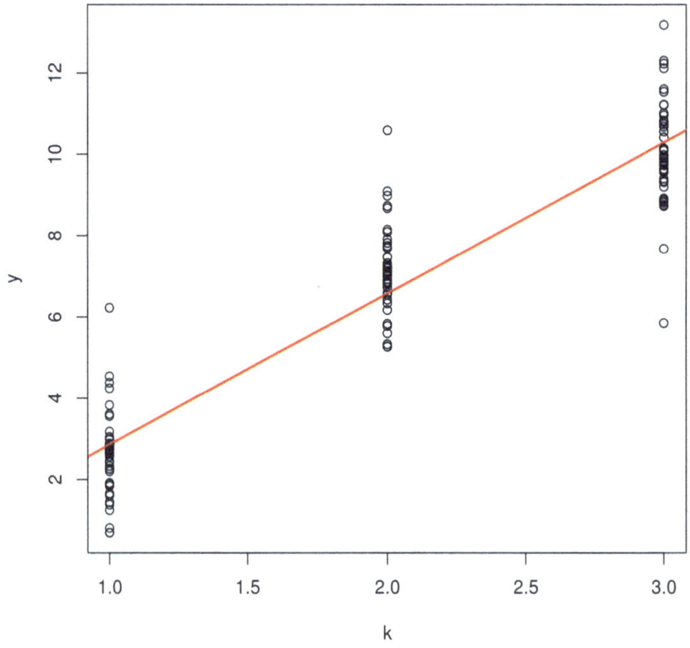

Fig. 4 Regression result for the pyramid structure, where k denotes the distance and y denotes $\ln(|\mathcal{N}_\kappa(W)|/|W|)$. The red line shows the regression result $y = -0.8461 + 3.7160k$ with p-value less than $2e^{-16}$

4.2 *On the Congressional Representatives*

Now let us study the congressional representatives. We have special interest in this case because it is a typical situation requiring appropriate representatives in a *hidden* social network (exists in each country). Our approach is to use statistics to find the relationship between the number R of congressional representatives and the population n, then compare it with the result derived by our model.

We studied 192 countries using data from [10]. The regression estimates the number of congressional representatives $R = 0.3323 \times n^{0.3922}$ (Fig. 5) for the population n. This matches a previous study [1]. For a simple, easy-to-remember expression, we propose

$$R = \frac{1}{3}n^{2/5}. \tag{5}$$

Notice that $1/3 < 2/5 < 5/9$. Therefore we may say our model well explains the number of congressional representatives.

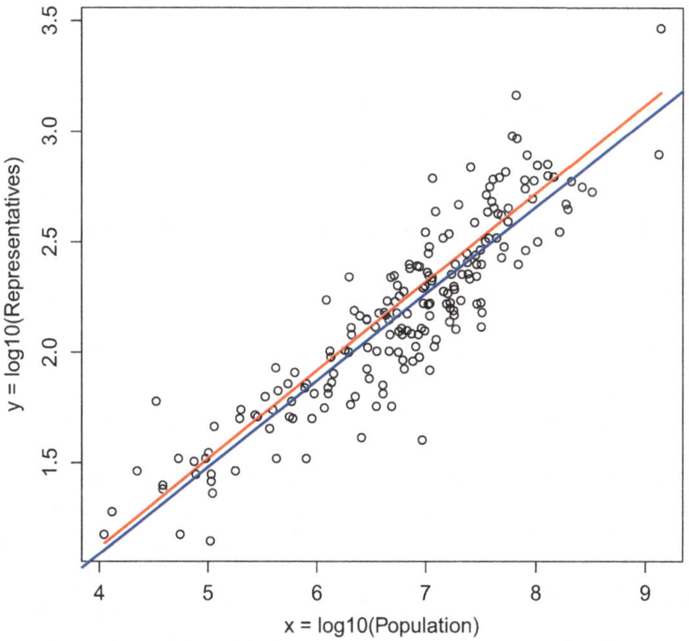

Fig. 5 The relationship between population and the number of congressional representatives for the 192 countries (from right to left, China, India, USA, ..., Nauru, Tuvalu). The regression estimates the number of congressional representatives by $0.3323\times$ population $^{0.3922}$ (blue line) with a p-value $< 2e^{-16}$. The red line shows $y = \frac{1}{3}x^{2/5}$

It is interesting to apply this scaling to other networks. An example is a Japanese government published report on the number of data analysts required (http://www.mext.go.jp/b_menu/shingi/gijyutu/gijyutu23/002/shiryo/_icsFiles/afieldfile/2015/11/19/1364662_002.pdf, in Japanese), which shows a pyramid structure (page 7) saying the supposed total number is 555555 and the number of representatives is 55. If we apply Eq. (5), we can have

$$R = \frac{1}{3} \times 555555^{2/5} \approx 66.$$

This is a much preciser than the estimation $555555^{1/3} = 82$ by the cube root law (on the other hand, there is no exact expression for the square root law thus we cannot compare with it). Therefore we can say our estimation Equation (5) provides good estimations for real cases.

Return back to the first question on the number of representative users of Facebook. Since it has about 2.13 billion active users (see https://newsroom.fb.com/company-info/), we can have a rough estimation:

$$\frac{1}{3} \times 2130000000^{2/5} \approx 1976.$$

5 Conclusion and Future Work

In this paper, we provided a novel, distance-based representing model for social networks and cases studies show the number of representatives scales to a γ-th power of the size of nodes, where $0 \leq \gamma < 1$ is a constant. This model is much simpler than existing studies and explains the real data very well comparing to the conventional cube root theory and square root theory. In fact, empirical studies suggest $1/3 \leq \gamma \leq 5/9$, and $\gamma = 2/5$ (Equation: $R = \frac{1}{3}n^{2/5}$) works well for the numbers of congressional representatives and some other applications.

As future works, we are working on more detailed analysis of the model and are applying it to study more networks. Since finding an exact λ and set W^* for a given network is NP-hard, and this paper only considered the scale of representative set, developing a fast and accurate approximate algorithm or heuristic to calculate the number and the set of representatives for a given network is a challenging task. For that purpose and for further generalization, combining with the Influence Maximization approach such as the Sketch-based algorithms (see [14] for more detail) and [23] to provide a more detailed calculation is interesting.

Acknowledgement This research was supported by JSPS KAKENHI Grant Number 18K11182.

References

1. Auriol, E., Gary-Bobo, R.J.: On the optimal number of representatives. Public Choice **153**(3–4), 419–445 (2012)
2. Barabasi, A.-L.: Network Science. Cambridge University Press, Cambridge (2016). http://networksciencebook.com/. Accessed 27 Sept 2019
3. Bhagat, S., Burke, M., Diuk, C., Filiz, I.O., Edunov, S.: Three and a half degrees of separation (2016). https://research.fb.com/three-and-a-half-degrees-of-separation/. Retrieved on April 2019
4. Brooks, L., Philips, J., Sinitsyn, M.: The cabals of a few or the confusion of a multitude: the institutional trade-off between representation and governance. Am. Econ. J. Econ. Pol. **3**, 1–24 (2011)
5. Clauset, A., Moore, C., Newman, M.E.J.: Hierarchical structure and the prediction of missing links in networks. Nature **453**, 98–101 (2008)
6. CNET News: Sorry, Facebook friends: our brains can't keep up. https://www.cnet.com/news/sorry-facebook-friends-our-brains-cant-keep-up/. Accessed 15 April 2018
7. Dunbar, R.I.M.: Neocortex size as a constraint on group size in primates. J. Hum. Evol. **22**(6), 469–493 (1992) https://doi.org/10.1016/0047-2484(92)90081-J
8. Godefroy, R., Klein, N.: Parliament shapes and sizes. Econ. Inq. (2018). https://doi.org/10.1111/ecin.12584
9. Haynes, T.W.: Stephen Hedetniemi, Peter Slater, Fundamentals of Domination in Graphs. Marcel Dekker, New York (1998)
10. Inter-Parliamentary Union (IPU): Annual report 2017. https://www.ipu.org/file/4474/download?token=YvWKNbJ3. Retrieved 26 Oct 2018
11. Jacobs, K., Otjes, S.: Explaining the size of assemblies. A longitudinal analysis of the design and reform of assembly sizes in democracies around the world. Elect. Stud. **40**, 280–292 (2015)
12. Kempe, D., Kleinberg, J., Tardos, E.: Maximizing the Spread of Influence through a Social Network. In: Proceedings of the 9th ACM SIGKDD, pp. 137–146 (2003)
13. Leskovec, J., Krevl, A.: SNAP datasets: stanford large network dataset collection (2014). http://snap.stanford.edu/data.
14. Li, Y., Fan, J., Wang, Y., Tan, K.L.: Influence maximization on social graphs: a survey. IEEE Trans. Knowl. Data Eng. **30**(10), 1852–1872. https://ink.library.smu.edu.sg/sis_research/3981
15. Lu, C., Yu, J.X., Li, R.H., Wei, H.: Exploring hierarchies in online social networks. IEEE Trans. Knowl. Data Eng. **28**(8), 2086–2100 (2016)
16. Mones. E., Vicsek, L., Vicsek, T.: Hierarchy measure for complex networks. PLoS One **7**(3), e33799 (2012). https://doi.org/10.1371/journal.pone.0033799
17. Newman, M.E.J.: Network: An Introduction. Oxford Press, Oxford (2010)
18. Penrose, L.S.: The elementary statistics of majority voting. J. R. Stat. Soc. **109**(1), 53–57 (1946)
19. Ranjbar-Sahraei, B., Bou Ammar, H., Tuyls, K., Weiss, G.: On the prevalence of hierarchies in social networks. Soc. Netw. Anal. Min. **6**, 58 (2016)
20. Rossi, R.A., Ahmed, N.K.: The network data repository with interactive graph analytics and visualization. In: Proceedings of the 29th AAAI Conference on Artificial Intelligence (2015). http://networkrepository.com
21. Stigler, G.J.: The Sizes of legislatures. J. Legal Stud. **5**(1), 17–34 (1976)
22. Taagepera, R.: The size of national assemblies. Soc. Sci. Res. **1**(4), 385–401 (1972)
23. Tang, J., Tang, X., Yuan, J.: Influence maximization meets efficiency and effectiveness: a hop-based approach. In: Proceedings of the ASONAM'17, pp. 64-41 (2017)
24. Zhao, L.: Majority dominating and democratic number: a proposal to define the democracy of a social network. In: Proceedings of the 4th Internatinal Conference on Operations Research and Statistics (ORS 2016), Singapore, pp. 66–67 (2016)
25. Zhao, L.: Finding small dominating sets in large-scale networks. In: Dehmer, M., et al. Big Data of Complex Networks. Chapman and Hall/CRC, Boca Raton (2016)

NeXLink: Node Embedding Framework for Cross-Network Linkages Across Social Networks

Rishabh Kaushal, Shubham Singh, and Ponnurangam Kumaraguru

Abstract Users create accounts on multiple social networks to get connected to their friends across these networks. We refer to these user accounts as user identities. Since users join multiple social networks, therefore, there will be cases where a pair of user identities across two different social networks belong to the same individual. We refer to such pairs as Cross-Network Linkages (CNLs). In this work, we model the social network as a graph to explore the question, *whether we can obtain effective social network graph representation such that node embeddings of users belonging to CNLs are closer in embedding space than other nodes, using only the network information*. To this end, we propose a modular and flexible node embedding framework, referred to as *NeXLink*, which comprises of three steps. First, we obtain local node embeddings by preserving the local structure of nodes within the same social network. Second, we learn the global node embeddings by preserving the global structure, which is present in the form of common friendship exhibited by nodes involved in CNLs across social networks. Third, we combine the local and global node embeddings, which preserve local and global structures to facilitate the detection of CNLs across social networks. We evaluate our proposed framework on an augmented (synthetically generated) dataset of 63,713 nodes & 817,090 edges and real-world dataset of 3338 Twitter-Foursquare node pairs. Our approach achieves an average Hit@1 rate of 98% for detecting CNLs across social networks and significantly outperforms previous state-of-the-art methods.

R. Kaushal (✉)
Department of Information Technology, Indira Gandhi Delhi Technical University for Women, Delhi, India

Precog Research Lab, Indraprastha Institute of Information Technology, Delhi, India
e-mail: rishabhk@iiitd.ac.in

S. Singh · P. Kumaraguru
Precog Research Lab, Indraprastha Institute of Information Technology, Delhi, India
e-mail: shubham12101@iiitd.ac.in; pk@iiitd.ac.in

N. Masuda et al. (eds.), *Proceedings of NetSci-X 2020: Sixth International Winter School and Conference on Network Science*, Springer Proceedings in Complexity,
https://doi.org/10.1007/978-3-030-38965-9_5

1 Introduction

Online Social Networks (OSNs) are popular platforms on the Internet, helping users to connect with their friends, enabling them to view and share information. OSNs offer different types of content to its users. For instance, YouTube offers videos, Instagram offers images, while Facebook and Twitter offers a mix of text, images, and videos. OSNs also offer different types of friend network to its users. LinkedIn provides access to the professional network while others like Facebook, help in connecting to personal friends. With the presence of these multiple social networks, it is evident that users join more than one social network to avail these several benefits offered by OSNs. In this scenario, it is of great interest to find user identities across multiple social networks belonging to the same individual, which we refer to as *cross-network linkage* and refer these identities as *linked identities*. User behaviors exhibited through these linked identities across multiple OSNs help in building a collective digital footprint [12]. Users' popularity and friendships trends [23] and influence [3] across OSNs can be better understood using such digit footprints. For an adversary to launch social engineering attacks [1], this helps in harvesting information about users based on their activities in multiple OSNs. In digital marketing, it helps to know and identify your customer [11] for the targeted advertisement.

Given the immense importance of finding linked identities, we propose a solution based on the construction of effective graph representations. The goal is to learn node embeddings in a social graph such that nodes with similar characteristics are represented by similar node embedding vectors. In the context of our problem, we ask the question *whether we can obtain effective social network graph representation such that node embeddings of users belonging to CNLs are closer in embedding space than other nodes*. In other words, as depicted in Fig. 1, the goal is to propose an embedding framework that transforms nodes into embedding vectors

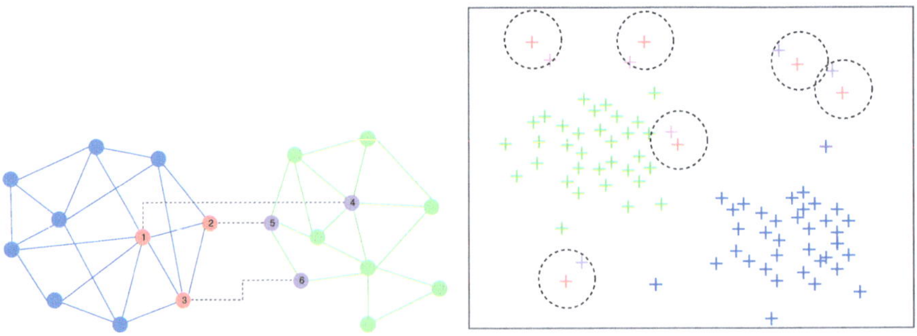

Fig. 1 Our proposed NeXLink framework learns node embeddings from two social networks (represented as graphs, on the left side) with few cross-network linkages. On the right side, we depict embedding space in which nodes corresponding to user identities belonging to same individual are closer than other nodes

such that nodes present in linked identities are closer in embedding space than other nodes. To this end, we propose a three-step *NeXLink* framework that learns node representations to detect CNLs across social networks. In the first step, the local structure of nodes within the same network is preserved. In social networks, these local structures would comprise of friendship relation or follow-followee relation maintained by user identities. In particular, we learn node embeddings of nodes within the same network using the normalized edge weights so that nodes that are structurally near to each other, their corresponding embeddings are also close in embedding space. In the second step, the global structure of nodes connected across multiple networks is preserved. In social networks, these global structures would comprise of cross-platform linkages which represent user identities across social networks belonging to the same individual. These linked identities are expected to exhibit a number of common friends across social networks. In particular, we learn the node embeddings of nodes that are part of Cross-Network Linkages (CNLs) by biasing the random walk in proportion to the common friendship. As a result, node embeddings of nodes that are part of CNLs with more common friends are closer in embedding space. In the third step, we directly leverage the node embeddings to evaluate their efficacy in the detection of cross-network linkages across social networks. The code and data of our work are available at the GitHub repository.[1] We evaluate our proposed approach of the NeXLink framework on two datasets. The first dataset is an augmented dataset synthetically generated using the Facebook social network [18] comprising of 63,713 nodes (users) and 817,090 edges. Our approach works well in all possible augmentations of the Facebook dataset achieving an average Hit@1 rate of 98%, which means that the probability of hitting on the correct cross-network linkage across social networks is 98%. Further, our approach outperforms the state-of-the-art prior approaches of node representations namely LINE [17] and DeepWalk [15] on synthetically generated graphs, which we refer to as augmented dataset. The second dataset comprises of a real-world dataset of Twitter-Foursquare social networks [24] comprising of 3338 nodes (user) pairs. We find that except for Hit@1 rate, our approach better than the state-of-the-art prior approaches of user identity linkages namely IONE [9] and REGAL [6] in Hit@5 rates and above. The key contributions of our work are as below.

- We propose a modular and flexible NeXLink framework as a two-step optimization process that preserves local structure within the same network and preserves global structure manifested in the form of cross-network linkages.
- We extensively evaluate our framework on two datasets, one augmented dataset obtained from Facebook and other real-world dataset comprising of Twitter-Foursquare node pairs. Our framework works well on the synthetically generated dataset and outperforms prior node representation approaches (LINE and DeepWalk) and identity linkage approaches (IONE and REGAL).

[1]Code and dataset of our work can be found at: https://github.com/precog-iiitd/nexlink-netscix-2020.

2 Related Work

Recently, there are a few prior works that have addressed the problem of user identity linkage using the network embedding approach whose aim is to learn a low dimensional representation for a given node in a graph. We categorize these prior methods in the field of network embedding into two main categories, as explained below.

Problem Independent Approaches These works only aim to learn generic low-dimensional representations without focusing on user linkage problems. The objective is to learn effective node representations in low dimensions. Tang et al. [17] propose a framework for network embedding in large graphs to preserve node structures of nodes which are directly connected (first-order node proximity) and connected at a distance of two hops (second-order node proximity). Perozzi et al. [15] leverage the notion of the skip-gram model in language modeling to perform truncated random walks in order to learn latent representations of nodes in a graph. Wang et al. [19] preserve the first and second-order node proximity using a semi-supervised deep learning model. Grover et al. [2] extend the notion of a random walk by introducing biased walks in node neighborhood to learn feature representations of the node in a network. Xu et al. [22] propose two embeddings for each node that capture the structural proximity of nodes as well as the semantic similarity, which they express in terms of common interests. Liang et al. [8] propose a dynamic user and word embedding model (DUWE) that monitors over some time, the relationship between user and words to model their embeddings. Liu et al. [10] present a self-translation network embedding (STNE) framework that is a sequence-to-sequence framework taking into consideration both content and network features of the node.

Problem Dependent Approaches These learn low-dimensional embedding focusing on specific problem, which in our case is to detect cross-network linkages representing user identities across social networks. Liu et al. [9] propose an input-output node embedding (IONE) framework to align user identities across social networks belonging to the same person by learning node representations that preserve follower-followee relationships. Man et al. [13] introduce a framework referred to as PALE, which predicts anchor links via embeddings. First, it converts a social network into a low dimensional node representation. They follow it up by learning a matching function that is supervised by known anchor links. Heimann et al. [6] explain the REGAL framework, which stands for representation learning-based graph alignment based on the cross-network matrix factorization method. Wang et al. [20] propose LHNE mode referred to as linked heterogeneous network embedding which creates an unified framework to leverage structure and content posted by users for learning node representations. Xie et al. [21] use the concept of factoid embedding, which is an unsupervised approach to perform user identity linkage. Our proposed approach outperforms some of these existing approaches, as explained later in this paper.

3 Proposed Approach

In this section, we discuss our proposed NeXLink framework for effective representation and detection of cross-network linkages across social networks. We consider two social networks X and Y as two undirected graphs $G_X(V_X, E_X)$ and $G_Y(V_Y, E_Y)$, where V_X & V_Y represent the nodes (users) of graphs and E_X & E_Y represent the edges. An edge between nodes u_i and u_j indicates friendship relation between users u_i and u_j. We divide the set of node pairs (u_i^X, u_j^Y) across social networks X and Y into two types, namely, cross-network linkages, denoted by $CNL(V_X, V_Y)$ and other pairs are denoted by $NCNL(V_X, V_Y)$. Nodes u_i^X and u_j^Y belonging to social networks X and Y are referred to as cross-network linkage if u_i^X and u_j^Y belong to the same individual and the pair $(u_i^X, u_j^Y) \in CNL(V_X, V_Y)$ else $(u_i^X, u_j^Y) \in NCNL(V_X, V_Y)$. Further, it may be observed in Fig. 2, that the two users represented as two nodes u_i^X and u_j^Y have a^X, b^X and c^X as friends in social network X and same friends a^Y, b^Y and c^Y in social network Y. We refer to such familiar friends as common friendship and leverage this behavior in learning node representations in our NeXLink framework. Besides familiar friends, each node also has some friends which are specific to one social network only. In Fig. 2, nodes d^X and e^X are friends of u_i^X in only social network X whereas nodes f^Y and g^Y are friends of u_j^Y in only social network Y. We note that above formulations for undirected graphs are also applicable in case of directed graphs, in which case the friendship relation would get replaced by follow-followee relation using directed edges.

3.1 Problem Statement

Given two graphs $G_X(V_X, E_X)$ and $G_Y(V_Y, E_Y)$ as input, we define cross-network linkage $CNL(G_X, G_Y)$ as the set of user identity pairs across these two networks X and Y, which belong to the same person. Similarly, we denote all other user pairs which do not belong to the same person by $NCNL(G_X, G_Y$. The goal of network

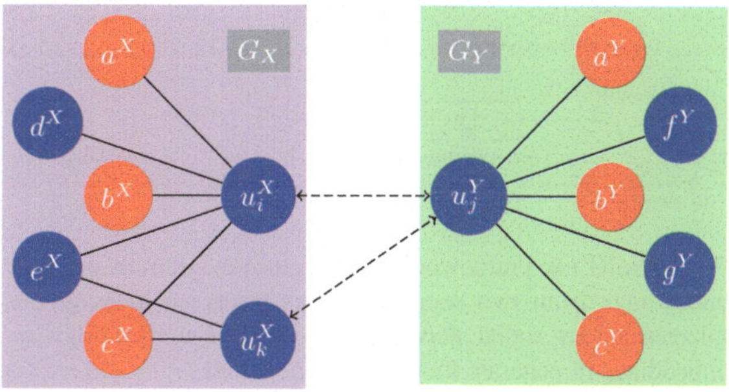

Fig. 2 Illustration of common neighbors of user identities u_i^X and u_j^Y belonging to networks G_X and G_Y. Since all neighbors are common, it is highly likely that u_i^X and u_j^Y belong to same individual than u_k^X and u_j^Y

embedding function (denoted by f_{emb}) is to transform each user identity $u_i^X \in V_X$ and $u_j^Y \in V_Y$ into low *d-dimensional* vectors z_i^X and z_j^Y such that if user identities u_i^X and u_j^Y belong to the same individual (i.e. they represent cross-network linkage), then their corresponding node embeddings z_i^X and z_j^Y are closer in embedding space else they are far apart.

$$\begin{aligned} z_i^X &= f_{emb}(u_i^X), \forall u_i^X \in V_X. \\ z_j^Y &= f_{emb}(u_j^Y), \forall u_j^Y \in V_Y. \\ &such\ that \\ &sim(z_i^X, z_j^Y) \gg sim(z_k^X, z_j^Y)\ and \\ &\exists\,(u_i^X, u_j^Y) \in CNL(V_X, V_Y) \wedge (u_k^X, u_j^Y) \in NCNL(V_X, V_Y). \end{aligned} \tag{1}$$

3.2 NeXLink Framework

The goal of our proposed NeXLink node embedding framework is to obtain representations of nodes in two networks X and Y such that node pairs participating in cross-network linkages have similar node embeddings than other node pairs. To achieve this goal, we follow a two-step approach, as depicted in Fig. 3. In the first step, structural similarities of nodes within or local in their respective networks are preserved independently of the other network. In the second step, similarities of nodes across (or global) the two networks are preserved using a common friendship

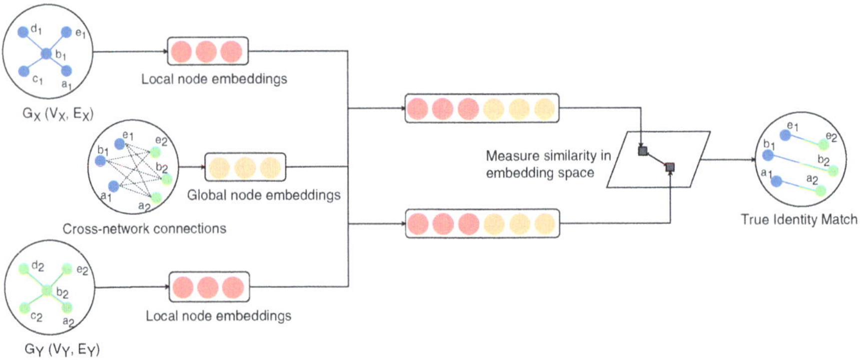

Fig. 3 NeXLink Framework. Architecture diagram of our proposed framework that learns node embeddings from two social networks (represented as graphs) to represent the cross-network linkages across social networks. Local node embeddings are concatenated with global node embeddings to generate final node embeddings

relationship. Given the two-step process, the embedding function f_{emb} can be broken down into two embedding functions, as shown below.

$$
\begin{aligned}
z_i^X &= f_{global}(u_i^X) \oplus f_{local}(u_i^X), \forall u_i^X \in V_X. \\
z_j^Y &= f_{global}(u_j^Y) \oplus f_{local}(u_j^Y), \forall u_j^Y \in V_Y.
\end{aligned} \tag{2}
$$

There can be different ways of combining the global and local node embeddings, however, it turned out that concatenation is the best operation $\oplus$ to combine local and global node embeddings, which we finally used in our proposed NeXLink framework. Further, we note that our proposed approach makes use of only the network structure in the two social networks. However, it can be easily extended to include other sources of information from content and profile information of users, which we leave for future work.

Step 1—Preserving Local Structure Within Social Networks We perform the first step on the intuition that directly connected user nodes within their respective social networks are likely to exhibit similar characteristics, based on the well established social behavioral principle of homophily [14]. Given two nodes u_i^X and u_k^X in same social network X, the goal is to define an encoding function f_{local} that takes these nodes as input and learns their *d-dimensional* embedding vectors $z_i^X \in R^d$ and $z_k^X \in R^d$. To learn z_i^X and z_k^X for all nodes in V_X, we rely upon the probabilistic approach. The empirical probability of the relationship between two nodes u_i^X and u_k^X within the same social network X can be defined as the normalized weight of edge $(w_{i,k}^X)$ between the nodes. Since we consider only the structural information of the network, therefore, for this work, we consider weights to have binary values 1 or 0, depending upon whether there is an edge or not, respectively. In general, the weight of the edge between nodes is intuitively proportional to the similarity between two nodes. Similarity, we can measure other criteria like content similarity. However, we consider only the network structure similarity in this work. We employ a well-established network embedding algorithm, LINE [17], to preserve the local structure.

Step 2—Preserving Global Structure Across Social Networks We propose the second step based on the intuition that user nodes with common friends (CF) across the social networks are likely to belong to the same person. The degree to which two nodes (users) u_i^X and u_j^Y on two social networks X and Y, respectively, having *common friendship*, is expressed as below.

$$
CF(u_i^X, u_j^Y) = \frac{N(u_i^X) \cap N(u_j^Y)}{N(u_i^X) \cup N(u_j^Y)} \tag{3}
$$

where $N(u_i^X)$ and $N(u_j^Y)$ represent the set of friends of ith user in network X and j^{th} user in network Y, respectively. Higher is the value of common friendship (CF), more likely the users u_i^X and u_j^Y would belong to the same person. Therefore, the goal of second encoding function f_{global} is to take u_i^X and u_j^Y as inputs and generate *d-dimensional* node embeddings vectors $z_{G,i}^X \in R^d$ and $z_{G,j}^Y \in R^d$, respectively by using supervisory information of common friendship between u_i^X and u_j^Y in networks X and Y, respectively, along with structural information. If u_i^X and u_j^Y have more common friends, their embedding vectors $z_{G,i}^X$ and $z_{G,j}^Y$ are expected to be closer in embedding space. We employ a well-established network embedding DeepWalk [15] algorithm to preserve the local structure.

4 Data

We evaluate our approach on two network datasets, one augmented, and another a real-world dataset.

4.1 Augmented Dataset

We use the Facebook friendship network dataset,[2] provided by Viswanath et al. [18], comprising 63,713 users and 817,090 edges. We create an undirected graph from the dataset and filter out the nodes that have a degree less than 5, reducing the graph to 40,711 nodes and 766,579 edges. We use this graph to create two subgraphs using a sampling algorithm proposed by Man et al. [13]. Given a graph $G(V, E)$, the algorithm takes two parameters, α_s, α_c and produces two subgraphs $G_X(V_X, E_X), G_Y(V_Y, E_Y)$. The parameter α_s represents how likely are the two subgraphs to retain the edges from the original graph, or the sparsity level. Similarly, the parameter α_c indicates the expected fraction of edges shared among the two subgraphs, or the overlap level. Table 1 shows the number of edges and nodes in the generated subgraphs for different values of α_s and α_c. Once we have the subgraphs, we need to generate node pairs which represent CNLs and NCNLSs across the two subgraphs, which we call as X-node pairs. To do so, we consider all the common nodes in both the graphs, $V_{CNL} = V_X \cap V_Y$, and call them as our CNL nodes, while we term others as NCNL nodes. Now, we take a CNL node and initiate a random walk of a variable length t in G_X, and later in G_Y. The random walks generate $2 \times t$ nodes from G_X and G_Y collectively, and these nodes are then paired with the CNL node to form node pairs.

---

[2] http://socialnetworks.mpi-sws.org/data-wosn2009.html.

Table 1 Statistics for the two datasets used for the evaluation

Graph	#Nodes	#Edges	#CNLs
Augmented dataset			
$G_X(\alpha_s = 0.5, \alpha_c = 0.5)$	40,558	383,463	39,061
$G_Y(\alpha_s = 0.5, \alpha_c = 0.5)$	40,563	382,380	
$G_X(\alpha_s = 0.5, \alpha_c = 0.9)$	40,562	383,360	40,458
$G_Y(\alpha_s = 0.5, \alpha_c = 0.9)$	40,547	383,528	
$G_X(\alpha_s = 0.9, \alpha_c = 0.5)$	40,602	422,295	40,418
$G_Y(\alpha_s = 0.9, \alpha_c = 0.5)$	40,708	689,481	
$G_X(\alpha_s = 0.9, \alpha_c = 0.9)$	40,709	689,856	40,705
$G_Y(\alpha_s = 0.9, \alpha_c = 0.9)$	40,709	690,103	
Real-world dataset			
Twitter	5120	130,575	1288
Foursquare	5313	54,233	

4.2 Real-World Dataset

Kong et al. [7] introduced a network dataset collected from Twitter and Foursquare social networks. The data collection process is described in [7, 25] and used in multiple social link prediction problems [9, 24, 26]. Since the dataset comprises two graphs on its own, we do not need to employ any sampling algorithm to generate subgraphs, and we present the statistical details about the dataset in Table 1. The cross-network linkages represent the users that have profiles on both the social networks. It is evident that such users are less in number in this real-world dataset, compared to the number of CNLs in our augmented dataset.

5 Experiments

We design our experiments to answer the following research questions:

RQ1 How do the values α_s and α_c affect the retrieval of a cross-network node match?

RQ2 How does the choice of second node embedding function f_{global} affect the cross-network node retrieval?

RQ3 How does our proposed NeXLink framework compare with other baselines on a real-world dataset?

We implement all our experiments using NetworkX [4] for graph functions and use OpenNE[3] to run network embedding implementations. To generate the $NCNL$ node pairs, we keep the depth of random walk, $t = 20$ throughout the experiments. When generating the embeddings for cross-network linkages, all

[3]https://github.com/thunlp/OpenNE.

embeddings functions treat node pairs as the edges of the cross-network graph, with CF values as the weights for cross-network edges. Given that our proposed NeXLink framework has two steps for the preservation of structure at the local and global level, we employ prior state-of-the-art node embedding methods at these steps. We typically employ LINE [17] to preserve local structure and consider only first-order proximity calculated over first-order nodes and run over 50 epochs, with early stopping. We do not use second-order proximity since that is taken care of in the second step of our NeXLink framework. We employ various node embedding methods (LINE [17] and DeepWalk[15]) to preserve the global structure in the second step of our NeXLink framework. However, as we explain in this section, it turns out that node2vec [2] when employing common friendship across social networks gives the best results. In node2vec, we set the parameters as $p = 1$ and $q = 2$ which, as mentioned by the authors, are more suited towards preserving structural equivalence. All embedding functions yield 128D embeddings. We evaluate our approach to measure how effectively can node embeddings preserve the CNLs in lower dimensional space, and how closely do network embeddings for CNL lie in that space. In order to compute closeness, we measure the cosine similarity over the node embeddings. When querying for a node u_i^X from the CNL pair (u_i^X, u_j^Y), we count a hit if the matching node embedding z_j^Y for node u_j^Y is present in a set of k node embeddings, ordered on their similarity. To measure accuracy, we calculate a ratio of hits over number of queries and term it as *Hit-Rate@k*. *Hit-Rate@k* is defined as:

$$Hit(u_i^X) = \begin{cases} 1, & if\ z_j^Y \in \{z_1^Y, z_1^Y, ..., z_k^Y\} \\ 0, & otherwise \end{cases}$$

$$Hit - Rate@k = \frac{\sum_{i=0}^{N_{CNL}} Hit(u_i^X)}{N_{CNL}} \tag{4}$$

We choose $k = [1, 5, 10, 20, 50]$ for all the experiments to evaluate our approach.

5.1 Effect of Sparsity and Overlap levels

The α_s and α_c values affect the Common Friendship (CF) values for the CNL nodes, and since the second embedding function is trained to preserve the CF property across networks, we see significant differences in the performances with respect to the difference in α_s and α_c values. We start by employing LINE [17] to learn the local as well as cross-network similarity structure over the four subgraph configurations, as mentioned in Sect. 4, and present our results in Fig. 4a. We observe that the X-node pairs with $\alpha_s = 0.5$ and $\alpha_c = 0.9$ values achieve the highest Hit-Rate@k for all values of k, starting from 0.75 at $k = 1$, and up to 0.96 at $k = 50$. The X-node pairs with $\alpha_s = 0.9$ and $\alpha_c = 0.5$ values achieve the

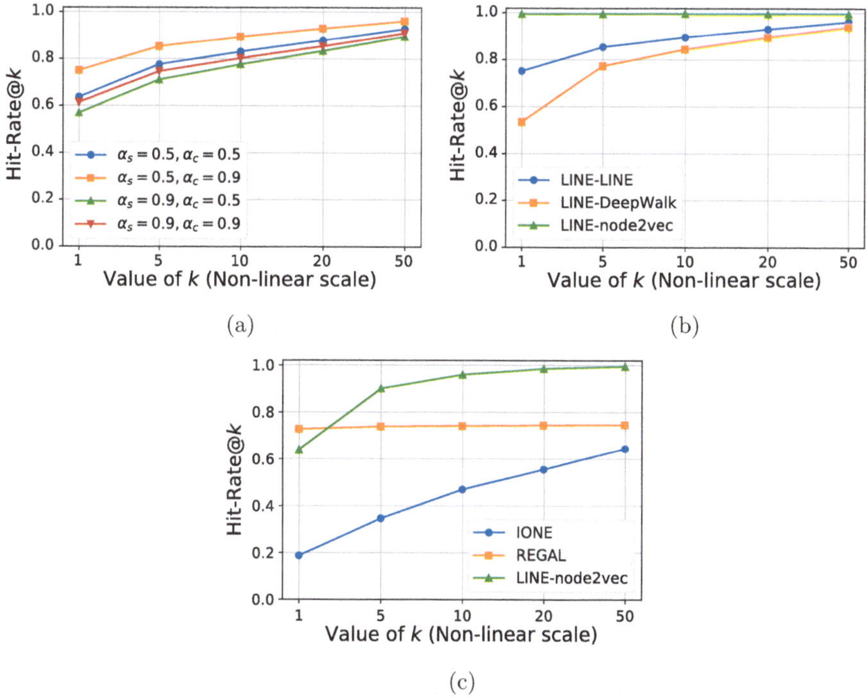

Fig. 4 Results of the three experiments for our research questions (RQ1-RQ3). (**a**) Comparison of Hit-Rate@k values for different sparsity (α_s) and overlap (α_c) levels. (**b**) Comparison of Hit-Rate@k values for different cross-network node embeddings. (**c**) Comparison of Hit-Rate@k values for the baselines and NeXLink (LINE-node2vec) over the real-world dataset

lowest Hit-Rate@k values with 0.57 at $k = 1$ and 0.89 at $k = 50$. We attribute this behavior to the fact that less number of edges and more the overlap between the two subgraphs help the embeddings to capture structural similarities with less noise.

5.2 *Effect of Cross-Network Node Embedding*

We study the role of different network embedding techniques in our proposed NetXLink framework help to preserve CNLs and their impact on the performance of the detection of CNLs across social networks. LINE [17] is suitable for a majority of the number of graphs which preserve local network structure through first-order proximity, which makes it an ideal choice of node embedding method for our within-network embeddings. Along with LINE, we use DeepWalk [15] over *X-node pairs* to get cross-network embeddings, as it uses the structural information about inter-connected nodes by performing truncated random walks to learn latent representations of nodes in a graph, which in our case would be CNLs across networks. Similarly, we employ node2vec [2] which proposes a flexible notion of node neighborhood by designing a biased random walk to learn feature representations of graph nodes. Figure 4b shows the results of our experiments with different node embeddings. The LINE-DeepWalk performs relatively low at $k = 1$, but reaches closer to the Hit-Rate@k of LINE-LINE at higher values of k. It can be explained as the DeepWalk algorithm uses a random walk to sample neighbors of a node to gather the structural information, however, it doesn't take into the account the weights of the edges, because of which it can not leverage the CF values for cross-network links. LINE-LINE performs relatively well as it preserves the first-order proximity proportional to the CF values and achieves a Hit-Rate@1 of about 0.75. However, using the bias parameters from node2vec to better represent structural equivalence, we gain a significant advantage over LINE-LINE and LINE-DeepWalk to get a Hit-Rate@k of around 0.99 for most of the k values. By biasing the walk towards detecting cross-linkages and weighting the transition probabilities towards the CF values, LINE-node2vec gives an optimal representation of cross-linkages that are placed closer to each other in the embedding space.

5.3 *Comparison with the Baselines*

Finally, we evaluate our best performing combination of LINE-node2vec in the NeXLink framework with competing baselines. Along with the structural information, REGAL [6] allows using attribute information for node similarity. However, when comparing with our approach, we only use the structural information from the real-world dataset, described in Sect. 4.2. We also compare our approach with IONE [9] that takes two network graphs as input and produces node embeddings based on the follower and followee relationship among the nodes. We employ our best performing LINE-node2vec technique and elaborate on its performance on the real-world dataset. Figure 4c illustrates the performance of the baselines, as compared to our approach. Given the evaluation of IONE uses the same dataset, we were to successfully reproduce their results, as mentioned in their work [9]. However, it still underperforms when compared to the other approaches. REGAL achieves the highest Hit-Rate@1 as it uses node degrees to capture structural similarities, and node degrees partially contribute to the CF values. However, it still fails to leverage the essential CF values completely, as one of its limitations is not being able to take the edge weights into account. Therefore, its performance stagnates at higher k values. In contrast, LINE-node2vec starts below REGAL at $k = 1$, but achieves higher Hit-Rate@k values with the increase in k. LINE-node2vec learns both within-graph and cross-graph structural features from the real-world dataset and effectively represents the similarities in low-embedding space.

6 Limitations, Discussions and Future Work

While developing NeXLink, we identify some of the limitations of our approach. Firstly, we only include structural information indicating standard connections in the two networks, to learn node representations. We can utilize more rich features to gain more comprehensive node representations. Secondly, an essential step in our approach is to create cross-network pairs, which we accomplish using random walks. We can evaluate more efficient ways to sample the cross-network pairs. And last, the two significant limitations of node embeddings are (a) the need to define an objective function, based on which we learn the embeddings, and (b) node embedding models are transductive, which means that it is not possible to generate the embeddings for the nodes that we do not see during the training. To this end, we can consider the use of graph neural networks [5, 16].

In this work, we propose our *NeXLink* framework for effective representation of cross-network linkages across social networks. Our framework works by preserving the local structure of nodes within the same social network and global structure manifested in the form of common friends exhibited by nodes participating in cross-network linkages. We perform an extensive evaluation of our approach on two datasets, one of which we augment from Facebook social network, and the other comprises of Twitter-Foursquare node pairs. Given that NeXLink framework is flexible, we explored numerous state-of-the-art node embedding algorithms and found that LINE-node2vec performs the best when provided with supervisory information of common friendship. It performs with average Hit@1 rate of 98% across all configurations of the augmented dataset. Further our approach outperforms state-of-the-art node representation algorithms LINE and DeepWalk for representing cross-network linkages across the social networks. This can be primarily attributed to the fact that our approach preserves local and global cross-network links more effectively than these previous approaches which are specifically targeted to perform well on single networks. Our framework works better than other state-of-the-art node embedding approaches like IONE and REGAL for identity linkage on a real-world dataset. This is because our framework performs biased walks in accordance with the common friendship metric for cross-network links.

As future work, we can include node attributes derived from user profile configuration and user content in the NeXLink framework and their impact on performance measured. At the algorithmic level, deep learning-based approaches for node embedding would also be a right direction to explore.

References

1. Edwards, M., Larson, R., Green, B., Rashid, A.: Panning for gold: automatically analysing online social engineering. Comput. Secur. **39**, 396–405 (2013)
2. Grover, A., Leskovec, J.: node2vec: Scalable feature learning for networks. In: Proceedings of the 22nd ACM SIGKDD International Conference on Knowledge Discovery and Data Mining, pp. 855–864. ACM, New York (2016)

3. Guo, L., Zhang, D., Cong, G., Wu, W., Tan, K.L.: Influence maximization in trajectory databases. IEEE Trans. Knowl. Data Eng. **29**(3), 627–641 (2016)
4. Hagberg, A., Swart, P., S Chult, D.: Exploring network structure, dynamics, and function using networkx. Technical report, Los Alamos National Laboratory (LANL), Los Alamos, NM (2008)
5. Hamilton, W., Ying, Z., Leskovec, J.: Inductive representation learning on large graphs. In: Advances in Neural Information Processing Systems, pp. 1024–1034 (2017)
6. Heimann, M., Shen, H., Safavi, T., Koutra, D.: Regal: Representation learning-based graph alignment. In: Proceedings of the 27th ACM International Conference on Information and Knowledge Management, pp. 117–126. ACM, New York (2018)
7. Kong, X., Zhang, J., Yu, P.S.: Inferring anchor links across multiple heterogeneous social networks. In: Proceedings of the 22nd ACM International Conference on Information & Knowledge Management, pp. 179–188. ACM, New York (2013)
8. Liang, S., Zhang, X., Ren, Z., Kanoulas, E.: Dynamic embeddings for user profiling in twitter. In: Proceedings of the 24th ACM SIGKDD International Conference on Knowledge Discovery & Data Mining, pp. 1764–1773. ACM, New York (2018)
9. Liu, L., Cheung, W.K., Li, X., Liao, L.: Aligning users across social networks using network embedding. In: IJCAI, pp. 1774–1780 (2016)
10. Liu, J., He, Z., Wei, L., Huang, Y.: Content to node: Self-translation network embedding. In: Proceedings of the 24th ACM SIGKDD International Conference on Knowledge Discovery & Data Mining, pp. 1794–1802. ACM, New York (2018)
11. Lu, C.T., Shuai, H.H., Yu, P.S.: Identifying your customers in social networks. In: Proceedings of the 23rd ACM International Conference on Conference on Information and Knowledge Management, pp. 391–400. ACM, New York (2014)
12. Malhotra, A., Totti, L., Meira Jr, W., Kumaraguru, P., Almeida, V.: Studying user footprints in different online social networks. In: Proceedings of the 2012 International Conference on Advances in Social Networks Analysis and Mining (ASONAM 2012), pp. 1065–1070. IEEE Computer Society, Silver Spring (2012)
13. Man, T., Shen, H., Liu, S., Jin, X., Cheng, X.: Predict anchor links across social networks via an embedding approach. In: IJCAI, vol. 16, pp. 1823–1829 (2016)
14. McPherson, M., Smith-Lovin, L., Cook, J.M.: Birds of a feather: Homophily in social networks. Ann. Rev. Sociol. **27**(1), 415–444 (2001)
15. Perozzi, B., Al-Rfou, R., Skiena, S.: Deepwalk: Online learning of social representations. In: Proceedings of the 20th ACM SIGKDD International Conference on Knowledge Discovery and Data Mining, pp. 701–710. ACM, New York (2014)
16. Scarselli, F., Gori, M., Tsoi, A.C., Hagenbuchner, M., Monfardini, G.: The graph neural network model. IEEE Trans. Neural Netw. **20**(1), 61–80 (2008)
17. Tang, J., Qu, M., Wang, M., Zhang, M., Yan, J., Mei, Q.: Line: large-scale information network embedding. In: Proceedings of the 24th International Conference on World Wide Web, pp. 1067–1077. International World Wide Web Conferences Steering Committee, Geneva (2015)
18. Viswanath, B., Mislove, A., Cha, M., Gummadi, K.P.: On the evolution of user interaction in facebook. In: Proceedings of the 2nd ACM Workshop on Online Social Networks, pp. 37–42. ACM, New York (2009)
19. Wang, D., Cui, P., Zhu, W.: Structural deep network embedding. In: Proceedings of the 22nd ACM SIGKDD International Conference on Knowledge Discovery and Data Mining, pp. 1225–1234. ACM, New York (2016)
20. Wang, Y., Feng, C., Chen, L., Yin, H., Guo, C., Chu, Y.: User identity linkage across social networks via linked heterogeneous network embedding. World Wide Web **22**, 2611–2632 (2019)
21. Xie, W., Mu, X., Lee, R.K.W., Zhu, F., Lim, E.P.: Unsupervised user identity linkage via factoid embedding. In: 2018 IEEE International Conference on Data Mining (ICDM), pp. 1338–1343. IEEE, Piscataway (2018)
22. Xu, L., Wei, X., Cao, J., Yu, P.S.: On exploring semantic meanings of links for embedding social networks. In: Proceedings of the 2018 World Wide Web Conference on World Wide

Web, pp. 479–488. International World Wide Web Conferences Steering Committee, Geneva (2018)

23. Zafarani, R., Liu, H.: Users joining multiple sites: friendship and popularity variations across sites. Inform. Fusion **28**, 83–89 (2016)
24. Zhang, J., Philip, S.Y.: Integrated anchor and social link predictions across social networks. In: IJCAI, pp. 2125–2132 (2015)
25. Zhang, J., Kong, X., Yu, P.S.: Transferring heterogeneous links across location-based social networks. In: Proceedings of the 7th ACM International Conference on Web Search and Data Mining, pp. 303–312. ACM, New York (2014)
26. Zhang, J., Yu, P.S., Zhou, Z.H.: Meta-path based multi-network collective link prediction. In: Proceedings of the 20th ACM SIGKDD International Conference on Knowledge Discovery and Data Mining, pp. 1286–1295. ACM, New York (2014)

Improved Algorithm for Neuronal Ensemble Inference by Monte Carlo Method

Shun Kimura and Koujin Takeda

Abstract Neuronal ensemble inference is one of the significant problems in the study of biological neural networks. Various methods have been proposed for ensemble inference from their activity data taken experimentally. Here we focus on Bayesian inference approach for ensembles with generative model, which was proposed in recent work. However, this method requires large computational cost, and the result sometimes gets stuck in bad local maximum solution of Bayesian inference. In this work, we give improved Bayesian inference algorithm for these problems. We modify ensemble generation rule in Markov chain Monte Carlo method, and introduce the idea of simulated annealing for hyperparameter control. We also compare the performance of ensemble inference between our algorithm and the original one.

1 Introduction

In recent study of biological neural network, advanced recording technologies such as calcium imaging or functional magnetic resonance imaging enable us to obtain neuronal activity data from thousands of neurons simultaneously. Such activity data will reveal features of neural network. For instance, neurons in the same neuronal ensemble tend to fire synchronously [1, 2], therefore identification of ensembles is significant for understanding whole neural network structure and its dynamical behavior. In fact, there are some studies on neural network structure using ensemble information [3, 4].

Several statistical methods are known for ensemble identification in activity data. For instance, principal component analysis or singular value decomposition can identify ensembles [5]. Their advantage is that they can effectively reduce dimension

S. Kimura · K. Takeda (✉)
Department of Mechanical Systems Engineering, Graduate School of Science and Engineering, Ibaraki University, Ibaraki, Japan
e-mail: koujin.takeda.kt@vc.ibaraki.ac.jp

N. Masuda et al. (eds.), *Proceedings of NetSci-X 2020: Sixth International Winter School and Conference on Network Science*, Springer Proceedings in Complexity,
https://doi.org/10.1007/978-3-030-38965-9_6

and volume of activity data, which are very large in general. On the other hand, these methods require prior knowledge such as the number of ensembles. There is an alternative approach using graph theory, where neuronal activity is expressed as nodes in graph. From such graph, ensemble activity can be extracted by graph clustering method such as spectral clustering [6], while this method is basically applied to static data and neglects dynamical behavior.

One of the strategies to overcome above-mentioned problems is Bayesian modeling for neuronal activity. In recent work, generative model of ensemble activity was proposed by Bayesian inference framework [7]. Using Markov chain Monte Carlo (MCMC) method, this enables us to infer neuronal ensembles and their dynamical behavior. However, this requires large computational cost in MCMC, and the result sometimes gets stuck in bad local maximum solution of Bayesian inference.

In this work, for reduction of computational cost and bad local maximum problem, we propose an improved algorithm for neuronal ensemble identification. First, we change the update rule in MCMC for controlling the number of ensembles. Second, we introduce the idea of simulated annealing for hyperparameter control. We also compare our algorithm and the original one in terms of ensemble identification using synthetic data, and discuss the advantage of our method.

2 Theory

2.1 Bayesian Inference Model

Here we outline the framework of Bayesian inference. We basically follow the notation in the original paper [7]. See it for the detail. Each neuron has the label $i \in \{1, 2, \dots N\}$, and N is the total number of neurons. The variable $k \in \{1, 2, \dots M\}$ is the time step, and M is the size of time frame. The variable $\mu \in \{1, 2, \dots A\}$ is the label of neuronal ensemble, and A is the total number of ensembles. The ith neuron has neuronal membership label to an ensemble, $t_i \in \{1, 2, \dots, A\}$, and activity variable $s_{ik} \in \{0, 1\}$ at time k. The μth ensemble has ensemble activity variable $\omega_{k\mu} \in \{0, 1\}$ at time k. The state 1 for binary variable means active (firing) neuron or neuronal ensemble, and the state 0 is inactive.

With these variables, we give generative model for neuronal activity as the conditional joint probability,

$$P(\boldsymbol{t}, \boldsymbol{\omega}, \boldsymbol{s} | \boldsymbol{n}, \boldsymbol{p}, \boldsymbol{\lambda}) \propto \left(\prod_{i=1}^{N} n_{t_i} \right) \cdot \left(\prod_{\mu=1}^{A} \prod_{k=1}^{M} p_{\mu}^{\omega_{k\mu}} (1 - p_{\mu})^{1-\omega_{k\mu}} \right) \cdot \left(\prod_{i=1}^{N} \prod_{k=1}^{M} [\lambda_{t_i}(\omega_{kt_i})]^{s_{ik}} [1 - \lambda_{t_i}(\omega_{kt_i})]^{(1-s_{ik})} \right), \quad (1)$$

where boldface letter represents the set of variables (e.g. $\boldsymbol{t} = \{t_1, t_2, \ldots t_N\}$). The μth ensemble has activity rate p_μ, which means activity of ensemble. It also has assign probability n_μ, which describes how many neurons belong to this ensemble. Conditional activity rate λ_{t_i} of the ith neuron for given ensemble activity ω_{kt_i} is defined as

$$\lambda_{t_i}(\omega_{kt_i}) = P(s_{ik} = 1 \mid \omega_{kt_i}) \ \text{ for } \ \omega_{kt_i} \in \{0, 1\}. \tag{2}$$

Accordingly, inactivity rate is expressed as $1 - \lambda_{t_i}(\omega_{kt_i}) = P(s_{ik} = 0 \mid \omega_{kt_i})$. The parameter $\boldsymbol{\lambda}$ describes coherence or incoherence (=noise) between neuronal activity and ensemble activity. We assume that the priors of ensemble activity rate $\boldsymbol{p}$ and conditional activity rate $\boldsymbol{\lambda}$ are beta distribution (denoted by Beta), while the prior of assign probability $\boldsymbol{n}$ is Dirichlet distribution (by Dir), namely

$$P(p_\mu) = \text{Beta}\left(\alpha_\mu^{(p)}, \beta_\mu^{(p)}\right), \tag{3}$$

$$P(\lambda_\mu(z)) = \text{Beta}\left(\alpha_{z,\mu}^{(\lambda)}, \beta_{z,\mu}^{(\lambda)}\right), \tag{4}$$

$$P(n_1, \cdots, n_A) = \text{Dir}\left(\alpha_1^{(n)}, \cdots, \alpha_A^{(n)}\right), \tag{5}$$

where $\alpha_\mu^{(p)}, \beta_\mu^{(p)}, \alpha_{z,\mu}^{(\lambda)}, \beta_{z,\mu}^{(\lambda)}, \alpha_\mu^{(n)}$ ($z \in \{0, 1\}, \mu \in \{1, 2, \ldots, A\}$) are hyperparameters. The relation among variables/parameters in this generative model is represented graphically in Fig. 1a.

For improvement of inference accuracy, we analytically integrate out the set of model parameters $\{\boldsymbol{n}, \boldsymbol{p}, \boldsymbol{\lambda}\}$. Integration over these parameters leads to the joint probability as

$$\begin{aligned}
P(\boldsymbol{t}, \boldsymbol{\omega}, \boldsymbol{s}) &= \int d\boldsymbol{n} d\boldsymbol{p} d\boldsymbol{\lambda} \ P(\boldsymbol{t}, \boldsymbol{\omega}, \boldsymbol{s} | \boldsymbol{n}, \boldsymbol{p}, \boldsymbol{\lambda}) P(\boldsymbol{n}, \boldsymbol{p}, \boldsymbol{\lambda}) \\
&\propto \int d\boldsymbol{n} d\boldsymbol{p} d\boldsymbol{\lambda} \left(\prod_{i=1}^{N} n_{t_i}\right) \cdot \text{Dir}\left(\alpha_1^{(n)}, \ldots, \alpha_A^{(n)}\right) \\
&\cdot \left(\prod_{\mu=1}^{A} \prod_{k=1}^{M} p_\mu^{\omega_{k\mu}} (1 - p_\mu)^{1-\omega_{k\mu}}\right) \cdot \left(\prod_{\mu=1}^{A} \text{Beta}\left(\alpha_\mu^{(p)}, \beta_\mu^{(p)}\right)\right) \\
&\cdot \left(\prod_{i=1}^{N} \prod_{k=1}^{M} [\lambda_{t_i}(\omega_{kt_i})]^{s_{ik}} [1 - \lambda_{t_i}(\omega_{kt_i})]^{(1-s_{ik})}\right) \\
&\cdot \left(\prod_{\mu=1}^{A} \prod_{z \in \{0,1\}} \text{Beta}\left(\alpha_{z,\mu}^{(\lambda)}, \beta_{z,\mu}^{(\lambda)}\right)\right)
\end{aligned}$$

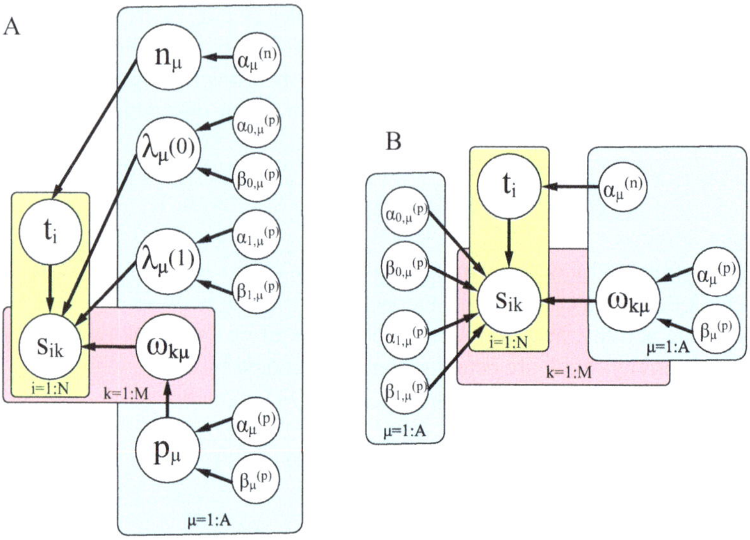

Fig. 1 Graphical representation of the relation among variables/parameters in Bayesian inference model: (**a**) the full model and (**b**) the model after integrating out of $\boldsymbol{n}$, $\boldsymbol{p}$, $\boldsymbol{\lambda}$

$$= \left(\frac{\mathcal{B}(\alpha_1^{(n)} + G_1,\ \alpha_2^{(n)} + G_2, \ldots,\ \alpha_A^{(n)} + G_A)}{\mathcal{B}(\alpha_1^{(n)}, \alpha_2^{(n)}, \ldots, \alpha_A^{(n)})} \right)$$

$$\cdot \prod_{\mu=1}^{A} \left\{ \frac{B(H_\mu, \bar{H}_\mu)}{B(\alpha_\mu^{(p)}, \beta_\mu^{(p)})} \prod_{z \in \{0,1\}} \frac{B(T_\mu^{z1}, T_\mu^{z0})}{B(\alpha_{z,\mu}^{(\lambda)}, \beta_{z,\mu}^{(\lambda)})} \right\}. \tag{6}$$

Note that $B(\cdot, \cdot)$ is beta function, and $\mathcal{B}$ is defined by gamma functions as

$$\mathcal{B}(x_1, \cdots, x_A) \equiv \frac{\prod_{k=1}^{A} \Gamma(x_k)}{\Gamma(\sum_{k=1}^{A} x_k)}. \tag{7}$$

In addition, we introduce several variables,

$$G_\mu = \sum_{i=1}^{N} \delta_{\mu,t_i}, \quad H_\mu = \alpha_\mu^{(p)} + \sum_{k=1}^{M} \omega_{k\mu}, \quad \bar{H}_\mu = \beta_\mu^{(p)} + \sum_{k=1}^{M} (1 - \omega_{k\mu}),$$

$$T_\mu^{z1} = \alpha_{z,\mu}^{(\lambda)} + \sum_{k=1}^{M} \left(\sum_{i \in \boldsymbol{\mu}} \delta_{z,\omega_{k\mu}} \delta_{1,s_{ik}} \right), \quad T_\mu^{z0} = \beta_{z,\mu}^{(\lambda)} + \sum_{k=1}^{M} \left(\sum_{i \in \boldsymbol{\mu}} \delta_{z,\omega_{k\mu}} \delta_{0,s_{ik}} \right), \tag{8}$$

where δ_{ij} is Kronecker delta function, boldface $\boldsymbol{\mu}$ is the set of neurons in the μth ensemble, and $z \in \{0, 1\}$. All variables are defined for the μth ensemble, and their meanings are as follows: G_μ counts the number of neurons in the ensemble. H_μ and $\bar{H}_\mu$ indicate frequencies of active and inactive states, respectively. T_μ^{z1} and T_μ^{z0} measure coherence between ensemble activity and neuronal activity under the same superscript numbers, and incoherence (=noise) under different superscript numbers. The relation among variables/parameters in the model after integrating out of parameters $\{\boldsymbol{n}, \boldsymbol{p}, \boldsymbol{\lambda}\}$ is represented in Fig. 1b.

The posterior $P(\boldsymbol{t}, \boldsymbol{\omega}|\boldsymbol{s})$ can be constructed from this model. With this posterior, we can infer membership $\boldsymbol{t}$ and ensemble activity $\boldsymbol{\omega}$ from neuronal activity variable $\boldsymbol{s}$, which is obtained experimentally.

2.2 *Improvement of Inference Algorithm for the Number of Neuronal Ensembles*

With the scheme mentioned above, we can obtain neuronal ensembles and ensembles activity by Bayesian inference. However, the number of possible neuronal states in all neurons is huge, therefore direct Bayesian inference is infeasible.

For computational cost problem, we employ MCMC to evaluate maximum of posterior distribution. In the previous work [7] collapsed Gibbs sampling is used, where the number of ensembles A is unknown and we also need to evaluate it. For inference of A, Dirichlet process (DP) [8] is introduced, and we can vary A till convergence of DP. However, we need to start with *large* A as initial condition in DP. If we start with small A, we will get stuck at the solution of very few ensembles or without ensemble structure, which is supposed to be bad local maximum solution of Bayesian inference. Hence, this method still requires large computational cost at early stage of MCMC for successful inference, because the cost is proportional to A at a given MCMC stage. (See Algorithm 1.)

To cope with this problem, we propose a novel method. The differences from the previous work are summarized as follows:

- In our method, when new ensemble is created for increasing A in DP, *multiple* neurons can move to new ensemble *simultaneously*, while single neuron can move to new ensemble in the original. Introduction of such simultaneous move will make new ensemble hard to vanish.
- We apply the idea of *simulated annealing* to transient probability to new ensemble in DP.

As shown later, our method enables us to infer appropriate ensemble structure *without starting large* A.

We give the detail of our algorithm in the following. In our MCMC, we first update ensemble activity $\boldsymbol{\omega}$, then update ensemble membership label $\boldsymbol{t}$ by collapsed Gibbs sampling. This process is the same as the original. Next, we employ DP in order to increase/decrease the number of ensembles. Suppose that we have A ensembles in the intermediate stage of MCMC. Destination ensemble of the ith neuron after MCMC update, denoted by t_i^*, is drawn from the probability,

$$q_i(t_i^*) = \begin{cases} \dfrac{G_{t_i^*}^{(\backslash i)}}{q_\alpha^{[\gamma]} + N - 1} & \text{for } t_i^* = 1, \ldots, A, \\ \dfrac{q_\alpha^{[\gamma]}}{q_\alpha^{[\gamma]} + N - 1} & \text{for } t_i^* = A + 1. \end{cases} \tag{9}$$

The backslash \ denotes the removal of a specific element, and $G_{t_i^*}^{(\backslash i)}$ means the number of neurons in the t_i^*th ensemble, where the ith neuron is not counted. Note that $\sum_{\mu=1}^{A} G_\mu^{(\backslash i)} = N - 1$ and $q_i(t_i^*)$ satisfies $\sum_{\mu=1}^{A+1} q_i(t_i^*) = 1$. The parameter $q_\alpha^{[\gamma]}$ is proportional to transient probability to the new $(A+1)$th ensemble. As mentioned before, we apply the idea of simulated annealing to DP. In our method, the transient parameter at the γth MCMC stage $q_\alpha^{[\gamma]}$ decays exponentially as

$$q_\alpha^{[\gamma]} = q_\alpha^{[0]} e^{-\frac{\gamma}{\tau}}, \tag{10}$$

where τ is decay constant. The idea of Eq. (10) is that the number of ensembles A is changed frequently at early MCMC stage for exploring appropriate A, while the change is suppressed at late stage for convergence. If new ensemble is accepted in DP, we must generate new ensemble activity $\boldsymbol{\omega}$ and hyperparameters of new ensemble. We give hyperparameters of the new $(A + 1)$th ensemble as arithmetic average of already-existing hyperparameters, because new hyperparameter should have the same scale as others for appropriate convergence of MCMC.

$$\begin{aligned} \alpha_{A+1}^{(p)} &= \frac{1}{A}\sum_{\mu=1}^{A} \alpha_\mu^{(p)}, \quad \beta_{A+1}^{(p)} = \frac{1}{A}\sum_{\mu=1}^{A} \beta_\mu^{(p)}, \\ \alpha_{z,A+1}^{(\lambda)} &= \frac{1}{A}\sum_{\mu=1}^{A} \alpha_{z,\mu}^{(\lambda)}, \quad \beta_{z,A+1}^{(\lambda)} = \frac{1}{A}\sum_{\mu=1}^{A} \beta_{z,\mu}^{(\lambda)} \quad \text{for } z \in \{0, 1\}, \\ \alpha_{A+1}^{(n)} &= \frac{1}{A}\sum_{\mu=1}^{A} \alpha_\mu^{(n)}. \end{aligned} \tag{11}$$

In addition, if some already-existing ensembles become empty (=no neuron) after update, we delete these ensembles and their hyperparameters.

Now we consider the case that the membership label $\boldsymbol{t}^0 = \{t_1^0, t_2^0, \ldots, t_N^0\}$ may be updated to new one $\boldsymbol{t}^* = \{t_1^*, t_2^*, \ldots, t_N^*\}$ in MCMC. In this update, the ratio of conditional probabilities between $\boldsymbol{t}^0$ and $\boldsymbol{t}^*$ is calculated from Eq. (6), which is necessary for acceptance rule of MCMC update,

$$\frac{P(\boldsymbol{t}^*, \boldsymbol{\omega}, \boldsymbol{s})}{P(\boldsymbol{t}^0, \boldsymbol{\omega}, \boldsymbol{s})} = \frac{\prod_{\mu=1}^{A} \left\{ \Gamma(G_\mu + \alpha_\mu) \prod_{z=\{0,1\}} B(T_\mu^{z1}, T_\mu^{z0}) \right\} |_{\boldsymbol{t}=\boldsymbol{t}^*}}{\prod_{\mu=1}^{A} \left\{ \Gamma(G_\mu + \alpha_\mu) \prod_{z=\{0,1\}} B(T_\mu^{z1}, T_\mu^{z0}) \right\} |_{\boldsymbol{t}=\boldsymbol{t}^0}}$$
$$\cdot \left[\frac{\Gamma(\sum_{\mu=1}^{A+1} \alpha_\mu^{(n)})}{\Gamma(\sum_{\mu=1}^{A} \alpha_\mu^{(n)})} \cdot \frac{\Gamma(G_{A+1} + \alpha_{A+1}^{(n)})}{\Gamma(\alpha_{A+1}^{(n)})} \right.$$
$$\left. \cdot \frac{B(H_{A+1}, \bar{H}_{A+1}) \cdot \prod_{z=\{0,1\}} B(T_{A+1}^{z1}, T_{A+1}^{z0})}{B(\alpha_{A+1}^{(p)}, \beta_{A+1}^{(p)}) \cdot \prod_{z=\{0,1\}} B(\alpha_{z,A+1}^{(\lambda)}, \beta_{z,A+1}^{(\lambda)})} \right]. \quad (12)$$

The factor in the square bracket is the contribution from transient neurons to the new $(A+1)$th ensemble. If there is no neuron to the new ensemble, the factor in the square bracket vanishes because the denominator and the numerator cancel out.

For Metropolis-Hastings update rule, we also need to define proposal distribution from the t_i^0th ensemble to the t_i^*th, $Q_i(t_i^*|t_i^0)$, and its reverse process $Q_i(t_i^0|t_i^*)$ for the ith neuron. These probabilities are calculated to satisfy detailed balance condition as

$$Q_i(t_i^*|t_i^0) = \begin{cases} \dfrac{G_{t_i^*}^{(\setminus i)}}{q_\alpha^{[\gamma]} + N - 1} & \text{for } t_i^* = 1, \ldots, A, \\ \dfrac{q_\alpha^{[\gamma]}}{q_\alpha^{[\gamma]} + N - 1} & \text{for } t_i^* = A+1, \end{cases} \quad (13)$$

$$Q_i(t_i^0|t_i^*) = \frac{G_{t_i^0}^{(\setminus i)}}{N-1}, \quad (14)$$

for the γth MCMC stage, where Eq. (13) is the same as (9). If multiple neurons move simultaneously, we must consider the product of the probabilities above for all transient neurons. As a result, acceptance rate from the membership label $\boldsymbol{t}^0$ to $\boldsymbol{t}^*$ is written as

$$a(\boldsymbol{t}^*, \boldsymbol{t}_0) = \min \left\{ 1, \frac{P(\boldsymbol{t}^*, \boldsymbol{\omega}, \boldsymbol{s})}{P(\boldsymbol{t}^0, \boldsymbol{\omega}, \boldsymbol{s})} \frac{Q(\boldsymbol{t}^0|\boldsymbol{t}^*)}{Q(\boldsymbol{t}^*|\boldsymbol{t}^0)} \right\}, \quad (15)$$

$$\text{where} \quad Q(\boldsymbol{t}^0|\boldsymbol{t}^*) = \prod_{i=1}^{N} Q_i(t_i^0|t_i^*), \quad Q(\boldsymbol{t}^*|\boldsymbol{t}^0) = \prod_{i=1}^{N} Q_i(t_i^*|t_i^0). \quad (16)$$

Finally, we update hyperparameters of $\{\boldsymbol{p}, \boldsymbol{\lambda}, \boldsymbol{n}\}$ for remaining ensembles. For hyperparameter update, we introduce learning rate $\varepsilon^{[\gamma]}$, where γ is the stage of MCMC update as in (10), to control the effect of simulated annealing. Here we use sigmoid function for $\varepsilon^{[\gamma]}$ because it is bounded and smooth,

$$\varepsilon^{[\gamma]} = \frac{1}{1 + e^{-\frac{\gamma}{\tau}}}. \tag{17}$$

The decay constant τ is the same as in (10). Note that we do not need to introduce additional hyperparameter for the learning rate. Following the update rule in the original [7], hyperparameters should be updated with learning rate $\varepsilon^{[\gamma]}$ as

$$\begin{aligned}
\tilde{\alpha}_{\mu}^{(p)} &= \alpha_{\mu}^{(p)} + \varepsilon^{[\gamma]} \left(\sum_{k=1}^{M} \omega_{k\mu} \right), \quad \tilde{\beta}_{\mu}^{(p)} = \beta_{\mu}^{(p)} + \varepsilon^{[\gamma]} \left(\sum_{k=1}^{M} (1 - \omega_{k\mu}) \right), \\
\tilde{\alpha}_{z,\mu}^{(\lambda)} &= \alpha_{z,\mu}^{(\lambda)} + \varepsilon^{[\gamma]} \left(\sum_{k=1}^{M} \left(\sum_{i \in \boldsymbol{\mu}} \delta_{z,\omega_{k\mu}} \delta_{1,s_{ik}} \right) \right), \\
\tilde{\beta}_{z,\mu}^{(\lambda)} &= \beta_{z,\mu}^{(\lambda)} + \varepsilon^{[\gamma]} \left(\sum_{k=1}^{M} \left(\sum_{i \in \boldsymbol{\mu}} \delta_{z,\omega_{k\mu}} \delta_{0,s_{ik}} \right) \right), \\
\tilde{\alpha}_{\mu}^{(n)} &= \alpha_{\mu}^{(n)} + \varepsilon^{[\gamma]} G_{\mu},
\end{aligned} \tag{18}$$

where tilde means updated hyperparameter. By introducing learning rate, hyperparameter update is suppressed at early stage of MCMC, when the number of ensembles A is frequently changed instead.

To conclude, we summarize our MCMC process as the pseudo code in Algorithm 1. The symbol $\boldsymbol{\omega}_{\setminus k\mu}$ means the set of parameters $\boldsymbol{\omega}$ excepting $\omega_{k\mu}$ for describing collapsed Gibbs sampling.

3 Experiment

3.1 Generative Model for Synthetic Data

Before discussion on utility of our algorithm, we summarize how to generate synthetic neuronal activity data for our numerical experiment. In our generative model, neuronal activities are closely related to ensemble activities as in Eq. (2), and the relation between neuronal/ensemble activities is characterized by conditional activity rate $\boldsymbol{\lambda}$. Hence, to generate synthetic data, we first divide all neurons into ensembles. Next we generate ensemble activity data $\boldsymbol{\omega}$ based on ensemble activity parameter $\boldsymbol{p}$. In the last we determine activity of each neuron $\boldsymbol{s}$ by conditional activity rate $\boldsymbol{\lambda}$.

The algorithm of synthetic data generation is summarized as the pseudo code in Algorithm 2. Detail can be found in the original work as well [7].

Algorithm 1 Inference of ensembles activity and the number of ensembles

```
initialize ω and t
while the number of ensembles A converges do
    for each ensemble μ ∈ [1, A], k ∈ [1, M] do
        draw ω_kμ ~ P(ω_kμ = 1|t, ω_\kμ, s)
    end for
    for each neuron i ∈ [1, N] do
        draw destination ensemble t_i* ~ q(t_i*) in Eq. (9)
        if t_i* = A + 1 then
            A → A + 1
            give new hyperparameters as Eq. (11)
        end if
    end for
    for each neuron i ∈ [1, N] do
        draw t ~ a(t*, t^0) in Eq. (16)
    end for
    for each ensemble μ ∈ [1, A] do
        if G_μ = 0 then
            delete the μth ensemble and its hyperparameters
        end if
    end for
    update hyperparameters as Eq. (18)
end while
```

Algorithm 2 Generation of synthetic neuronal activity data

```
set all ω and s to be 0
for each ensemble μ ∈ [1, A], k ∈ [1, M] do
    draw ω_kμ ~ P(ω_kμ)
end for
for each neuron i ∈ [1, N] do
    deal the ith neuron to an ensemble
end for
for each neuron i ∈ [1, N], k ∈ [1, M] do
    if ω_kt_i = 1 then
        draw s_ik ~ P(s_ik|ω_kt_i = 1)
    else
        draw s_ik ~ P(s_ik|ω_kt_i = 0)
    end if
end for
```

3.2 *Numerical Validation*

To validate our method, we conduct numerical experiment for neuronal ensemble inference. In the experiment, we first generate synthetic data by Algorithm 2, then we extract the information of ensembles by Algorithm 1. We illustrate a sample of activity matrix $\boldsymbol{s}$ in Fig. 2a, which is obtained by Algorithm 2. In this sample we have 10 ensembles and 100 neurons, where each ensemble has equally 10 neurons. The vertical axis shows the neuron label, which is sorted by neuronal membership label $\boldsymbol{t}$. We can easily see the structure of 10 ensembles, however we should note

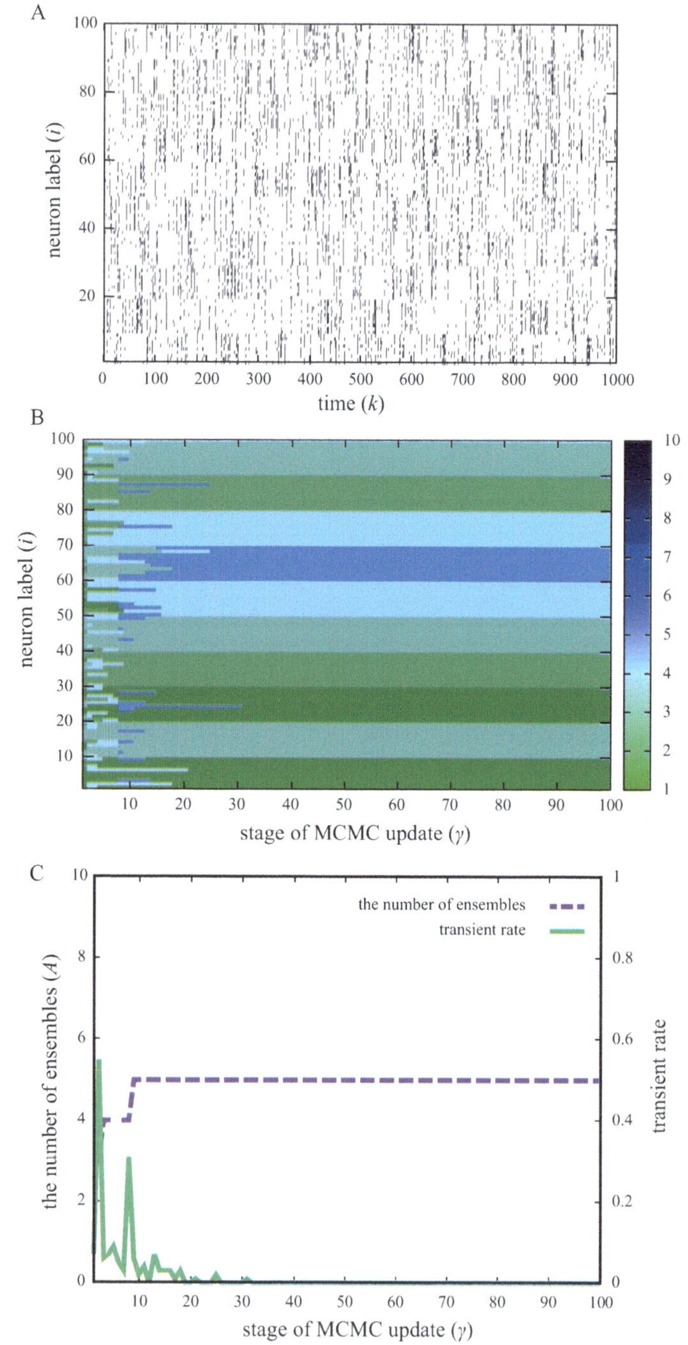

Fig. 2 Dynamical behavior of Algorithm 1 in MCMC: (**a**) Synthetic data of neuronal activity with 100 neurons, 10 ensembles and 1000 time steps. Black = active, white = inactive. (**b**) Behavior of ensemble membership label. (**c**) Behavior of the number of ensembles (broken) and transient rate (solid)

Table 1 The condition of synthetic data generation

Parameter	Value
The number of neurons	$N = 100$
The number of ensembles	$A = 10$
Ensemble activity rate	$p_\mu = 0.1 \ (\forall \mu)$
Conditional activity rate	$\lambda_\mu(0) = 0.01, \lambda_\mu(1) = 0.6 \ (\forall \mu)$

that it cannot be seen if neuron labels are randomly permuted. The parameters for synthetic data generation are summarized in Table 1. All ensembles/neurons are generated with the same ensemble activity rate $\boldsymbol{p}$ and conditional activity rate $\boldsymbol{\lambda}$.

For ensemble inference, we use the data in Fig. 2a as input activity $\boldsymbol{s}$. As shown in Algorithm 1, we repeatedly update ensemble activity $\boldsymbol{\omega}$, ensemble membership label $\boldsymbol{t}$, and hyperparameters till the number of transient neurons in DP becomes sufficiently small. In this experiment, we set initial number of ensembles $A = 3$, decay constant $\tau = 10$, initial transient parameter $q_\alpha^{[0]} = 100$ and hyperparameters $\alpha_\mu^{(p)} = 100, \beta_\mu^{(p)} = 100, \alpha_{z,\mu}^{(\lambda)} = 100, \beta_{z,\mu}^{(\lambda)} = 100, \alpha_\mu^{(n)} = 100$ for all μ, z. At initialization step, we randomly assign initial membership label to each neuron uniformly between 1 to A.

In Fig. 2b, we show a typical example of dynamical membership label behavior in MCMC by the heat map, where the horizontal axis indicates the stage of MCMC update. The colors in the heat map distinguish ensemble numbers $\{1, 2, \ldots, A\}$. The original ensemble structure is clearly obtained at most after 40 MCMC stages.

In Fig. 2c, we show the behavior of the number of ensembles A and transient rate in MCMC, which is calculated from the result in Fig. 2b. Transient rate is defined by

$$\text{transient rate} = \frac{1}{N} \sum_{i=1}^{N} \left(1 - \delta_{t_i^{[\gamma]}, t_i^{[\gamma-1]}}\right) \tag{19}$$

at the γth MCMC stage, namely the fraction of transient neurons between the $(\gamma - 1)$th and the γth MCMC stages. The broken line represents the number of ensembles and the solid line represents transient rate. From Fig. 2c, we find that the number of ensembles and ensemble membership label converge after 40 MCMC stages.

We should note that the number of final ensembles is 5, which is smaller than the ground-truth value 10. This is because some ensembles are merged in the final result. To manage this problem, we use our algorithm repeatedly with another initial membership label, which leads to different ensemble structure. Even in this case, we will obtain blockwise ensemble structure again like in Fig. 2b, while the number of ensemble is still smaller than 10. However, we should note that *other* ensembles are merged under different initial membership label. Therefore, we can obtain the original 10-ensemble structure exactly by combining several MCMC results with different initial membership label.

When we change conditional activity rate $\boldsymbol{\lambda}$, which controls coherence or noise, the ensemble inference becomes easy/hard. Even under hard condition or noisy case, we can still obtain nearly correct ensemble structure, and noise can be removed as much as possible. In addition, even if the sizes of ground-truth ensembles are not equal unlike Fig. 2a, we can infer correct ensemble structure.

3.3 Comparison with the Original Algorithm

We compare our algorithm and the original one in Ref. [7]. For comparison, we use the same synthetic data in Fig. 2a. In the original algorithm, we do not use simultaneous update rule for multiple neurons to new ensemble, nor simulated annealing idea for hyperparameters (or take the limit of $\tau \to +\infty$).

A typical example of dynamics is shown in Fig. 3. In Fig. 3a, we start with small number of ensembles ($A = 3$) like our algorithm in Sect. 3.2. However, we cannot obtain blockwise structure. All ensembles are merged into one at late MCMC stage. In Fig. 3b, we show the number of ensembles and transient rate in the result of Fig. 3a, which exhibits slow convergence of transient rate. We verify that this result does not depend on the initial condition such as ensemble membership label. From these results, we conclude that the original algorithm will get stuck in bad local maximum of Bayesian inference, when we start with small initial number of ensembles.

4 Discussion and Perspective

In this work, we proposed Bayesian inference algorithm with faster convergence. In our method, we introduced the update rule to new ensemble for multiple neurons and the idea of simulated annealing, to avoid bad local maximum solution of Bayesian inference. For simulated annealing, we introduced decay constant τ to control annealing schedules of transient probability and learning rate. As a consequence, we find that our method can successfully obtain blockwise neuronal ensemble structure by numerical experiment for synthetic data, even with small initial number of ensembles. We also compare our algorithm with the original one, and the result indicates our algorithm has advantage for finding correct ensemble solution.

Note that our method in this work focuses on neuronal ensemble identification, not for the detail of neural network structure like connection. However, we believe that our idea for ensemble identification will be helpful for understanding whole structure of biological neural network. Moreover, by further improvement of our method, we think that we can construct Bayesian inference framework for the detail of neural network structure.

Several issues are remained as future works. First, the additional hyperparameter τ may also be useful for finding hierarchical ensemble structure. If τ is set to be large or annealing schedule is slow, we can obtain finer ensemble structure, while it requires many MCMC stages till convergence. On the other hand, if τ is small

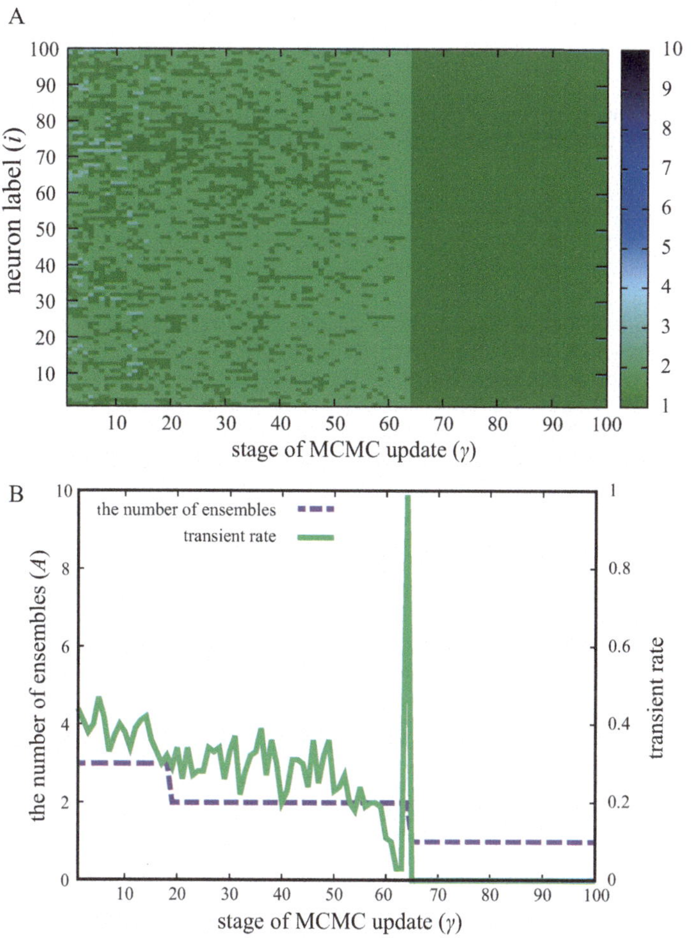

Fig. 3 Dynamical behavior of the original algorithm in Ref. [7] in MCMC: (**a**) Behavior of ensemble membership label. (**b**) Behavior of the number of ensembles (broken) and transient rate (solid)

we can obtain ensemble structure more faster. However, only large scale structure will be found and fine structure will be neglected. Even in this case, we can obtain fine structure if we use this method repeatedly, as mentioned in Sect. 3.2. We should investigate the role of the hyperparameter τ in more detail.

Second, we should apply our algorithm to real neuronal activity data, which we are planning at present. One of the problems for application is that real experimental data of neuronal activity is often continuous, not binary like our formulation. The natural idea for application to continuous data is to binarize real activity data, however this may neglect significant information in neuronal activity. Another idea is to generalize our formalism to continuous activity data, and for this idea we must modify generative model for continuous data. We should consider which strategy is more appropriate for application to real activity data.

Acknowledgements We appreciate the comments from Giovanni Diana and Yuishi Iwasaki. This work is supported by KAKENHI Nos. 18K11175, 19K12178.

References

1. Fries, P.: A mechanism for cognitive dynamics: neuronal communication through neuronal coherence. Trends Cogn. Sci. **9**, 474–480 (2005). https://doi.org/10.1016/j.tics.2005.08.011
2. Lopes-dos-Santos, V., Conde-Ocazionez, S., Nicolelis, M.A.L., Ribeiro, S.T., Tort, A.B.L.: Neuronal assembly detection and cell membership specification by principal component analysis. PLoS One **6**, 1–16 (2011). https://doi.org/10.1371/journal.pone.0020996
3. Friedrich, R.W., Habermann, C.J., Laurent, G.: Multiplexing using synchrony in the zebrafish olfactory bulb. Nat. Neurosci. **7**, 862–871 (2004). https://doi.org/10.1038/nn1292
4. Palva, J.M., Monto, S., Kulashekhar, S., Palva, S.: Neuronal synchrony reveals working memory networks and predicts individual memory capacity. Proc. Natl. Acad. Sci. **107**, 7580–7585 (2010). https://doi.org/10.1073/pnas.0913113107
5. Romano, S.A., Pietri, T., Pérez-Schuster, V., Jouary, A., Haudrechy, M., Sumbre, G.: Spontaneous neuronal network dynamics reveal circuit's functional adaptations for behavior. Neuron **85**, 1070–1085 (2015). https://doi.org/10.1016/j.neuron.2015.01.027
6. Sakuma, H., Teramoto, T., Kuge, S., Ishihara, T., Iwasaki, Y.: Large graph laplacian matrix and functional map of whole brain of C. elegans. In: Proceedings of 2016 International Symposium on Nonlinear Theory and Its Applications, pp. 696–699 (2016)
7. Diana, G., Sainsbury, T.T.J., Meyer, M.P.: Bayesian inference of neuronal assemblies. PLoS Comput. Biol. **15**, e1007481 (2019). https://doi.org/10.137/journal.pcbi.1007481
8. Neal, R.M.: Markov chain sampling methods for Dirichlet process mixture models. J. Comput. Graph. Stat. **9**, 249–265 (2000)

Testing for Network and Spatial Autocorrelation

Youjin Lee and Elizabeth L. Ogburn

Abstract Testing for dependence has been a well-established component of spatial statistical analyses for decades. In particular, several popular test statistics have desirable properties for testing for the presence of spatial autocorrelation in continuous variables. In this paper we propose two contributions to the literature on tests for autocorrelation. First, we propose a new test for autocorrelation in categorical variables. While some methods currently exist for assessing spatial autocorrelation in categorical variables, the most popular method is unwieldy, somewhat ad hoc, and fails to provide grounds for a single omnibus test. Second, we discuss the importance of testing for autocorrelation in data sampled from the nodes of a network, motivated by social network applications. We demonstrate that our proposed statistic for categorical variables can both be used in the spatial and network setting.

1 Introduction

In studies using spatial data, researchers routinely test for spatial dependence before proceeding with statistical analysis [8, 17, 20]. Spatial dependence is usually assumed to have an autocorrelation structure, whereby pairwise correlations between data points are a function of the geographic distance between the two observations [6, 26]. Because autocorrelation is a violation of the assumption of *independent and identically distributed* (i.i.d.) observations or residuals required by most standard statistical models and hypothesis tests [1, 17, 18], testing for spatial autocorrelation is a necessary step for valid statistical inference using spatial data.

Y. Lee (✉)
University of Pennsylvania, Philadelphia, PA, USA
e-mail: youjin.lee@pennmedicine.upenn.edu

E. L. Ogburn
Johns Hopkins Bloomberg School of Public Health, Baltimore, MD, USA
e-mail: eogburn@jhu.edu

N. Masuda et al. (eds.), *Proceedings of NetSci-X 2020: Sixth International Winter School and Conference on Network Science*, Springer Proceedings in Complexity,
https://doi.org/10.1007/978-3-030-38965-9_7

Taking temporal dependence into account is also widely practiced in time series settings. But other kinds of statistical dependence are routinely ignored. In many public health and social science studies, observations are collected from individuals who are members of one or a small number of social networks within the target population, often for reasons of convenience or expense. For example, individuals may be sampled from one or a small number of schools, institutions, or online communities, where they may be connected by ties such as being related to one another; being friends, neighbors, acquaintances, or coworkers; or sharing the same teacher or medical provider. If individuals in a sample are related to one another in these ways, they may not furnish independent observations, and yet most statistical analyses in the literature use i.i.d. data methods [16]. This failure to account for dependence can result in anticonservative inference: inflated false positive rates and artificially small p-values.

In the literature on spatial and temporal dependence, dependence is often implicitly assumed to be the result of latent traits that are more similar for observations that are close than for distant observations. This *latent variable dependence* [24] is likely to be present in many network contexts as well. In networks, ties often present opportunities to transmit traits or information from one node to another, and such direct transmission will result in *dependence due to direct transmission* [24] that is informed by the underlying network structure. In general, both of these sources of dependence result in positive pairwise correlations that tend to be larger for pairs of observations from nodes that are close in the network and smaller for observations from nodes that are distant in the network. Network distance is usually measured by geodesic distance, which is a count of the number of edges along the shortest path between two nodes. This is analogous to spatial and temporal dependence, which are generally thought to be inversely related to (Euclidean) distance.

Despite increasing interest in and availability of social network data, there is a dearth of valid statistical methods to account for network dependence. Although many statistical methods exist for dealing with dependent data, almost all of these methods are intended for spatial or temporal data—or, more broadly, for observations with positions in $\mathbb{R}^k$ and dependence that is related to Euclidean distance between pairs of points. The topology of a network is very different from that of Euclidean space, and many of the methods that have been developed to accommodate Euclidean dependence are not appropriate for network dependence. The most important difference is the distribution of pairwise distances which, in Euclidean settings, is usually assumed to skew towards larger distances as the sample grows, with the maximum distance tending to infinity with sample size n. In social networks, on the other hand, pairwise distances tend to be concentrated on shorter distances and may be bounded from above. However, as we elaborate in Sect. 2, methods that have been used to test for spatial dependence can be adapted and applied to network data.

The most popular tests for spatial autocorrelation use Moran's I statistic [23] and Geary's C statistic [13] for continuous random variables. In a companion paper, we show that Moran's I provides valid tests of network dependence whenever the dependence is inversely related to a measure of network distance [16].

For categorical random variables, however, available tests based on join count analysis [5] are unwieldy and fail to provide a single omnibus test of dependence. Categorical random variables are especially important in social network settings, where group affiliations are often of interest [15, 19, 35]. Join count analysis has been recently used for testing autocorrelation in categorical outcomes sampled from social network nodes (e.g. [21]). Farber et al. [9] proposed a more elegant test for categorical network data and explored its performance in data generated from linear spatial autoregression (SAR) models [14, 20], which are parametric models for network data [9, 12]. As far as we are aware, all of the previous work on testing for network dependence in categorical variables assumes that the data were generated from SAR models, and none of this previous work has considered the performance of autocorrelation tests for more general network settings. Although SAR models are often used to model network dependent data, there is very little evidence that most social network data truly conform to these models. In particular, these models cannot capture general forms of latent variable dependence or of dependence due to direct transmission.

In this paper we propose a new test statistic that generalizes Moran's I for categorical random variables. We demonstrate that both Moran's I and our new test for categorical data can be used to test for dependence among observations sampled from a single social network (or a small number of networks). We assume that any dependence is monotonically inversely related to the pairwise distance between nodes, but otherwise we make no assumptions about the structure of the dependence, and we do not require any parametric assumptions. These tests allow researchers to assess the validity of i.i.d statistical methods, and are therefore the first step towards correcting the practice of defaulting to i.i.d. methods even when data may exhibit network dependence.

2 Methods

2.1 Moran's I

Moran's I takes as input an n-vector of continuous random variables and an $n \times n$ weighted distance matrix $\mathbf{W}$, where entry w_{ij} is a non-negative, non-increasing function of the Euclidean distance between observations i and j. Moran's I is expected to be large when pairs of observations with greater w values (i.e. closer in space) have larger correlations than observations with smaller w values (i.e. farther in space). The choice of non-increasing function used to construct $\mathbf{W}$ is informed by background knowledge about how dependence decays with distance; it affects the power but not the validity of tests of independence based on Moran's I.

Let Y be a continuous variable of interest and y_i be its realized observation for each of n units ($i = 1, 2, \ldots, n$). Each observation is associated with a location, traditionally in space but we will extend this to networks. Let $\mathbf{W}$ be a weight matrix signifying closeness between the units, e.g. a matrix of pairwise Euclidean distances for spatial data or an adjacency matrix for network data. (The entries A_{ij} in the adjacency matrix **A** for a network are indicators of whether nodes i and j share a tie.) Then Moran's I is defined as follows:

$$I = \frac{\sum_{i=1}^{n} \sum_{j=1}^{n} w_{ij}(y_i - \bar{y})(y_j - \bar{y})}{S_0 \sum_{i=1}^{n} (y_i - \bar{y})^2 / n}, \tag{1}$$

where $S_0 = \sum_{i=1}^{n} (w_{ij} + w_{ji})/2$ and $\bar{y} = \sum_{i=1}^{n} y_i/n$. Under independence, the pairwise products $(y_i - \bar{y})(y_j - \bar{y})$ are each expected to be close to zero. On the other hand, under network dependence adjacent pairs are more likely to have similar values than non-adjacent pairs, and $(y_i - \bar{y})(y_j - \bar{y})$ will tend to be relatively large for the upweighted adjacent pairs; therefore, Moran's I is expected to be larger in the presence of network dependence than under the null hypothesis of independence.

2.2 New Methods for Categorical Random Variables

For a K-level categorical random variable, join count statistics compare the number of adjacent pairs falling into the same category to the expected number of such pairs under independence, essentially performing K separate hypothesis tests. As the number of categories increases, join count analyses become quite cumbersome. Furthermore, they only consider adjacent observations, thereby throwing away potentially informative pairs of observations that are non-adjacent but may still exhibit dependence. Finally, the K separate hypothesis tests required for a join count analysis are non-independent and it is not entirely clear how to correct for multiple testing. To overcome this last limitation, Farber et al. [10] proposed a single test statistic that combines the K separate join count statistics.

Instead of extending join count analysis, we propose a new statistic for categorical observations using the logic of Moran's I. This has two advantages over the proposal of [10]: it incorporates information from discordant, in addition to concordant, pairs, and it weights pairs according to their probability under the null, allowing more "surprising" pairs to contribute more information to the test. To illustrate, under network dependence adjacent nodes are more likely to have concordant outcomes—and less likely to have discordant outcomes—than they would be under independence. We operationalize independence as random distribution of the outcome across the network, holding fixed the marginal probabilities of each category. The less likely a concordant pair (under independence), the more evidence it provides for network dependence, and the less likely a discordant pair (under independence), the more evidence it provides against network dependence. Using this rationale, a test statistic should put higher weight on more unlikely observations. The following is our proposed test statistic:

$$\Phi = \{\sum_{i=1}^{n}\sum_{j=1}^{n} w_{ij}\{2\mathbf{I}(y_i = y_j) - 1\}/p_{y_i}p_{y_j}\}/S_0, \quad (2)$$

where $p_{y_i} = P(Y = y_i)$, $p_{y_j} = P(Y = y_j)$, and $S_0 = \sum_{i=1}^{n}(w_{ij} + w_{ji})/2$. The term $(2\mathbf{I}(y_i = y_j) - 1) \in \{-1, 1\}$ allows concordant pairs to provide evidence for dependence and discordant pairs to provide evidence against dependence. The product of the proportions p_{y_i} and p_{y_j} in the denominator ensures that more unlikely pairs contribute more to the statistic. As the true population proportion is generally unknown, $\{p_k : k = 1, \ldots, K\}$ should be estimated by sample proportions for each category.

The first and second moment of Φ are derived in Appendix 1. Asymptotic normality of the statistic Φ under the null can also be proven based on the asymptotic behavior of statistics defined as weighted sums under some constraints. For more details see Appendix 2. For binary observations, which can be viewed as categorical or continuous, our proposed statistic has the desirable property that the standardized version of Φ is equivalent to the standardized Moran's I. Tests can be derived based on the asymptotic normal distribution of Φ under the null, but tests based on the permutation distribution of Φ when node labels are permuted but the adjacency matrix is held fixed may have better performance in finite sample sizes.

2.3 *Choosing the Weight Matrix* W

Tests for spatial dependence take Euclidean distances (usually in $\mathbb{R}^2$ or $\mathbb{R}^3$) as inputs into the weight matrix $\mathbf{W}$. In networks, the entries in $\mathbf{W}$ can be comprised of any non-increasing function of geodesic (or other) distance, but for robustness we use the adjacency matrix $\mathbf{A}$ for $\mathbf{W}$, where A_{ij} is an indicator of nodes i and j sharing a tie. The choice of $\mathbf{W} = \mathbf{A}$ puts weight 1 on pairs of observations at a distance of 1 and weight 0 otherwise. In many spatial settings, subject matter expertise can facilitate informed choices of weights for $\mathbf{W}$ (e.g. [27, 33]), and if researchers have concrete information about how dependence decays with geodesic network distance then a more informed choice of $\mathbf{W}$ can improve the power of the test.

3 Simulations

In Sect. 3.1, we demonstrate the validity and performance of our new statistic, Φ, for testing spatial autocorrelation in categorical variables. In Sect. 3.2, we demonstrate the performance of Φ for testing for network dependence.

3.1 Testing for Spatial Autocorrelation in Categorical Variables

We replicated one of the data generating settings used by Farber et al. [10] and implemented permutation tests of spatial dependence using Φ. First, we generated a binary weight matrix $\mathbf{W}$ with entries w_{ij} indicating whether regions i and j are adjacent. The number of neighbors (q_i) for each site i was randomly generated through $q_i = 1 + \text{Binomial}(2(d-1), 0.5)$ for a fixed parameter d that controls the expected number of neighbors. We simulated 500 independent replicates of $n = 100$ observations under each of four different settings, varying the values of $d = 3, 5, 7, 10$. We then used $\mathbf{W}$ to generate a continuous, autocorrelated variable:

$$Y^* = (I_n - \rho \mathbf{W})^{-1} \boldsymbol{\epsilon}, \quad \epsilon = \{\epsilon_i \overset{i.i.d.}{\sim} N(0, 1) : i = 1, \ldots, n\},$$

where I_n is a $n \times n$ identity matrix and ρ controls the amount of dependence. When $\rho = 0$, Y^*'s are i.i.d. while positive ρ induces some dependence among Y^*'s informed by $\mathbf{W}$. Since Y^* is continuous, we applied cutoffs based on the $(0.25, 0.5, 0.75)$ quantiles of each simulated dataset to convert Y^* into categorical observations $\mathbf{Y} = (Y_1, Y_2, \ldots, Y_n)$ having $K = 4$ categories.

Figure 1 presents the simulation results. It shows that under the null ($\rho = 0$), the rejection rate is close to the nominal level of $\alpha = 0.05$ and that the power to detect dependence increases with ρ. Moreover, as the expected number of adjacent neighbors, d, increases, power tends to increase at fixed ρ when ρ is relatively small ($\rho < 0.6$). This relationship is reversed when $\rho \geq 0.6$. This can be explained by

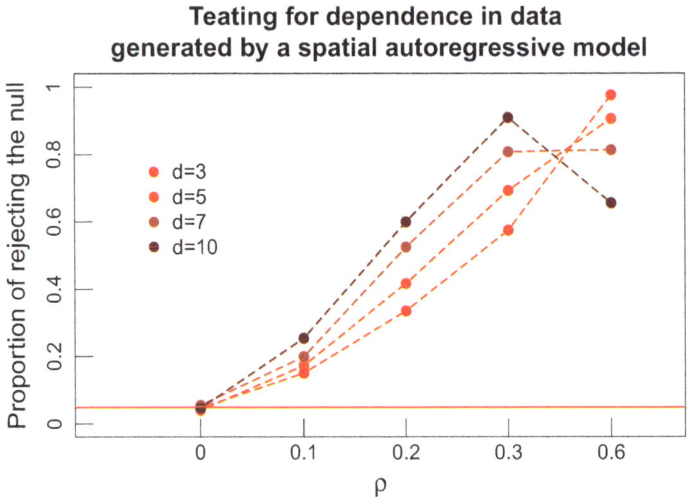

Fig. 1 Permutation tests based on Φ. Dependence increases as ρ increases, and the y-axis is the proportion of 500 independent simulations in which the test rejected the null hypothesis of independence

the fact that, when ρ and d are large relative to n, all of the data points tend to look similar to one another, leading to smaller contrasts between pairs of data points that are close and pairs that are distant, i.e. to weaker evidence for dependence. This is an inevitable feature of any test of dependence that does not rely heavily on a parametric data-generating model.

3.2 *Testing for Network Dependence*

To illustrate the performance of Φ, we simulated categorical outcomes Y associated with nodes in a single interconnected network and with dependence structure informed by the network ties. Y had five levels and marginal probabilities $(p_1, p_2, p_3, p_4, p_5) = (0.1, 0.2, 0.3, 0.25, 0.15)$; we seeded each node with independent outcomes and then induced dependence due to direct transmission by running a contagious process across the nodes over several time steps; details are provided in the Appendix 2. The number of time steps, t, indexes the amount of dependence induces, with $t = 0$ indexing i.i.d. observations. To demonstrate the consequences of using i.i.d. inference in the presence of dependence, as is currently standard practice for network data, we calculated simultaneous 95% confidence intervals for estimates of p_1 through p_5 (using the method proposed in [32]). We tested for network dependence using permutation tests based on Φ and report power as the percentage of 500 simulations in which the test rejected the null.

Table 1 summarizes the simulation results. As dependence increases, coverage rates of the 95% confidence intervals that were estimated under the i.i.d. assumption decrease, representing anticonservative inference. The power of Φ to reject the null simultaneously increases. These results indicate (a) that the common practice of using i.i.d. data for network data may be invalid, and (b) that tests based on Φ can operate as a good screening process for settings in which i.i.d. models are especially problematic.

The netdep R package for testing network dependence and generating network dependent observations is available through Github (https://github.com/youjin1207/netdep).

Table 1 Coverage rate of simultaneous 95% CIs, empirical power of tests of independence using asymptotic normality of Φ, and empirical power of permutation tests of independence based on Φ, under direct transmission for $t = 0, 1, 2, 3$

	95% CI coverage rate	% of p-values(z) ≤ 0.05	% of p-values (permutation) ≤ 0.05
$t = 0$	0.94	5.40	4.80
$t = 1$	0.81	39.40	36.20
$t = 2$	0.63	67.80	65.00
$t = 3$	0.43	85.40	83.40

The size of the tests is $\alpha = 0.05$

4 Applications

4.1 Spatial Data

In this section we apply Φ to spatial data on a categorical variable describing the race/ethnicity of populations immediately surrounding 473 U.S. power generating facilities [28]. We compare the results to standard analyses using join count statistics.

Figure 2a depicts the composition of the population living within a 100 km radius of each power generating facility, with the shade of each dot representing the proportion of the population falling into each race/ethnicity category (White/Hispanic/African American). We can apply Moran's I separately to data on each of the three categories, but Moran's I cannot provide a single aggregate test statistic. Figure 2b depicts the distributions of two alternative categorical summaries of the information from Fig. 2a: a 3-level variable indicating the most populous group in the area surrounding each facility, and a 4-level variable indicating whether more than 10% of the population is Hispanic and African American, respectively. Using each of these categorial variables, we can perform an omnibus test for dependence using Φ. We observe greater evidence of dependence in the second categorization (Φ:22.72) than the first categorization (Φ:9.17). This direct

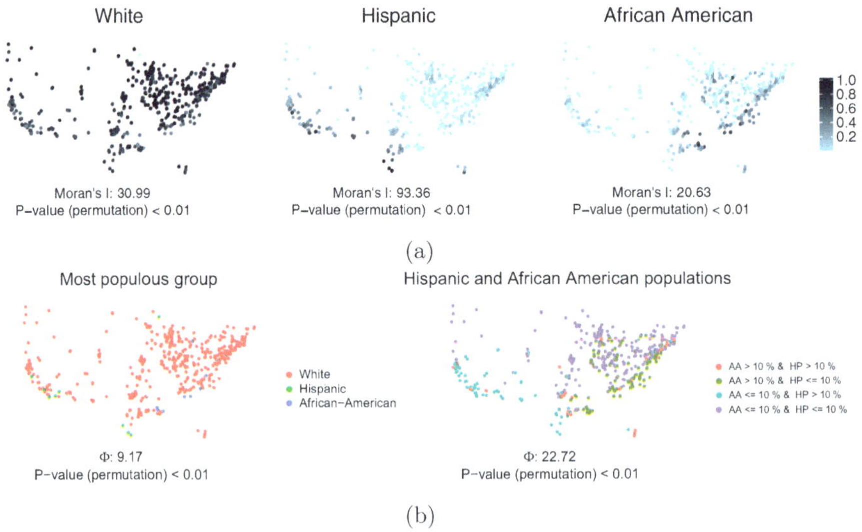

Fig. 2 Panel (**a**): Proportion of race/ethnicity groups around 473 power-producing facilities across the U.S. Applying Moran's I separately to each proportion, all of the tests reject the null hypothesis of independence at the $\alpha = 0.05$ level. Panel (**b**): Most populous group (left) and categories defined by having $\leq 10\%$ or $>10\%$ Hispanic or African American residents (right). Omnibus tests of dependence based on Φ reject the null hypothesis of independence at the $\alpha = 0.05$ level for both variables

Table 2 Permutation tests of dependence based on join count statistics applied to the most populous group

Most populous group	White	Hispanic	African–American
n	446	13	14
Join count statistic	212.63	0.97	0.77
p-value (permutation)	<0.01	<0.01	<0.01

Table 3 Permutation tests of dependence based on join count statistics applied to four different population categories, defined by having ≤ 10 or $>10\%$ Hispanic or African American residents

	AA $>$ 10%, HP $>$ 10%	AA $>$ 10%, HP $\leq$ 10%	AA $\leq$ 10%, HP $>$ 10%	AA $\leq$ 10%, HP $\leq$ 10%
n	52	106	98	217
Join-count statistic	7.07	26.63	30.30	69.20
p-value (permutation)	<0.01	<0.01	<0.01	<0.01

comparison is possible using Φ but would not be possible using join count statistics. The join count statistics for these two categorical variables are given in Tables 2 and 3. The statistics themselves count the frequency of concordant neighboring pairs for each category and standardize it; the p-values are derived from a permutation test that permutes the location of each observation while holding the values fixed. Join count analysis requires a notion of adjacency; we specified a neighborhood size of 15, meaning that observation j is considered to be adjacent to i if j is one of i's closet 15 neighbors in Euclidean distance.

4.2 Network Data

The Framingham Heart Study, initiated in 1948, is an ongoing cohort study of participants from the town of Framingham, Massachusetts that was originally designed to identify risk factors for cardiovascular disease. The study has grown over the years to include five cohorts. For decades, FHS has been one of the most successful and influential epidemiologic cohort studies in existence. It is arguably the most important source of data on cardiovascular epidemiology. It has been analyzed using i.i.d. statistical models (as is standard practice for cohort studies) in over 3400 peer-reviewed publications since 1950: to study cardiovascular disease etiology (e.g. [2, 7]), risks for developing obesity (e.g. [34]), factors affecting mental health (e.g. [29, 30]), and many other outcomes.

In addition to being a very prominent cohort study, more recently FHS has played a uniquely influential role in the study of social networks and social contagion. Researchers reconstructed the (partial) social network underlying the cohort and used this network to study social contagion and peer influence for a variety of outcomes in a series of highly influential papers [3, 4, 11]. However, even these

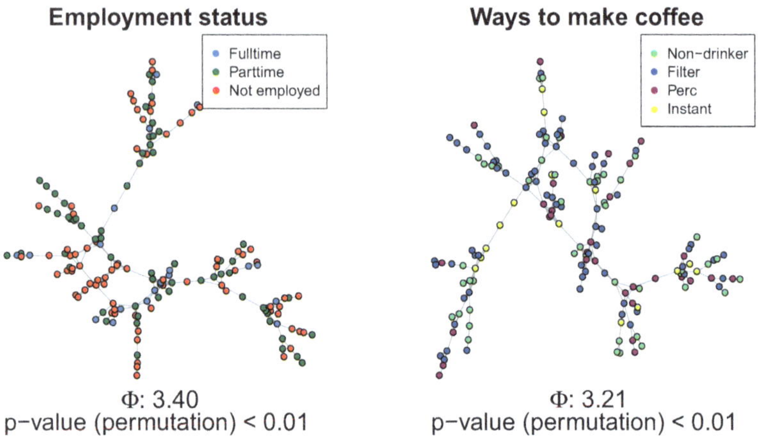

Fig. 3 Network dependence test for categorical variables with three levels (left) and four levels (right) using Φ

analyses use methods that assume independence across subjects [16, 22]. In a companion paper we test for dependence in continuous and binary variables in the FHS data, and discuss the implications of network dependence for the body of research that relies on i.i.d. analyses these data. Here we illustrate that dependence in these data may extend beyond continuous and binary variables to categorical variables, which previous methods would not have been able to ascertain. We analyzed $n = 1033$ subjects with 690 undirected social network ties from the Offspring Cohort at Exam 5, which was conducted between 1991 and 1995.

We tested for dependence in two different categorical random variables using Φ: employment status and preferred method of making coffee. Figure 3 shows the distribution of these two variables over the largest connected component of the network. We found significant evidence of network dependence for both variables, resulting p-value of <0.01 in both variables.

5 Concluding Remarks

In this paper, we proposed a simple test for dependence among categorical observations sampled from geographic space or from a network. We demonstrated the performance of our proposed test in simulations under both spatial and network dependence, and applied it to spatial data on U.S. power producing facilities and to social network data from the Framingham Heart Study.

Under network dependence, adjacent pairs are expected to exhibit the greatest correlations, and for robustness we used the adjacency matrix as the weight matrix for calculating the test statistic, thereby restricting our analysis to adjacent pairs;

if researchers have substantive knowledge of the dependence mechanism other weights may increase power and efficiency.

Researchers should be aware of the possibility of dependence in their observations, both when studying social networks explicitly and when observations are sampled from a single community for reasons of convenience. As we have seen in the classic Framingham Heart Study example, such observations can be dependence, potentially rendering i.i.d. statistical methods invalid. In a companion paper [16], we delve deeper into the consequences of assuming that observations are independent when they may in fact exhibit network dependence. That paper focuses on continuous and binary variables, but similar conclusions hold for the categorical variables that we addressed in this paper.

Acknowledgements Youjin Lee and Elizabeth Ogburn were supported by ONR grant N000141512343. The Framingham Heart Study is conducted and supported by the National Heart, Lung, and Blood Institute (NHLBI) in collaboration with Boston University (Contract No. N01-HC-25195 and HHSN268201500001I). This manuscript was not prepared in collaboration with investigators of the Framingham Heart Study and does not necessarily reflect the opinions or views of the Framingham Heart Study, Boston University, or NHLBI.

Appendix 1

Moments of Φ

Here we derive $\mu_\Phi := E[\Phi]$ and $E[\Phi^2]$, the first and second moments of Φ. Based on these moments, we can derive the variance of Φ, $\sigma^2_\Phi := E[\Phi^2] - \mu^2_\Phi$. When K is the number of categories and p_j is the proportion of Y in category j ($j = 1, 2, \ldots, K$),

$$
\begin{aligned}
\mu_\Phi &= \frac{1}{n(n-1)}\{n^2K(2-k) - nQ_1\},\\
E[\Phi^2] &= \frac{1}{S_0^2}\Bigg[\frac{S_1}{n(n-1)}(n^2Q_{22} - nQ_3)\\
&+ \frac{S_2-2S_1}{n(n-1)(n-2)}((K-4)K+4)n^3Q_1 + n(n((2K-4)Q_2-Q_{22})+2Q_3)\\
&+ \frac{S_0^2 - S_2 + S_1}{n(n-1)(n-2)(n-3)}\Big\{n(-4Q_3 + 2nQ_{22} - 6KnQ_2 + 12nQ_2\\
&- 3K^2n^2Q_1 + 14Kn^2Q_1 - 16n^2Q_1 + K^4n^3 - 4K^3n^3 + 4K^2n^3)\\
&- ((2K-4)n^2Q_2 + n^2(Kn(2Q_1 - KQ_1) - Q_{22}) + 2nQ_3)\Big\}\Bigg],
\end{aligned}
\tag{3}
$$

where $Q_m := \sum_{l=1}^{K} 1/p_l^m, (m = 1, 2, 3)$; $Q_{22} := \sum_{l=1}^{K}\sum_{u=1}^{K} 1/p_l p_u$; $S_0 = \sum_{i=1}^{n}\sum_{j=1}^{n}(w_{ij} + w_{ji})/2$; $S_1 = \sum_{i=1}^{n}\sum_{j=1}^{n}(w_{ij} + w_{ji})^2/2$; $S_2 = \sum_{i=1}^{n}(w_{i\cdot} + w_{\cdot i})^2$.

Asymptotic Distribution of Φ Under the Null

Shapiro and Hubert [31] proved the asymptotic normality of permutation statistics of the form H_n for i.i.d. random variables $Y_1, Y_2, \ldots, Y_n$ under some conditions:

$$H_n = \sum_{i=1}^{n} \sum_{j=1, j\neq i}^{n} d_{ij} h(Y_i, Y_j), \tag{4}$$

where $h(\cdot, \cdot)$ is a symmetric real valued function with $E[h^2(Y_i, Y_j)] < \infty$ and $\mathbf{D} := \{d_{ij}; i, j = 1, \ldots, n\}$ is a $n \times n$ symmetric, nonzero matrix of which all diagonal terms must be zero. In the context of Φ, $h(Y_i, Y_j) = \big(2I(Y_i = Y_j) - 1\big)/(p_{Y_i} p_{Y_j})$ and $\mathbf{D} = \mathbf{W}$. Requirements for asymptotic normality include $\sum_{i,j=1, j\neq i}^{n} d_{ij}^2 / \sum_{i=1}^{n} d_{i\cdot}^2 \to 0$ and $\max_{1\leq i\leq n} d_{i\cdot}^2 / \sum_{k=1}^{n} d_{k\cdot}^2 \to 0$ as $n \to 0$ for $d_{i\cdot} = \sum_{j=1}^{n} d_{ij}$. If we use the adjacency matrix for $\mathbf{W}$, this implies $\sum_{i,j=1,i\neq j}^{n} A_{ij} / \sum_{i=1}^{n} A_{i\cdot}^2 \to 0$ and $\max_{1\leq i\leq n} A_{i\cdot} / \sum_{i=1}^{n} A_{i\cdot}^2 \to 0$ where $A_{i\cdot}$ is the degree of node i. More details can be found in [31]; see also [25].

Appendix 2: Simulation of Categorical Observations Over Network

Direct Transmission Simulations

We specify the starting probability that each observation falls into one of K categories, $\{(p_1, p_2, \ldots, p_K) : \sum_{j=1}^{K} p_k = 1\}$. We then simulate initial outcomes from a multinomial distribution, and generate outcomes at subsequent time points iteratively:

$$Y_1^0, Y_2^0, \ldots, Y_n^0 \overset{i.i.d}{\sim} \text{Multinomial}\big((p_1, p_2, \ldots, p_K)\big)$$

$$Y_i^t = \begin{cases} Z_i^t \sim \text{Multinomial}\big((\hat{p}_{i1}^{t-1}, \hat{p}_{i2}^{t-1}, \ldots, \hat{p}_{iK}^{t-1})\big) & \text{with probability } q \\ Y_i^{t-1} & \text{with probability } 1-q \end{cases} \tag{5}$$

where $\hat{p}_{im}^{t-1} := \sum_{j=1}^{n} w_{ij} I(y_j^{t-1} = m) / \sum_{j=1}^{n} w_{ij};\ m = 1, .., K;\ 0 < q \leq 1$. At each time point, with probability q, a node's outcome is updated as a draw from a new multinomial with probabilities influenced by the proportion of adjacent nodes falling into each category at the previous time. The amount of influence from adjacent peers can be controlled by pre-specified maximum susceptibility probability $q_m (0 \leq q_m \leq 1)$, where $q \in [0, q_m]$, and we set $q_m = 0.4$.

References

1. Anselin, L., Bera, A.K., Florax, R., Yoon, M.J.: Simple diagnostic tests for spatial dependence. Reg. Sci. Urban Econ. **26**(1), 77–104 (1996)
2. Castelli, W.: Cholesterol and lipids in the risk of coronary artery disease–the Framingham heart study. Can. J. Cardiol. **4**, 5A–10A (1988)
3. Christakis, N.A., Fowler, J.H.: The spread of obesity in a large social network over 32 years. N. Engl. J. Med. **357**(4), 370–379 (2007)
4. Christakis, N.A., Fowler, J.H. The collective dynamics of smoking in a large social network. N. Engl. J. Med. **358**(21), 2249–2258 (2008)
5. Cliff, A.D., Ord, K.: Spatial autocorrelation: a review of existing and new measures with applications. Econ. Geogr. **46**(suppl.1), 269–292 (1970)
6. Cliff, A., Ord, K.: Testing for spatial autocorrelation among regression residuals. Geogr. Anal. **4**(3), 267–284 (1972)
7. D'Agostino, R.B., Vasan, R.S., Pencina, M.J., Wolf, P.A., Cobain, M., Massaro, J.M., Kannel, W.B.: General cardiovascular risk profile for use in primary care the Framingham heart study. Circulation **117**(6), 743–753 (2008)
8. Diniz-Filho, J.A.F., Bini, L.M., Hawkins, B.A.: Spatial autocorrelation and red herrings in geographical ecology. Glob. Ecol. Biogeogr. **12**(1), 53–64 (2003)
9. Farber, S., Páez, A., Volz, E.: Topology and dependency tests in spatial and network autoregressive models. Geogr. Anal. **41**(2), 158–180 (2009)
10. Farber, S., Marin, M.R., Páez, A.: Testing for spatial independence using similarity relations. Geogr. Anal. **47**(2), 97–120 (2015)
11. Fowler, J.H., Christakis, N.A.: Dynamic spread of happiness in a large social network: longitudinal analysis over 20 years in the Framingham heart study. BMJ **337**, a2338 (2008)
12. Fujimoto, K., Chou, C.P., Valente, T.W.: The network autocorrelation model using two-mode data: affiliation exposure and potential bias in the autocorrelation parameter. Soc. Networks **33**(3), 231–243 (2011)
13. Geary, R.C.: The contiguity ratio and statistical mapping. Inc. Stat. **5**(3), 115–146 (1954)
14. Griffith, D.A.: A linear regression solution to the spatial autocorrelation problem. J. Geogr. Syst. **2**(2), 141–156 (2000)
15. Kossinets, G., Watts, D.J.: Empirical analysis of an evolving social network. Science **311**(5757), 88–90 (2006)
16. Lee, Y., Ogburn, E.L.: Network dependence and confounding by network structure lead to invalid inference. arXiv preprint ArXiv:1908.00520 (2019)
17. Legendre, P.: Spatial autocorrelation: trouble or new paradigm? Ecology **74**(6), 1659–1673 (1993)
18. Lennon, J.J.: Red-shifts and red herrings in geographical ecology. Ecography **23**(1), 101–113 (2000)
19. Lewis, K., Kaufman, J., Gonzalez, M., Wimmer, A., Christakis, N.: Tastes, ties, and time: a new social network dataset using facebook.com. Soc. Networks **30**(4), 330–342 (2008)

20. Lichstein, J.W., Simons, T.R., Shriner, S.A., Franzreb, K.E.: Spatial autocorrelation and autoregressive models in ecology. Ecol. Monogr. **72**(3), 445–463 (2002)
21. Long, J., Harre, N., Atkinson, Q.D.: Social clustering in high school transport choices. J. Environ. Psychol. **41**, 155–165 (2015)
22. Lyons, R.: The spread of evidence-poor medicine via flawed social-network analysis. Stat. Polit. Policy **2**(1), 1–26 (2011)
23. Moran, P.A.: The interpretation of statistical maps. J. R. Stat. Soc. Ser. B Methodol. **10**(2), 243–251 (1948)
24. Ogburn, E.L.: Challenges to estimating contagion effects from observational data. arXiv preprint ArXiv:1706.08440 (2017)
25. O'Neil, K.A., Redner, R.A.: Asymptotic distributions of weighted U-statistics of degree 2. Ann. Probab. **21**, 1159–1169 (1993)
26. Ord, J.K., Getis, A.: Local spatial autocorrelation statistics: distributional issues and an application. Geogr. Anal. **27**(4), 286–306 (1995)
27. Overmars, K.P., De Koning, G., Veldkamp, A.: Spatial autocorrelation in multi-scale land use models. Ecol. Model. **164**(2), 257–270 (2003)
28. Papadogeorgou, G., Choirat, C., Zigler, C.M.: Adjusting for unmeasured spatial confounding with distance adjusted propensity score matching. Biostatistics **20**(2), 256–272 (2018)
29. Qiu, W.Q., Dean, M., Liu, T., George, L., Gann, M., Cohen, J., Bruce, M.L.: Physical and mental health of homebound older adults: an overlooked population. J. Am. Geriatr. Soc. **58**(12), 2423–2428 (2010)
30. Saczynski, J.S., Beiser, A., Seshadri, S., Auerbach, S., Wolf, P., Au, R.: Depressive symptoms and risk of dementia the Framingham heart study. Neurology **75**(1), 35–41 (2010)
31. Shapiro, C.P., Hubert, L., et al.: Asymptotic normality of permutation statistics derived from weighted sums of bivariate functions. Ann. Stat. **7**(4), 788–794 (1979)
32. Sison, C.P., Glaz, J.: Simultaneous confidence intervals and sample size determination for multinomial proportions. J. Am. Stat. Assoc. **90**(429), 366–369 (1995)
33. Smouse, P.E., Peakall, R.: Spatial autocorrelation analysis of individual multiallele and multilocus genetic structure. Heredity **82**(5), 561–573 (1999)
34. Vasan, R.S., Pencina, M.J., Cobain, M., Freiberg, M.S., D'Agostino, R.B.: Estimated risks for developing obesity in the Framingham heart study. Ann. Intern. Med. **143**(7), 473–480 (2005)
35. Weaver, I.S., Williams, H., Cioroianu, I., Williams, M., Coan, T., Banducci, S.: Dynamic social media affiliations among UK politicians. Soc. Networks **54**, 132–144 (2018)

Part II
Dynamics

Approximate Identification of the Optimal Epidemic Source in Complex Networks

S. Jalil Kazemitabar and Arash A. Amini

Abstract We consider the problem of identifying the source of a network epidemic from a complete snapshot of the infected nodes. We take a fully statistical approach and derive novel recursions to compute the Bayes optimal solution, under a heterogeneous susceptible-infected (SI) epidemic model. Our analysis is time and rate independent, and holds for general network topologies. We then provide two highly scalable algorithms for solving these recursions, a mean-field approximation and a greedy approach, and evaluate their performance on real and synthetic networks. Previous work on the problem has mostly focused on tree-like network topologies. Real networks are far from tree-like and an emphasis will be given to networks with high transitivity, such as social networks and those with communities. We show that on such networks, our approaches significantly outperform popular geometric and spectral centrality measures, most of which perform no better than random guessing.

1 Introduction

Modern transportation networks have had profound effects on geographical spread of infectious diseases [1, 2] giving rise to complicated epidemic evolutions [3]. These evolutions can be modeled as dynamic processes on transportation networks. The epidemic spread on networks can take other forms, such as outbreaks of foodborne diseases [4], intercontinental cascade of failures among financial institutions [5, 6], computer malware propagation on the internet and mobile networks [7, 8] spread of targeted fake news [9, 10] and rumors [11] on social media, especially during presidential elections [12–14]. In response to an adverse diffusion on a network, it is critical to trace back sources to enable appropriate prevention and containment of the spread [15]. Inferential methods have been developed to locate

S. J. Kazemitabar (✉) · A. A. Amini
University of California, Los Angeles, CA, USA
e-mail: sjalilk@ucla.edu; aaamini@ucla.edu

N. Masuda et al. (eds.), *Proceedings of NetSci-X 2020: Sixth International Winter School and Conference on Network Science*, Springer Proceedings in Complexity,
https://doi.org/10.1007/978-3-030-38965-9_8

the source of foodborne diseases [16, 17] and influenza pandemics [18, 19]. In the context of online social networks, the spread of misinformation can be limited by the identification of influential users [20, 21]. Source recovery can also be used to assess the power of diffusions in generating anonymity in network protocols [22].

The epidemic source identification problem has received considerable attention in the past decade. Given a snapshot of the infected nodes in a network, the task is to discover who has originated the epidemic. Since the seminal work of Shah and Zaman [23], numerous attempts have been made to address the question and its extensions [24–29]. By now, there are multiple methods that show satisfactory results in limited experimental setups or have proven guarantees in restricted network topologies [30]. However, identifying the source under general conditions still remains a difficult task. The problem of optimal recovery appears to be NP-hard in infection size [28, 31]. The theoretical guarantees for optimal and consistent recovery are restricted to regular infinite trees [23, 26], and as we show in this paper, the popular and well-cited methods are quite unreliable in a wide range of real networks.

Source identification has remained largely unsolved and poorly understood for real complex networks [30]. As we will show through experiments in Sect. 5, in real networks, even the optimal Bayes estimator applied to small infected sets has difficulty narrowing down to the true source. It is thus important to recover as much information from the likelihood of the model as possible. We develop techniques for computing the full likelihood of the infection, as opposed to identifying the most likely sample-path [26]. Moreover, we fully exploit the information from the boundary of the infection set, in addition to the structure inside the infected subgraph. We develop all these ideas without restricting the structure of the network to trees. Our framework also easily extends to the case where there are multiple infecting sources (Appendix 1).

In this paper, we develop statistical algorithms that outperform the state-of-the-art in a wide range of network topologies. Our contributions are distinct in several ways:

1. Our methods are parameter-free, meaning that they do not require knowing the duration of the epidemic or how fast it grows [25, 32].
2. We show that the exact maximum likelihood estimator (MLE) of the source—or equivalently the Bayes optimal solution under uniform prior—can be written as a dynamic programming (DP), with easily computable coefficients based on the adjacency matrix of the network.
3. We develop two schemes to approximate the DP: an efficient greedy elimination (GE), and a novel mean-field approximation (MFA) of the likelihood, computed by solving a linear system. MFA and GE both perform well in naturally occurring networks, extend directly to heterogeneous infection probabilities, and are scalable, while competing methods fail to succeed in general topology.
4. Our approximations are more disciplined than existing approaches. They do not impose restrictions on the topology of the network. Nor do they appeal to the

partial likelihood of the candidate infecting sets. This is in contrast to the use of spanning trees to deal with general topologies [23, 33] or the *path-based approaches* that rely on the likelihood of individual paths from potential sources to the infected set [26].

We will show that when applied to real networks, both approximation schemes (MFA and GE) outperform various geometric and spectral approaches, most of which perform no better than random guessing. We also show that even for basic models of real networks, e.g., models with community structure, most existing methods dramatically fail. The improvement in performance is most significant for the networks with many cycles, including social networks that are known to have high transitivity. In terms of computational efficiency, both the greedy and mean-field approximations are superior to the state-of-the-art likelihood-based and spectral approaches and comparable to centrality-based methods. In addition, the mean-field algorithm is easily parallelizable through standard linear algebraic routines and can be used to tackle very large-scale epidemics on real networks.

Related Work Most of the existing literature on the source identification problem are based on a SIR dynamic where the infection spreads with an exponential rate proportional to the number of infected neighbors. All nodes are *susceptible* to the infection and once *infected* may *recover* with a fixed exponential rate [34]. Moreover, the spread of infection through edges are mutually independent. Different variations of SIR may assume that no recovery is possible (SI) or the recovered is not immune to iterated infections (SIS).

Shah and Zaman [23] considered the SI dynamics and proposed the Rumor Centrality (RC), which counts the *permitted permutations*, a.k.a. infection paths, inside the infected subgraph. Their linear time algorithm is an optimal estimator in regular trees and enjoys strong theoretical properties in such idealized settings [35]. Zhou and Ying [26] consider SIR dynamics on a tree and show that the most likely infection path is rooted at a Jordan center (JC) of the infected set O, that is, a node with minimal eccentricity (i.e., minimal maximum distance to other nodes). It has been shown [26, 35] that in regular trees, eccentricity ranking generates, with high probability, a confidence set containing the true source, whose size does not grow with the infection size.

The Dynamic Message Passing (DMP) was proposed in [25] as an approximation of the maximum likelihood estimator in discrete SIR epidemics, by approximating the probability of an infected set, as the product of the marginal probabilities of infection for each node (i.e., a form of pseudo-likelihood). Despite compelling performance, DMP is computationally intensive and impractical for large networks with moderately dense structures, even for small infection sets. A spectral algorithm, called Dynamical Age (DA) was introduced in [24], based on how sensitive the maximum eigenvalue of the Laplacian matrix is to the elimination of each node in the infection set. The algorithm was mainly developed to discover the initial node in a growing preferential attachment model. Another spectral method for the discrete SI model is proposed in [29].

2 Source Detection in SI Epidemics

We consider a continuous-time heterogeneous susceptible-infected (SI) epidemic [34] with rate of infection β, on a static weighted (directed) network $G(V, E)$ with known edge set E and $V = [n]$. At time zero, all nodes but the source are in the susceptible state. Infection is a terminal state and susceptible nodes are exposed to the infection at an exponential rate proportional to the number of their infected neighbors. More precisely, given that nodes I are infected at some time t, we run exponential clocks $T_j \sim \text{Exp}(\beta \, \text{vol}(I, j))$ for all $j \in I^c$ and the first to expire determines the next infected node: If $\boldsymbol{j}^* = \text{argmin}_j \, T_j$, then the dynamics move to the infected set $I \cup \{\boldsymbol{j}^*\}$ at time $t + T_{\boldsymbol{j}*}$. It is clear that the contagion will eventually spread through the entire graph.

The infection source or patient zero, denoted as $\boldsymbol{i}_*$, is unknown. What we observe is a snapshot of the contagion at some time t, meaning the entire set of infected nodes at that time, which we denote by O. The objective is to find $\boldsymbol{i}_* \in O$ or form a confidence set for $\boldsymbol{i}_*$ with desired false exclusion probability. Our focus here will be on the single source setting, but the analysis is extensible to the multi-source setting (cf. Sect. 5).

Notation We write $A \in [0, 1]^{n \times n}$ for the weighted (asymmetric) adjacency matrix of the network and $\text{vol}(I, J) := \sum_{i \in I, j \in J} A_{ij}$ for the volume of a cut in the network between subsets $I, J \subset [n]$ of nodes. For singleton subsets, we often drop the braces, e.g., $\text{vol}(I, j) := \text{vol}(I, \{j\})$ and $O \setminus j = O \setminus \{j\}$.

2.1 *Time and Rate Invariant Analysis*

We start by examining the probability of observing a particular set of infected nodes given a starting source. Let us introduce a parameter-free formulation of the problem (i.e. not dependent on rate β and time t) that will be the foundation for our analysis of the continuous SI dynamics.

Suppose that, at some point in time, the infection reaches $I \subset [n]$. Let $O \subset [n]$ be some superset of I. We are interested in computing $\rho_{I \to O}$, the chance that all the nodes in O are infected before any node outside. More precisely, let

$$\rho_{I \to O} := \mathbb{P}\big(O \text{ is infected before } O^c \mid I \text{ is infected}\big). \tag{1}$$

We refer to $\rho_{I \to O}$ as the *transition probabilities*. Note that these transition probabilities are independent of the infection source. Given that in a snapshot of the contagion, nodes I are infected, $\rho_{I \to O}$ determines how likely it is that in some future snapshot, O is the set of infected nodes. The Markov property of (continuous-time) SI dynamics allows us to define $\rho_{I \to O}$ without reference to the source, or the time of the first snapshot. We will also show that these probabilities do not require the knowledge of the infection rate or the time of the second snapshot.

2.2 Statistical Inference

Given the observed (random) infected set O, the function $I \mapsto \rho_{I\to O}$ is *the likelihood* of the model. Writing $L_O(I) := \rho_{I\to O}$ for this likelihood, we observe that $L_O(I) = 0$ for all I not contained in O. So, we can restrict $L(\cdot)$ to all subsets of O. When dealing with the single-source setup, we restrict the parameter space to $I = \{i\}$ and with some abuse of notation write $\rho_{i\to O}$ for $\rho_{\{i\}\to O}$, and $L_O(i) = \rho_{i\to O}, i \in [n]$ for the likelihood.

We can further consider a Bayesian setup by putting a uniform prior on the source (i.e., uniform over $[n]$). The Bayesian setup allows us to consider various notions of optimality by changing the loss function. Letting $\boldsymbol{i}_*$ be the random initial source, we have a joint distribution on $(\boldsymbol{i}_*, O)$. Then the posterior probability that the source is i, given that we observed infected nodes O is

$$p_i := \mathbb{P}(\boldsymbol{i}_* = i \mid O) = \frac{\rho_{i\to O}}{\sum_{j\in O}\rho_{j\to O}} 1\{i \in O\}.$$

Therefore, the maximum a posteriori (MAP) estimate of the source is $\boldsymbol{i}^*_{\text{MAP}} = \operatorname{argmax}_i \rho_{i\to O}$ which minimizes the probability of error. That is, i^*_{MAP} minimizes $\mathbb{P}(\hat{\boldsymbol{i}} \neq \boldsymbol{i}_*)$ for any estimator $\hat{\boldsymbol{i}} = \hat{\boldsymbol{i}}(O)$. In some applications, the graph geodesic distance (d_G) to the source determines the error of estimation. In that case, the Bayes optimal estimator is $\boldsymbol{i}^*_{\text{dist}} = \operatorname{argmin}_i \sum_{j\in O} \text{dist}_G(i, j)\, \rho_{j\to O}$. It is not hard to see that $\boldsymbol{i}^*_{\text{dist}}$ minimizes $\mathbb{E}[d_G(\hat{\boldsymbol{i}}, \boldsymbol{i}_*)]$ among all possible estimators $\hat{\boldsymbol{i}}$.

A third choice is to output a ranking instead of a single source. In this case, an estimator is formally a permutation $\hat{\boldsymbol{\sigma}} = \hat{\boldsymbol{\sigma}}_O$ on $[n]$, suppressing the dependence on O for simplicity. We can then consider the *rank loss* $\ell(\hat{\boldsymbol{\sigma}}, \boldsymbol{i}_*) = \hat{\boldsymbol{\sigma}}(\boldsymbol{i}_*)$, and we call the associated risk the *expected (source) rank* $= \mathbb{E}\hat{\boldsymbol{\sigma}}(\boldsymbol{i}_*)$. The corresponding optimal Bayes estimator is obtained by minimizing the posterior risk:

$$\hat{\boldsymbol{\sigma}}^* := \underset{\sigma:[n]\to[n]}{\operatorname{argmin}}\ \mathbb{E}[\sigma(\boldsymbol{i}_*) \mid O].$$

Noting that $\mathbb{E}[\sigma(\boldsymbol{i}_*) \mid O] = \sum_i \sigma(i)\, p_i$, the optimal estimator in this case is the ranking that sorts p_i into descending order, i.e., $\hat{\boldsymbol{\sigma}}^*(j_i) = i$ where $p_{j_1} \geq p_{j_2} \geq \cdots \geq p_{j_n}$.

Remark 1 The distance loss might be suitable in some applications, but in general it is a poor measure if the goal is to reveal the actual source. This is especially true in small world networks, including most social networks, where the expected distance between any pair of nodes is small. On the other extreme, in terms of the precision in recovering the source, is the zero-one loss which is too stringent. The rank loss can be considered a more robust version of the zero-one loss, and it will be our main evaluation measure.

3 Exact Likelihood Computation

The Bayesian estimators introduced in Sect. 2.2 require us to evaluate the posterior probabilities (p_i), or equivalently the likelihood values $\rho_{j\to O}$ for all $j \in O$. The main difficulty of the source identification problem is that computing the likelihood is itself challenging. We now develop exact equations that allow us to recursively compute the likelihood values $L_O(I)$ for all subsets $I \subset O$.

Dynamic Programming To begin, note that $\rho_{O\to O} = 1$ for any $O \subset [n]$. In addition, $\rho_{I\to O} = 1$ whenever O corresponds to a connected component of G. We develop two dynamic programming expressions for $\rho_{I\to O}$ for general $I \subset O$:

Proposition 1 *For $I \subset O \subset [n]$, the probabilities $\rho_{I\to J}$ defined in* (1) *satisfy the forward program*

$$\rho_{I\to O} = \sum_{j\in O\setminus I} \frac{\mathrm{vol}(I, j)}{\mathrm{vol}(I, I^c)}\, \rho_{I\cup j\to O} \tag{2}$$

and the backward program

$$\rho_{I\to O} = \sum_{j\in O\setminus I} \rho_{I\to O\setminus j} \frac{\mathrm{vol}(O\setminus j, j)}{\mathrm{vol}(O\setminus j, (O\setminus j)^c)}. \tag{3}$$

In the forward programming (2), j effectively iterates over the boundary of I in O, as $\mathrm{vol}(I, j) = 0$ if j is outside that boundary. Therefore, the running time of the forward programming benefits from the sparsity of the network. Unlike the forward programming, the iteration over j in (3) cannot be restricted to a smaller set. A corollary of Proposition 1 is that the transition probabilities $\rho_{I\to J}$ are not affected by the rate and the duration of the infection.

Let us now observe some connection with the *path-based analysis*. A permitted permutation or an infection path starting at a node $\boldsymbol{i}_*$, refers to a permutation σ of nodes with $\sigma_1 = \boldsymbol{i}_*$, and such that σ_{k+1} is connected to at least one node in $\{\sigma_1, \ldots, \sigma_k\}$, for all $k \in [|\sigma| - 1]$. Notice that the probability of observing a given infection path is

$$\mathbb{P}\big(\text{path } \sigma \text{ observed} \mid \sigma_1 = \boldsymbol{i}_*\big) = \prod_{k=1}^{|\sigma|-1} \frac{\mathrm{vol}(\sigma_{[k]}, \sigma_{k+1})}{\mathrm{vol}(\sigma_{[k]}, \sigma_{[k]}^c)} \tag{4}$$

where $\sigma_I := (\sigma_i \mid i \in I)$. One can obtain the transition probability $\rho_{\{\boldsymbol{i}_*\}\to O}$ by summing (4) over all infection paths σ such that $\sigma_1 = \boldsymbol{i}_*$ and $\{\sigma_1, \ldots, \sigma_k\} = O$. Our recursive representation is novel, avoids these explicit summations, and will be key in deriving approximation schemes for $\rho_{I\to O}$ in Sect. 4.

Path-based approaches such as Jordan center [26] forgo computing the complete likelihood (i.e., avoid summing the odds of all infection paths) and instead find the most probable path, that is, one that maximizes (4) in a spanning tree. In contrast, Eqs. (2) and (3) compute the complete likelihood of the infection set, which has the following advantages over the path-based likelihood: It fully exploits the structure of the graph inside the infection set, not just a spanning tree or a permitted permutation of nodes in the infected subgraph. Moreover, it takes into account the boundary of the infected subgraph via $\mathrm{vol}(I, I^c)$.

Remark 2 Some previous papers, such as [25, 32], considered the discrete-time susceptible-infected dynamic. In that setup, the rate and time parameters are intertwined with the transition probabilities in a way that it is hard or infeasible to disentangle them. Therefore, the authors proposed to take β and t as inputs or estimate the probabilities for multiple candidates for the infection time. In this sense, our approach studies a more realistic model with less adverse consequences for estimation.

4 Approximations

We now provide two approximations to the likelihood function $L_O(I)$ based on the exact dynamic programming developed in Proposition 1.

Greedy Elimination (GE) We can obtain a singleton source set $I = \{i\}$ that maximizes $\rho_{I \to O}$ with greedy elimination of elements in O. The algorithm we propose is based on the backward recursion (3) and is detailed in Algorithm 1. We start with $O_0 := O$ and consider all maximal proper subsets of O_0 that induce a connected subgraph of G. Among those, we choose the one that maximizes the transition probability to O_0, i.e. $\rho_{O_0 \setminus j \to O_0} = \mathrm{vol}(O_0 \setminus j, j) / \mathrm{vol}(O_0 \setminus j, (O_0 \setminus j)^c)$. Suppose that $O_1 := O_0 \setminus j^*$ is the maximizer. Next, we iterate the same procedure for O_1 and so forth, until we reach a singleton set $I := O_{|O|-1}$. The procedure has an $O(k^2 m)$ runtime where $k = |O|$and m is the number of edges in the infected subgraph, G_O.

GE has a Bayesian justification. Let $\widetilde{O}_k$ be the random infected set after k steps. Suppose that we want to find the MAP for $\widetilde{O}_{k-1}$ given $\widetilde{O}_k$. The Bayesian posterior probability is

$$\mathbb{P}(\widetilde{O}_{k-1} = O \backslash j \mid \widetilde{O}_k = O) \propto \rho_{O \backslash j \to O} \cdot P(\widetilde{O}_{k-1} = O \backslash j).$$

Whenever $G_{O \backslash j}$ is connected, the prior is positive. GE finds a proxy for MAP through maximizing the evidence and ensuring the prior is positive.

Algorithm 1 has similarities with finding the most likely path from a source to the observed snapshot. Chang et al. [36] propose a similar path-based search called GSBA. They start from each node in O and approximate the most likely path and use it as a proxy to the most likely source. Algorithm 1, however, does this greedy search in a backward fashion.

Algorithm 1 Greedy elimination

Input: Graph $G([n], E)$ and $O \subset [n]$.
Output: $\boldsymbol{i}^*_{\mathrm{GE}} \in O$.
1: $O_0 := O$
2: **for** $i := 0$ to $|O| - 2$ **do**
3: $\quad O'_i := \{j \in O_i : G_{O_i \setminus j}$ remains connected$\}$
4: $\quad j^* := \operatorname{argmax}_{j \in O'_i} \dfrac{\mathrm{vol}(O_i \setminus j, j)}{\mathrm{vol}(O_i \setminus j, (O_i \setminus j)^c)}$.
5: $\quad O_{i+1} := O_i \setminus j^*$.
6: **end for**
7: $\boldsymbol{i}^*_{\mathrm{GE}} :=$ the single element in $O_{|O|-1}$.

Algorithm 2 Mean-field approximation

Input: Graph $G([n], E)$ and $O \subset [n]$.
Output: $\boldsymbol{i}^*_{\mathrm{MFA}} \in O$.
1: Compute S, z as defined in (7).
2: $\widehat{\boldsymbol{b}} := S^{-1} z$.
3: $\boldsymbol{i}^*_{\mathrm{MFA}} := \operatorname{argmax}_{j \in O} \widehat{\boldsymbol{b}}_j$.

Mean-Field Approximation (MFA) We now approximate $\rho_{I \to O}$ by the mean-field technique. The idea is to treat the set function $I \mapsto \rho_{I \to O}$ as if it was a distribution (or measure) on O and approximate it by the product of its marginals. Fix a subset $O \subset [n]$. For any $I \subset O$, let $\boldsymbol{x}^I = (x^I_j)_{j \in O}$ be the binary representation of I, i.e. $x^I_j = 1\{j \in I\}$ for any $j \in O$. We find α_0 and $(b_j)_{j \in O}$ such that

$$\widehat{\rho}_{I \to O} = \alpha_0 \prod_{j \in O} b_j^{x^I_j - 1} \tag{5}$$

is a good approximation to $\rho_{I \to O}$ for all $I \subset O$, in the sense of minimizing the quadratic deviation from the solution of the recursion (2). First note that $\alpha_0 = 1$ since $\rho_{O \to O} = 1$. Next, we plug-in $\widehat{\rho}_{I \to O}$ into the forward recursion, to get

$$\mathrm{vol}(I, I^c)\, \widehat{\rho}_{I \to O} - \sum_{j \in O \setminus I} \mathrm{vol}(I, j)\, \widehat{\rho}_{I \cup \{j\} \to O} = 0.$$

Dividing both sides by $\prod_{j \in O \setminus I} b_j$ gives $\mathrm{vol}(I, I^c) - \sum_{j \in O \setminus I} \mathrm{vol}(I, j)\, b_j = 0$. These equations in general cannot be satisfied exactly for all $I \subset O$. Instead, letting $\boldsymbol{b} = (b_j)_{j \in O}$, we solve the following least-squares problem:

$$\widehat{\boldsymbol{b}} \in \operatorname*{argmin}_{\boldsymbol{b}} \sum_{I : I \subset O} \Big(\mathrm{vol}(I, I^c) - \sum_{j \in O \setminus I} \mathrm{vol}(I, j)\, b_j \Big)^2 = \operatorname*{argmin}_{\boldsymbol{b}} \| Q\boldsymbol{b} - \boldsymbol{r} \|_2^2 \tag{6}$$

where $Q \in \mathbb{R}^{(2^{|O|}-1)\times|O|}$ and $\boldsymbol{r} \in \mathbb{R}^{(2^{|O|}-1)\times 1}$ are defined as follows:

$$Q_{I,j} = 1\{j \notin I\}\ \mathrm{vol}(I, j),\ \forall I \subset O,\ j \in O, \qquad \boldsymbol{r}_I = \mathrm{vol}(I, I^c),\ \forall I \subset O.$$

The solution of (6) satisfies the normal equations $Q^T Q\widehat{\boldsymbol{b}} = Q^T\boldsymbol{r}$. The following proposition shows that $Q^T Q$ and $Q^T\boldsymbol{r}$ can be computed efficiently. Let A be the adjacency matrix of the network.

Proposition 2 *The solution $\widehat{\boldsymbol{b}}$ of* (6) *satisfies the linear system $S\widehat{\boldsymbol{b}} = z$ with S and z given by*

$$\begin{aligned} S &= \Xi\big(A_{OO} \odot A_{OO}^T + A_{OO}^T A_{OO} - A_{OO} \odot (\boldsymbol{u}\mathbf{1}^T) \\ &\quad - A_{OO}^T \odot (\mathbf{1}\boldsymbol{u}^T) + \boldsymbol{u}\boldsymbol{u}^T\big) \in \mathbb{R}^{|O|\times|O|}, \\ z &= (\mathbf{1}^T\boldsymbol{u} + 2\mathbf{1}^T\boldsymbol{v})\boldsymbol{u} - 2\boldsymbol{v} \odot \boldsymbol{u} + 2A_{OO}\,\boldsymbol{v} + (\boldsymbol{u} - \boldsymbol{u}^{out}) \odot \boldsymbol{u} \\ &\quad + \big((A_{OO} + A_{OO}^T) \odot A_{OO}^T\big)\mathbf{1} + A_{OO}^T(\boldsymbol{u}^{out} - \boldsymbol{u}) \end{aligned} \tag{7}$$

where $\boldsymbol{u} := A_{OO}^T\mathbf{1}$, $\boldsymbol{u}^{out} := A_{OO}\mathbf{1}$, and $\boldsymbol{v} = A_{OO^c}\mathbf{1}$. Here $\odot$ is the element-wise matrix product, $\Xi(\cdot)$ is a matrix operator that returns the same matrix with double the diagonal entries, and $\mathbf{1}$ is the vector of all ones.

See Appendix 2 for the proof. Proposition 2 shows that the mean-field approach reduces to solving a linear system of equations in $|O|$ variables, a task with much better computational complexity than solving the original recursion. Both S and z can be computed in at most $O(k^2)$ time, where $k = |O|$. In the cases where A is sparse (which often the case for real networks), S will be a rank-one perturbation of a sparse matrix (both A_{OO} and $A_{OO}^T A_{OO}$ will be sparse), hence solving the resulting system is often much faster than the worst-case, i.e., faster than $O(k^3)$.

Remark 3 MFA and GE utilize the forward and backward programs ((2), (3)), respectively. We have tried to apply linearization to the backward program and greedy inclusion to the forward program. However, the former does not go through as smoothly and the latter leads to a sub-par method. Whether one can utilize both recursions simultaneously to achieve a better performance is open.

5 Simulations

The methods proposed in this paper, the Greedy Elimination (GE) and the Mean Field Approximation (MFA), show superior performance in source identification, compared to popular procedures, while having comparable runtimes. In this section, we make a comparison based on these two measures (source identification ability and runtime) on real and synthetic networks. As discussed in Sect. 2.2, we consider ranking estimators (i.e., those that output a permutation of the nodes according

Table 1 Network statistics

Network	Internet	Power	Wiki vote	UCSC68	UC64	DC-SBM
n	10670	4941	7066	8979	6810	1962
Mean degree	4	3	29	50	46	66
Max. degree	2312	19	1065	454	660	897
Clust. coeff.	0.01	0.10	0.13	0.17	0.19	0.30

to their likelihood of being the source) and focus on the rank loss. If the method does not return a ranking, we tweak it to do so. We evaluate the methods based on the expected rank, $\mathbb{E}[R]$, where R is the rank of the true source among the list of candidates (cf. Sect. 2.2). The expectation is taken with respect to the variation in choosing the true source, which is drawn at random from the entire network. We normalize the expected rank to get a number in [0, 1], with zero corresponding to perfect recovery, i.e., we use $(\mathbb{E}[R] - 1)/n$.

We consider a variety of real and simulated networks. Our selection includes an Internet Autonomous System [37, 38], US west-coast power grid [39], two Facebook-100 networks [40, 41], called UC64 and UCSC68, and a Wikipedia voting network [42]. In addition, we present our results on a number of synthetic networks that are well studied in the literature, including regular trees, random trees, and degree-corrected stochastic block models (DC-SBM) [43].

Table 1 summarizes the statistics on the largest connected component of these networks. The regular tree is of degree 3 and depth 10. The random tree has 500 nodes. For the DC-SBM network, we generate from a 3-community planted partition version, i.e., $\mathbb{E}[A_{ij}] = \theta_i \theta_j P_{ij}$ where $P_{ij} = 0.5$ if nodes i and j are in the same community and $P_{ij} = 0.02$ if they are in different communities. The degree parameters θ_i are generated from a rescaled Pareto distribution with $\alpha = 2$ and threshold $= 1$.

The results are illustrated in Figs. 1 and 2. The methods we consider besides the optimal Bayes solution (BO), the MFA, and the GE are the Rumor Centrality (RC), the Degree Centrality (DC), the Jordan Center (JC) and the Dynamical Age (DA). Our selection of the methods loosely follows the methods surveyed in [30]. Each curve shows the performance of one method for different values of the infection size, $2 \leq |O| \leq 300$. Each point is an average over 500 infection paths rooted at random sources. To avoid an unreasonable computation time, we skip the BO for the infected sets of size greater than 10. The BO curve serves as the benchmark for the best achievable performance. Note that even the optimal solution needs to output a large set to catch the source, signifying the inherent difficulty of the problem.

Rumor and Jordan centralities perform optimally on regular trees in Fig. 1a, as predicted by the theory [23, 26], although the network here is not exactly an *infinite* tree. Notice that RC, JC, and BO overlap for infection sizes not exceeding the depth of the tree. Degree centrality also turns out to be a close competitor in this figure. Moving to other networks, however, these popular methods do not perform better than random guessing. For all the three, the expected relative rank is close to 0.5,

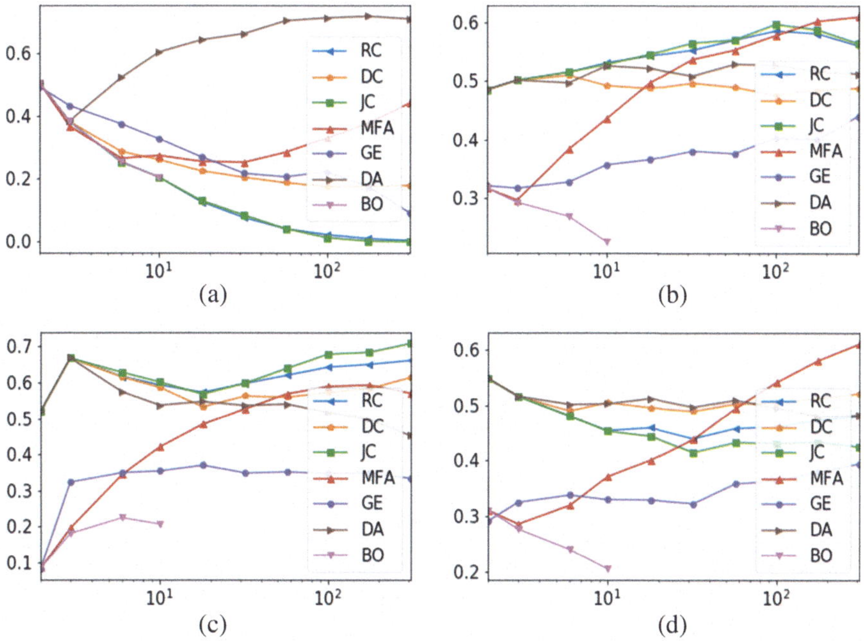

Fig. 1 Plots of the expected relative rank versus the infection size for low-transitivity networks. (**a**) Regular tree. (**b**) Random tree. (**c**) Internet AS. (**d**) US west power grid

even in a random tree. The plots in this section show that, despite their popularity, the RC and JC are quite unreliable for source recovery.

DA tends to perform well only when the infection size is sufficiently large. In some of the networks, i.e., in Fig. 2b–d, it is a close competitor to GE and MFA, while still behind them with a margin. DA also performs close to GE in the "Internet AS" network (Fig. 1c).

Among our proposed methods, MFA outperforms RC, JC, DC, and DA in Fig. 2a–d. MFA ranks the true source, on average, in the top 30%. The networks with superior MFA performance have the highest transitivity (a.k.a. clustering coefficient) in Table 1, that is, many triangles among triples of nodes. Transitivity has been studied extensively and distinguishes human social networks from random trees and less cyclical networks, such as the water distribution systems and traffic networks. In this sense, MFA is suitable for rumor source detection in social networks.

GE is the global winner, except in regular trees (Fig. 1a). We were surprised that a greedy algorithm had such a widespread success. GE not only performs well in highly transitive networks, but also outperforms RC, JC, DC, and DA on random trees (Fig. 1b) and less transitive networks (Fig. 1c, d).

Figure 3 illustrates the runtimes (on the log scale) for a single run on the UC64 network. Degree centrality is the fastest, followed by RC, MFA, GE, JC, and DA. The first three have comparable speed and scale quite well. In contrast, although the

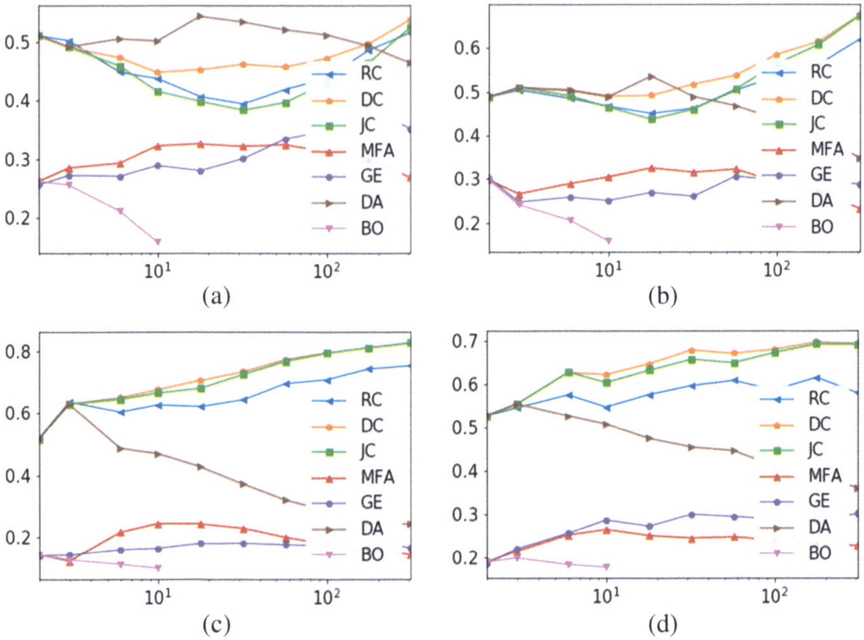

Fig. 2 Plots of the expected relative rank versus the infection size for high-transitivity networks. (**a**) UCSC68. (**b**) UC64. (**c**) Wiki vote. (**d**) DC-SBM

Fig. 3 Runtime in seconds

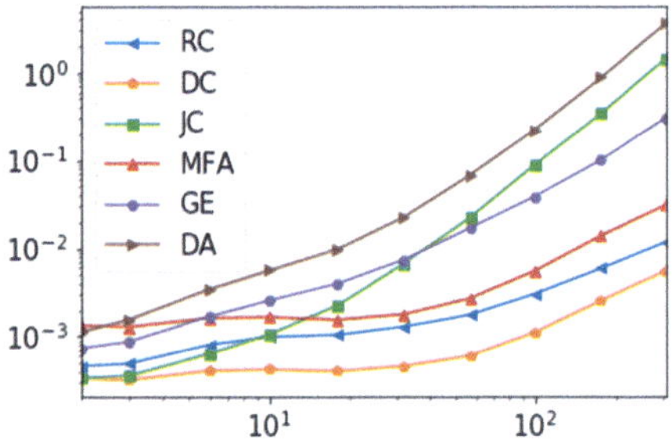

runtime for JC starts as low as that of RC, it accelerates past GE as the infection size grows. DA and JC do not scale well and GE follows them by a margin. It is worth noting that [44] gives a linear-time implementation of JC on trees that we have not tested here. BO is removed from this plot since its runtime grows exponentially with the infection size.

Based on these results, we advocate for the use of GE as the main tool for identifying sources of epidemics, regardless of the network topology or the nature of the epidemics (rumor propagation, disease contagion, etc.). MFA should be applied with caution. It is superior in social (transitive) networks, and attractive for its simplicity, scalability, and the potential for parallelism.

Acknowledgement We would like to thank Mason A. Porter for providing the Facebook-100 dataset.

Appendix 1: Multi-Source Extension

The inference problem discussed in Sect. 2.2 immediately extends to the multi-source situations. Consider the case were more than one independent source, denoted by $\boldsymbol{I}_*$, initiate the infection dynamics. Due to the Markovian nature of the dynamics, the infection path that leads to some set I does not influence the value of $\rho_{I\to O}$. Hence, Proposition 1 also describes the likelihood of the transition from the source set $\boldsymbol{I}_*$ to a snapshot O.

If we know that there are s original sources, e.g. $|\boldsymbol{I}_*| = s$, with a uniform prior on the patient zeros, the Bayesian solution would be characterized by the optimization

$$I^*_{\mathrm{MAP}} = \underset{I\subset O,\ |I|=s}{\operatorname{argmax}} \ \rho_{I\to O} \tag{8}$$

To compute this MAP estimate, we can still use the DP solution in Proposition 1, but we do not need to compute $\rho_{I\to O}$ for $|I| < s$. Thus, the multi-source problem is in a sense "easier", especially when $s \approx |O|$, since one can terminate the recursion earlier (i.e., the case $s = 1$ is the hardest).

Appendix 2: Proofs

Proof of Proposition 1

Let us first recall a known fact about the exponential distribution:

Lemma 1 *Let $T_i \sim \mathrm{Exp}(\beta_i)$ be a collection of independent exponential variables. Then,*

$$\mathbb{P}\Big(T_i < \min_{j\neq i} T_j\Big) = \frac{\beta_i}{\sum_j \beta_j}.$$

For a proof of Lemma 1, see [45]. The forward programming (2) is an application of the law of total probability in the following sense: The event that nodes in $O\setminus I$ are infected before any other node in I^c splits into sub-events that each node in $O\setminus I$ is infected before those in O^c and we have

$$\rho_{I\to O} = \sum_{j\in O\setminus I} \rho_{I\to I\cup j}\cdot \rho_{I\cup j\to O}$$

where we have also used the Markov property of SI dynamics to split the probabilities on the RHS into the products. The ratio in (2) corresponds to the transition probability from I to $I \cup j$, that is $\rho_{I \to I \cup j}$. Indeed, given that I is infected, we run exponential clocks $T_j \sim \text{Exp}(\beta \,\text{vol}(I, j))$ and the first to expire determines the next infected node. By Fact 1, this happens for any node $j \in I^c$ with probability $\propto_j \beta \,\text{vol}(I, j)$. Thus,

$$\rho_{I \to I \cup j} = \frac{\beta \,\text{vol}(I, j)}{\sum_{j'} \beta \,\text{vol}(I, j')} = \frac{\text{vol}(I, j)}{\text{vol}(I, I^c)}.$$

This proves the forward programming. The backward programming, on the other hand, connects $\rho_{I \to O}$ to $\rho_{I \to O \setminus j}$ and is proved similarly. Basically, the event of visiting O can be divided into sub-events based on the last node in O that is infected.

Proof of Proposition 2

We prove the following alternative expressions for $S = (S_{jj'})^{|O| \times |O|}$ and $z = (z_j)^{|O|}$,

$$S_{jj'} := \begin{cases} d^{in}_{O \setminus j'}(j) d^{in}_{O \setminus j}(j') + \sum_{i \in O} A_{ij} A_{ij'} & j \neq j' \\ 2\left[d^{in}_O(j)^2 + \sum_{i \in O} A^2_{ij}\right] & j = j' \end{cases}$$

$$z_j := \left[\text{vol}(O_{\setminus j}) + 2\,\text{vol}(O \setminus j, (O \setminus j)^c)\right] d^{in}_O(j)$$
$$+ \sum_{i \in O} (d^{out}_{O_{\setminus j}}(i) - d^{in}_{O_{\setminus j}}(i)) A_{ij} + 2 \sum_{i \in O} d^{out}_{(O \setminus j)^c}(i)\, A_{ij}.$$

Here, $d^{out}_O(i) := \sum_{j \in O} A_{ij}$ is the out-degree of node i in O, $d^{in}_O(i) := \sum_{j \in O} A_{ji}$ is the in-degree of node i in O, and $\text{vol}^{(2)}(i, j) := \sum_{r \in O} A_{ir} A_{rj}$ is the number of paths of length 2 between nodes i and j that pass through O. It is not hard to verify that these expressions are equivalent to the matrix form presented in (2).

Recall that $\text{vol}(I, I^c) = \sum_{i,k} A_{ik} 1\{i \in I, k \notin I\}$ and similarity $\text{vol}(I, j) = \sum_r A_{rj} 1\{r \in I\}$. Here, the indices, i, k and r run over all nodes in the network, i.e. $i, k, r \in [n]$. We have

$$(Q^T \boldsymbol{r})_j = \sum_{I \subset O} 1\{j \notin I\} \,\text{vol}(I, j) \cdot \text{vol}(I, I^c)$$
$$= \sum_{I \subset O \setminus \{j\}} \text{vol}(I, I^c) \cdot \text{vol}(I, j)$$

$$= \sum_{I \subset O \setminus \{j\}} \sum_{i,k,r} A_{ik} A_{rj} \, 1\{i \in I, \, k \notin I, \, r \in I\}$$

$$= \sum_{i,k,r} A_{ik} A_{rj} \gamma_{ikr}$$

where the last equality follows by interchanging the order of summations and defining

$$\gamma_{ikr} := \sum_{I \subset O \setminus \{j\}} 1\{i \in I, \, k \notin I, \, r \in I\}$$

If i or r do not belong to $O \setminus \{j\}$, or $k \in \{i, r\}$, then $\gamma_{ikr} = 0$. Thus, it what follows assume that $i, r \in O_{\setminus j} := O \setminus \{j\}$ and $k \notin \{i, r\}$. Then,

$$\gamma_{ikr} = 0 \begin{cases} 2^{|O|-4} & i \neq r, \; k \in O_{\setminus j} \\ 2^{|O|-3} & i = r, \; k \in O_{\setminus j} \\ 2^{|O|-3} & i \neq r, \; k \notin O_{\setminus j} \\ 2^{|O|-2} & i = r, \; k \notin O_{\setminus j} \end{cases}$$

To see the second equality, note that we are counting subsets of the set $O \setminus \{j\}$ (of cardinality $|O| - 1$) that contain or exclude certain elements. For example, when k, i, r are pairwise distinct, and $k \in O \setminus \{j\}$, looking at the binary representation of I, we have two ones in the positions i and r and a zero in position k, and the rest of $|O| - 1 - 3$ positions are free to be zero or one.

In what follows, i and r range over $O \setminus \{j\}$ (otherwise $\gamma_{ikr} = 0$). Also, condition $k \notin \{i, r\}$ can be replaced with $k \neq r$, since the $k \neq i$ is implicitly enforced by $A_{ik} = 0$ if $k = i$ (no self-loops). We have

$$(Q^T \boldsymbol{r})_j = \sum_{i,r} \sum_{k \neq r} A_{ik} A_{rj} \big[2^{|O|-4} (1 + 1\{i = r\}) 1\{k \in O_{\setminus j}\}$$

$$+ 2^{|O|-3} (1 + 1\{i = r\}) 1\{k \notin O_{\setminus j}\} \big]$$

$$= 2^{|O|-4} \sum_{i,r} d^{out}_{O \setminus \{j,r\}}(i) A_{rj} (1 + 1\{i = r\})$$

$$+ 2^{|O|-3} \sum_{i,r} d^{out}_{(O_{\setminus j})^c}(i) A_{rj} (1 + 1\{i = r\})$$

where in the second term, we used the fact that if $k \notin O_{\setminus j}$ then we automatically have $k \neq r$ since r ranges over $O_{\setminus j}$. We have

$$\sum_r d^{out}_{O\setminus\{j,r\}}(i)A_{rj} = \sum_r (d^{out}_{O_{\setminus j}}(i) - A_{ir})A_{rj}$$
$$= d^{out}_{O_{\setminus j}}(i)d^{in}_{O_{\setminus j}}(j) - \mathrm{vol}^{(2)}_{O_{\setminus j}}(i,j)$$

where $\mathrm{vol}^{(2)}_{O_{\setminus j}}(i,j) := \sum_{r\in O_{\setminus j}} A_{ir}A_{rj}$ is the number of paths of length two between i and j in $O_{\setminus j}$. Note that $\mathrm{vol}^{(2)}_{O_{\setminus j}}(i,j) = \mathrm{vol}^{(2)}_{O}(i,j)$ and similarly $d_{O_{\setminus j}}(j) = d_O(j)$ since $A_{jj} = 0$. Thus,

$$\sum_{i,r} d^{out}_{O\setminus\{j,r\}}(i)\, A_{rj}\big(1 + 1\{i = r\}\big) = \sum_i \Big[d^{out}_{O_{\setminus j}}(i)d^{in}_{O}(j) - \mathrm{vol}^{(2)}_{O}(i,j) + d^{out}_{O_{\setminus j}}(i)A_{ij}\Big]$$
$$= \sum_i d^{out}_{O_{\setminus j}}(i)d^{in}_{O}(j) + (d^{out}_{O_{\setminus j}}(i) - d^{in}_{O_{\setminus j}}(i))A_{ij}$$
$$= \mathrm{vol}(O_{\setminus j})d^{in}_{O}(j) + \sum_i (d^{out}_{O_{\setminus j}}(i) - d^{in}_{O_{\setminus j}}(i))A_{ij}$$

where $\mathrm{vol}(O_{\setminus j}) = \mathrm{vol}(O_{\setminus j}, O_{\setminus j})$ and the third equality follows since we have

$$\sum_{i\in A} \mathrm{vol}^{(2)}_{A}(i,j) = \sum_{i\in A}\sum_{r\in A} A_{ir}A_{rj} = \sum_{r\in A} d^{in}_{A}(r)A_{rj}$$

which was used with $A = O_{\setminus j}$. Similarly, we have

$$\sum_{i,r} d^{out}_{(O\setminus j)^c}(i)A_{rj}(1 + 1\{i = r\}) = \sum_i d^{out}_{(O\setminus j)^c}(i)\big(d^{in}_{O\setminus j}(j) + A_{ij}\big)$$
$$= \mathrm{vol}(O\setminus j, (O\setminus j)^c)\, d^{in}_{O}(j)$$
$$+ \sum_i d^{out}_{(O\setminus j)^c}(i)\, A_{ij}$$

It follows that

$$(Q^T\mathbf{r})_j = 2^{|O|-4}\Big[\mathrm{vol}(O_{\setminus j})d^{in}_{O}(j) + \sum_i (d^{out}_{O_{\setminus j}}(i) - d^{in}_{O_{\setminus j}}(i))A_{ij}$$
$$+ 2\,\mathrm{vol}(O\setminus j, (O\setminus j)^c)\, d^{in}_{O}(j) + 2\sum_i d^{out}_{(O\setminus j)^c}(i)\, A_{ij}\Big].$$

Calculating Q^TQ Let us first take $j \neq j'$. Then, similar to the previous argument,

$$(Q^TQ)_{jj'} = \sum_{I\subset O\setminus\{j,j'\}} \mathrm{vol}(I, j)\ \mathrm{vol}(I, j')$$

$$= \sum_{I \subset O \setminus \{j,j'\}} \sum_{i,r} A_{ij}\, A_{rj'} 1\{i \in I,\ r \in I\}$$
$$= \sum_{i,r} A_{ij}\, A_{rj'} \beta_{ir}$$

where we have defined

$$\beta_{ir} := \sum_{I \subset O \setminus \{j,j'\}} 1\{i \in I,\ r \in I\}$$
$$= 2^{|O|-4} 1\{i \neq r\} + 2^{|O|-3} 1\{i = r\}$$
$$= 2^{|O|-4} \big(1 + 1\{i = r\}\big)$$

assuming $i, r \in O \setminus \{j, j'\}$, otherwise $\beta_{ir} = 0$. Thus, restricting summations over indices $i, r \in O \setminus \{j, j'\}$

$$(Q^T Q)_{jj'} = 2^{|O|-4} \Big[\sum_{i,r} A_{ij}\, A_{rj'} + \sum_{i} A_{ij} A_{ij'} \Big]$$
$$= 2^{|O|-4} \Big[d^{in}_{O\setminus j'}(j) d^{in}_{O\setminus j}(j') + \sum_{i} A_{ij} A_{ij'} \Big].$$

Now consider the case $j = j'$. Then,

$$(Q^T Q)_{jj} = \sum_{I \subset O \setminus \{j\}} \mathrm{vol}(I, j)^2$$
$$= \sum_{I \subset O \setminus \{j\}} \sum_{i,r} A_{ij}\, A_{rj} 1\{i \in I,\ r \in I\}$$
$$= \sum_{i,r} A_{ij}\, A_{rj}\, 2^{|O|-3} \big(1 + 1\{i = r\}\big),$$

assuming $i, r \in O \setminus j$. It follows that

$$(Q^T Q)_{jj} = 2^{|O|-3} \Big[\sum_{i,r} A_{ij}\, A_{rj} + \sum_{i} A_{ij}^2 \Big]$$
$$= 2^{|O|-3} \big[d^{in}_{O}(j)^2 + \sum_{i} A_{ij}^2 \big].$$

References

1. Cliff, A., Haggett, P.: Time, travel and infection. Br. Med. Bull. **69**(1), 87–99 (2004)
2. Cohen, M.L.: Changing patterns of infectious disease. Nature **406**(6797), 762 (2000)
3. Colizza, V., Barrat, A., Barthélemy, M., Vespignani, A.: The role of the airline transportation network in the prediction and predictability of global epidemics. Proc. Natl. Acad. Sci. **103**(7), 2015–2020 (2006)
4. Slutsker, L., Altekruse, S.F., Swerdlow, D.L.: Foodborne diseases: emerging pathogens and trends. Infect. Dis. Clin. **12**(1), 199–216 (1998)
5. Elliott, M., Golub, B., Jackson, M.O.: Financial networks and contagion. Am. Econ. Rev. **104**(10), 3115–3153 (2014)
6. Acemoglu, D., Ozdaglar, A., Tahbaz-Salehi, A.: Systemic risk and stability in financial networks. Am. Econ. Rev. **105**(2), 564–608 (2015)
7. Kondakci, S.: Epidemic state analysis of computers under malware attacks. Simul. Model. Pract. Theory **16**(5), 571–584 (2008)
8. Fleizach, C., Liljenstam, M., Johansson, P., Voelker, G.M., Mehes, A.: Can you infect me now? Malware propagation in mobile phone networks. In: Proceedings of the 2007 ACM workshop on Recurring malcode, pp. 61–68. ACM, New York (2007)
9. Shao, C., Ciampaglia, G.L., Varol, O., Flammini, A., Menczer, F.: The spread of fake news by social bots (2017), pp. 96–104. Preprint. arXiv: 1707.07592
10. Shao, C., Ciampaglia, G.L., Varol, O., Yang, K.C., Flammini, A., Menczer, F.: The spread of low-credibility content by social bots. Nat. Commun. **9**(1), 4787 (2018)
11. Friggeri, A., Adamic, L., Eckles, D., Cheng, J.: Rumor cascades. In: Eighth International AAAI Conference on Weblogs and Social Media (2014)
12. Shin, J., Jian, L., Driscoll, K., Bar, F.: Political rumoring on twitter during the 2012 US presidential election: rumor diffusion and correction. New Media Soc. **19**(8), 1214–1235 (2017)
13. Jin, Z., Cao, J., Guo, H., Zhang, Y., Wang, Y., Luo, J.: Detection and analysis of 2016 US presidential election related rumors on twitter. In: International Conference on Social Computing, Behavioral-Cultural Modeling and Prediction and Behavior Representation in Modeling and Simulation, pp. 14–24. Springer, Berlin (2017)
14. Allcott, H., Gentzkow, M.: Social media and fake news in the 2016 election. J. Econ. Perspect. **31**(2), 211–36 (2017)
15. World Health Organization: Foodborne Disease Outbreaks: Guidelines for Investigation and Control, pp. 41–43. World Health Organization, Geneva (2008)
16. Manitz, J., Kneib, T., Schlather, M., Helbing, D., Brockmann, D.: Origin detection during food-borne disease outbreaks – a case study of the 2011 EHEC/HUS outbreak in Germany. PLoS Curr. **6** (2014)
17. Horn, A.L., Friedrich, H.: Locating the source of large-scale outbreaks of foodborne disease. J. R. Soc. Interface **16**(151), 20180624 (2019)
18. Shen, Z., Cao, S., Wang, W.X., Di, Z., Stanley, H.E.: Locating the source of diffusion in complex networks by time-reversal backward spreading. Phys. Rev. E **93**(3), 032301 (2016)
19. Pei, X., Jin, Z., Zhang, W., Wang, Y.: Detection of infection sources for avian influenza a (H7N9) in live poultry transport network during the fifth wave in China. IEEE Access **7**, 155759–155778 (2019)
20. Pei, S., Muchnik, L., Andrade Jr., J.S., Zheng, Z., Makse, H.A.: Searching for superspreaders of information in real-world social media. Sci. Rep. **4**, 5547 (2014)
21. Kitsak, M., Gallos, L.K., Havlin, S., Liljeros, F., Muchnik, L., Stanley, H.E., Makse, H.A.: Identification of influential spreaders in complex networks. Nat. Phys. **6**(11), 888 (2010)
22. Bojja Venkatakrishnan, S., Fanti, G., Viswanath, P.: Dandelion: redesigning the Bitcoin network for anonymity. Proc. ACM Meas. Anal. Comput. Syst. **1**(1), 22 (2017)
23. Shah, D., Zaman, T.: Rumors in a network: who's the culprit? IEEE Trans. Inf. Theory **57**(8), 5163–5181 (2011)

24. Fioriti, V., Chinnici, M.: Predicting the sources of an outbreak with a spectral technique (2012). Preprint. arXiv:1211.2333
25. Lokhov, A.Y., Mézard, M., Ohta, H., Zdeborová, L.: Inferring the origin of an epidemic with a dynamic message-passing algorithm. Phys. Rev. E **90**(1), 012801 (2014)
26. Zhu, K., Ying, L.: Information source detection in the sir model: a sample-path-based approach. IEEE/ACM Trans. Networking **24**(1), 408–421 (2016)
27. Luo, W., Tay, W.P.: Identifying multiple infection sources in a network. In: 2012 Conference Record of the Forty Sixth Asilomar Conference on Signals, Systems and Computers (ASILOMAR), pp. 1483–1489. IEEE, Piscataway (2012)
28. Nguyen, H.T., Ghosh, P., Mayo, M.L., Dinh, T.N.: Multiple infection sources identification with provable guarantees. In: Proceedings of the 25th ACM International on Conference on Information and Knowledge Management, pp. 1663–1672. ACM, New York (2016)
29. Prakash, B.A., Vreeken, J., Faloutsos, C.: Spotting culprits in epidemics: how many and which ones? In: 2012 IEEE 12th International Conference on Data Mining (ICDM), pp. 11–20. IEEE, Piscataway (2012)
30. Jiang, J., Wen, S., Yu, S., Xiang, Y., Zhou, W.: Identifying propagation sources in networks: state-of-the-art and comparative studies. IEEE Commun. Surv. Tutorials **19**(1), 465–481 (2017)
31. Lappas, T., Terzi, E., Gunopulos, D., Mannila, H.: Finding effectors in social networks. In: Proceedings of the 16th ACM SIGKDD International Conference on Knowledge Discovery and Data Mining, pp. 1059–1068. ACM, New York (2010)
32. Antulov-Fantulin, N., Lančić, A., Šmuc, T., Štefančić, H., Šikić, M.: Identification of patient zero in static and temporal networks: robustness and limitations. Phys. Rev. Lett. **114**(24), 248701 (2015)
33. Paluch, R., Lu, X., Suchecki, K., Szymański, B.K., Hołyst, J.A.: Fast and accurate detection of spread source in large complex networks. Sci. Rep. **8**(1), 2508 (2018)
34. Kiss, I.Z., Miller, J.C., Simon, P.L., et al.: Mathematics of Epidemics on Networks. Springer, Cham (2017)
35. Khim, J., Loh, P.L.: Confidence sets for the source of a diffusion in regular trees. IEEE Trans. Netw. Sci. Eng. **4**(1), 27–40 (2017)
36. Chang, B., Zhu, F., Chen, E., Liu, Q.: Information source detection via maximum a posteriori estimation. In: 2015 IEEE International Conference on Data Mining (ICDM), pp. 21–30. IEEE, Piscataway (2015)
37. University of Oregon Route Views Project: Online data and reports. http://www.routeviews.org
38. Leskovec, J., Kleinberg, J., Faloutsos, C.: Graphs over time: densification laws, shrinking diameters and possible explanations. In: Proceedings of the Eleventh ACM SIGKDD International Conference on Knowledge Discovery in Data Mining, pp. 177–187. ACM, New York (2005)
39. Watts, D.J., Strogatz, S.H.: Collective dynamics of 'small-world' networks. Nature **393**(6684), 440 (1998)
40. Traud, A.L., Kelsic, E.D., Mucha, P.J., Porter, M.A.: Comparing community structure to characteristics in online collegiate social networks. SIAM Rev. **53**(3), 526–543 (2011)
41. Traud, A.L., Mucha, P.J., Porter, M.A.: Social structure of Facebook networks. Physica A **391**(16), 4165–4180 (2012)
42. Leskovec, J., Huttenlocher, D., Kleinberg, J.: Signed networks in social media. In: Proceedings of the SIGCHI Conference on Human Factors in Computing Systems, pp. 1361–1370. ACM, New York (2010)
43. Karrer, B., Newman, M.E.: Stochastic blockmodels and community structure in networks. Phys. Rev. E **83**(1), 016107 (2011)
44. Luo, W., Tay, W.P., Leng, M.: How to identify an infection source with limited observations. IEEE J. Sel. Top. Sign. Proces. **8**(4), 586–597 (2014)
45. Ross, S.M.: Introduction to Probability Models. Academic, Cambridge (2014)

Maxwell's Demon: Controlling Entropy via Discrete Ricci Flow over Networks

Romeil Sandhu and Ji Liu

Abstract In this work, we propose to utilize discrete graph Ricci flow to alter network entropy through feedback control. Given such feedback input can "reverse" entropic changes, we adapt the moniker of Maxwell's Demon to motivate our approach. In particular, it has been recently shown that Ricci curvature from geometry is intrinsically connected to Boltzmann entropy as well as functional robustness of networks or the ability to maintain functionality in the presence of random fluctuations. From this, the discrete Ricci flow provides a natural avenue to "rewire" a particular network's underlying geometry to improve throughout and resilience. Due to the real-world setting for which one may be interested in imposing nonlinear constraints amongst particular agents to understand the network dynamic evolution, controlling discrete Ricci flow may be necessary (e.g., we may seek to understand the entropic dynamics and curvature "flow" between two networks as opposed to solely curvature shrinkage). In turn, this can be formulated as a natural control problem for which we employ feedback control towards discrete Ricci-based flow and show that under certain discretization, namely Ollivier-Ricci curvature, one can show stability via Lyapunov analysis. We conclude with preliminary results with remarks on potential applications that will be a subject of future work.

1 Introduction

In the current technological world, we increasingly depend upon the reliability, robustness, quality of service and timeliness of exceedingly large interconnected dynamical systems including those of power distribution, biological, transportation, and communication [1]. Over the past 20 years, we have witnessed a dramatic rise of information in which the analysis of such systems invariably present challenging "big data" complexity issues. For example, in transferring resources and

R. Sandhu (✉) · J. Liu
Stony Brook University, Stony Brook, NY, USA
e-mail: romeil.sandhu@stonybrook.edu

N. Masuda et al. (eds.), *Proceedings of NetSci-X 2020: Sixth International Winter School and Conference on Network Science*, Springer Proceedings in Complexity,
https://doi.org/10.1007/978-3-030-38965-9_9

information, a key requirement is the ability to adapt and reconfigure in response to structural and dynamic changes while avoiding disruption of service. In turn, exploiting functional properties such as robustness and heterogeneity (redundancy) are key to maintaining control and avoiding shotgun-based solutions during "black swan" events in which the continuous failing of interacting agents may result in catastrophic system failure.

As such, we have previously developed fundamental relationships between network functionality [2, 3] and certain topological and geometric properties of the corresponding graph [4, 5] to show that the geometric notion of curvature (a measure of "flatness") is positively correlated with network entropy and system's robustness or its ability to adapt to dynamic changes [6, 7]. This can be seen in Fig. 1. In this regard, network curvature may relate to anomaly detection, congestion in communication, to drug resistance [6, 8]. On the other hand, network entropy has often been chosen as a measure of network functional robustness [2, 9]. From this, if one is able to define such statistical properties over the graph that are proxies for functionality, then a natural progression would be to define corresponding theoretics in order to alter the networks behavior through such properties and for which in this note, we consider curvature and entropy. To this end, we focus on developing the necessary conditions to control network (curvature) entropy through the discrete Ricci flow. This flow in the graph setting has been proposed for congestion management, managing systemic risk [7], simulating biological resistance [6], as well as a generalized tool for network comparison [8, 10]. This said, the discrete Ricci flow for networks presents notable issues in that it not only reduces regions of negative curvature, but also reduces areas of highly positive curved regions. In the context of inducing network fragility (or vice versa), this may not be suitable as *increases* in negatively curved regions curvature relates to increases in entropy and subsequently network robustness. Further, in understanding network dynamics, one may want to "drive" the discrete flow between two networks [8, 10] as well as in augmented fashion for which one "pins down" the flow on regions considered "undruggable." The remainder of this note is outlined as follows: the next section provides preliminaries in motivating the theoretical need of understanding geometry

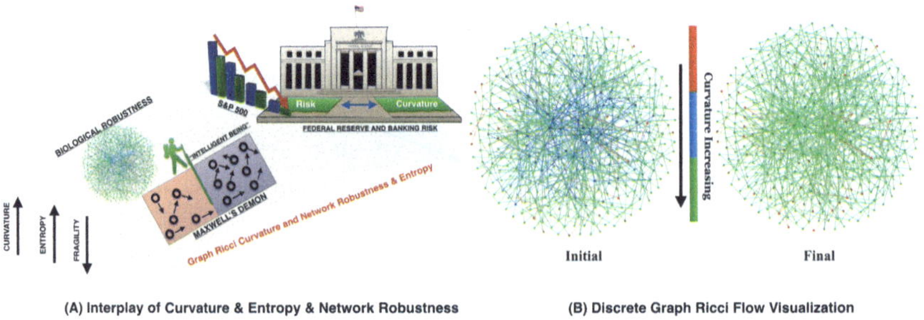

Fig. 1 Motivated by Maxwells "Demon", this work focuses on altering network entropy via Ollivier-Ricci flow whereby the "intelligent being" is a feedback operator. (**a**) Interplay of curvature and entropy on varied domains. (**b**) Example of discrete Ricci flow on a graph

as it pertains to functionality. From this, Sect. 3 lays the foundation of our framework for which we present the corresponding control laws and prove stability in the sense of Lypanuv. Then, Sect. 4 presents preliminary results on synthetic networks for illustration of theory. We conclude with a summary and future work towards applications in Sect. 5.

2 Preliminaries: Entropy and Curvature

To illustrate how geometry elucidates the functional behavior of a dynamical system, let us revisit optimal mass transport (OMT) [11]. The first notion of OMT was proposed by Gaspar Monge in 1781 with the concern of finding the minimal transportation cost for moving a pile of soil from one site to another. The modern formulation, given by Kantorovich, has been ubiquitously used in fields of econometrics, fluid dynamics, to shape analysis [11, 12] and recently, has received a renewed mathematical interest. More formally, let (X, μ_0) and (Y, μ_1) be two probability spaces and let $\pi(\mu_0, \mu_1)$ denote the set of all couplings on $X \times Y$ whose marginals are μ_0 and μ_1. As such, the Kantorovich costs seeks to minimize $\int c(x, y) d\pi(x, y)\ \forall \pi \in \pi(\mu_0, \mu_1)$ where $c(x, y)$ is the cost for transporting one unit of mass from x to y. The cost originally defined in a distance form on a metric space leads to the L^p Wasserstein distance as follows:

$$W_p(\mu_0, \mu_1) := \left(\inf_{\mu \in \pi(\mu_0, \mu_1)} \int \int d(x, y)^p d\mu(x, y)) \right)^{\frac{1}{p}}. \tag{1}$$

From this, let us begin considering M to be a Riemannian manifold such that

$$\mathcal{P} := \{\mu \geq 0 : \int \mu \text{ dvol}(M) = 1\} \qquad \mathcal{T}_\mu \mathcal{P} := \{\eta : \int \eta \text{ dvol}(M) = 0\} \tag{2}$$

as the space of probability densities and the tangent space at a given point μ, respectively. Due to the work of Benamou and Brenier [12], one can naturally compute the geodesic (*in the Wasserstein sense*) between two densities $\mu_0, \mu_1 \in \mathcal{P}$ as the below optimal control problem:

$$\inf_{\mu, g} \left\{ \int \int_0^1 \mu(t, x) \|\nabla g(t, x)\| dt \text{dvol}(M) \right.$$
$$\text{subject to} \quad \frac{\partial u}{\partial t} + \text{div}(\mu \nabla g) = 0$$
$$\left. \mu(0, .) = \mu_0, \quad \mu(1, .) = \mu_1 \right\}, \tag{3}$$

which leads us to give $\mathcal{P}$ a Riemannian structure due to the work of Jordan et al. [13]. From this, we can now consider Boltzmann entropy as

$$H(\mu_t) := \int_M \log \mu_t \mathrm{dvol}(M), \tag{4}$$

where the dependency on x has been dropped for convenience and we consider a family of densities evolving over time. Taking the second variation with respect to time t in the Wasserstein sense (i.e., rather than the Euclidean norm) and noting that, by construction, $\eta := \frac{\partial \mu}{\partial t}|_t = 0$, we have

$$\begin{aligned} \frac{d^2}{dt^2} H(\mu_t)|_{t=0} &= \langle Hess(H)(\eta), \eta \rangle_W \\ &= - \int_M \langle \nabla g_\eta, \nabla \Delta g_\eta \rangle + \frac{1}{2} \Delta \big(\|\nabla g_\eta\|^2 \big) \mu_0 \mathrm{dvol}(M), \end{aligned} \tag{5}$$

where μ_0 and g_η satisfy (3). Using the Bochner formula [14], which relates harmonic functions on a Riemannian manifold to Ricci curvature (herein denoted as "*Ric*"), we can further assume $Ric \geq kI$ as quadratic forms where k is a constant and I is the identity matrix. Then, due to Sturm [15] as well as Lott and Villani [11, 22], one can show that the $Hess(H)$ is k-convex:

$$H(\mu_t) \leq tH(\mu_0) + (1-t)H(\mu_1) - \phi(k, t, \mu_0, \mu_1) \forall t \in [0, 1] \tag{6}$$

where the right hand portion $\phi(.)$ can be shown to be $\phi(k, t, \mu_0, \mu_1) = \frac{k}{2} t(1 - t) W_2(\mu_0, \mu_1)^2$ allowing for k-convexity. That is, changes in entropy and curvature are positively correlated, i.e., $\Delta H \times \Delta Ric \geq 0$. Furthermore, through the Fluctuation Theorem [2], one may relate network robustness R to entropy; i.e., $\Delta H \times \Delta R \geq 0$ as well as Ricci curvature $\Delta Ric \times \Delta R \geq 0$—see [2, 6, 7] for details.

3 Proposed Framework

In this section, we propose a feedback based approach to control discrete Ricci flow over graphs due to a discretization by Ollivier [4, 5] which is discussed next.

3.1 Open-Loop View: Discrete Ollivier-Ricci Flow

While Ricci curvature relates to functionality, we require a discrete definition for networks. Here, we focus on the Ollivier formulation [4] given its relationship to the

Wasserstein distance, but refer to the reader to several works in this open problem area of varying discretizations including, but not limited to, Forman curvature [8, 16], Bakery Emery [17] as well as recent comparisons [17, 18]. This said, we can define Ollivier-Ricci curvature between any two nodes x and y as:

$$\kappa(x, y) := 1 - W_1(\mu_x, \mu_y)/d(x, y). \tag{7}$$

This definition, motivated by coarse geometry, is applicable to the graph setting whereby the geodesic distance $d(x, y)$ is given by the hop metric. From this, we can define the Ollivier-Ricci flow with an initial condition $\mu_0(x, y) = \phi_0(x, y)$ as:

$$\frac{d}{dt}\mu_t(x, y) : = -\kappa(x, y)\mu_t(x, y) \tag{8}$$

where $\mu_t(x, y)$ (with an abuse of notion) is the normalized edge weights, i.e., $\mu_t(x, y) \in [0, 1]$. Here, we can treat this flow as an open-loop control problem [19, 21] for which the "dynamics" to be controlled is Ollivier-Ricci curvature $\kappa(x, y)$. In particular, motivated philosophically by Maxwell's Demon [20], we seek to characterize an "intelligent being" to control entropy via discrete Ricci flow.

3.2 *Control Law Construction and Existence*

To begin developing our control-based approach, let us redefine the above flow as a closed-loop problem with the following form given as:

$$\frac{d}{dt}\mu_t(x, y) = \big[-\kappa(x, y) + \psi(\mu_t, \mu^*)\big]\mu_t(x, y) \tag{9}$$

$$\mu_0(x, y) = \phi_0(x, y)$$

where $\lim_{t\to\infty}\mu_t(x, y) \rightarrow \mu^*(x, y)$ and where $\psi(\mu, \mu^*)$ is the control law whereby the system is stable in the sense of Lypanuv (e.g., inputs "near" equilibrium stay or decay towards equilibrium). To understand the above in the context of this paper, we first note that it can be seen that the discrete Ricci Flow can be viewed as "rewiring" network functionality (to some steady state configuration) by altering the geometry or edge weights over the particular network. This is also indirectly correlated to altering network entropy through previous established relationships in Sect. 2. Thus, if we are to assume there exists a particular intelligent being whose "demonic" (discrete) input is able to combat such thermodynamically favorable "directions", then one can characterize such an action via $\psi(\mu_t, \mu^*)$ in the general sense and therefore, such action must be considered to be stable (i.e., the system does not implode due to the demon). To show how such actions can be performed in a stable manner, we assume $\mu^*(x, y)$ is ideal and for which there exists no error; i.e., we want the flow to converge entirely to such an end (network) point. To do so, let us further define the point-wise and total error as

$$\delta_t(x, y) := \mu_t(x, y) - \mu^*(x, y) \tag{10}$$

$$\Sigma_t(\delta_t) := \frac{1}{2}\sum_x \sum_y ||\delta_t(x, y)||^2. \tag{11}$$

Given this, we are now able to show the existence of a regulatory control.

Theorem 3.1 *The control law that stabilizes the closed-loop system in Eq. (9) from $\mu_t(x, y)$ to $\mu^*(x, y)$ is given by:*

$$\psi(\mu, \mu^*) = \beta_t^2(x, y)\delta_t(x, y) \tag{12}$$

where $\beta^2(x, y) \geq 2$ and $\delta_t(x, y)$ is given by Eq. (11).

Proof Let us first note that $\delta_t(x, y)$ is bounded, i.e., $-1 \leq \delta_t(x, y) \leq 1$ and that $-2 \leq \kappa(x, y) \leq 1$. From this, we choose Σ_t as the candidate Lyapunov function and differentiate it with respect t which yields the following:

$$\begin{aligned}
\frac{d\Sigma_t}{dt} &= \sum_{x,y} \delta_t(x, y) \cdot \frac{\partial \delta_t(x, y)}{\partial t} \\
&= \sum_{x,y} \delta_t(x, y) \cdot \Big[\frac{\partial}{\partial t}\mu_t(x, y) - \underbrace{\frac{\partial}{\partial t}\mu^*(x, y)}_{0}\Big] \\
&= \sum_{x,y} \delta_t(x, y) \cdot \big[-\kappa(x, y) - \beta_t^2(x, y)\delta_t(x, y)\big]\mu_t(x, y) \\
&\leq \sum_{x,y} \big[|\kappa(x, y)|\delta_t^2(x, y) - \beta_t^2(x, y)\delta_t^2(x, y)\big]\mu_t(x, y) \\
&\leq \sum_{x,y} \delta_t^2(x, y)\big[2 - \beta_t^2(x, y)\big]\mu_t(x, y) \\
&\leq 0
\end{aligned}$$

We note that while the above control law is due to the reliance on bounds of Ollivier-Ricci curvature, this will not hold for other discretizations such as Forman curvature [8, 10]. We have also assumed that one not only has an ideal representation of the corresponding network configuration $\mu^*(x, y)$, but the input is error-free and there are no modifications by a "demonic" operator during the entropic (Ricci) flow, e.g., impose node constraints. This is discussed in the next section.

3.3 "Non-Perfect Demonic" Input

We are now ready to define an estimator and observer-like framework for which an input may begin to control graph curvature and subsequently control network entropy. This can be akin to the thought experiment proposed by James Maxwell for which the "demon" seeks to violate the second law of thermodynamics, namely alter entropy [20]. Here, we assume there exists error from **both** the demon (and end targeted) state as well as the chosen (Ollivier-Ricci) flow model. As such, let us define $\hat{\mu}_t^*(x, y)$ as the estimate of the ideal knowledge $\mu^*(x, y)$ with corresponding error terms associated with the demon and the model as

$$\hat{\delta}_t(x, y) := \mu_t(x, y) - \hat{\mu}_t^*(x, y) \qquad \text{(Type I Error)}$$

$$\gamma_t(x, y) := \hat{\mu}_t^*(x, y) - \lambda_t(x, y) \qquad \text{(Type II Error)}$$

where $\lambda_t(x, y) := \sum_{l=0}^{l=k} \epsilon_t^k(x, y)$ and $\epsilon_t^k(x, y) := \pm p$ (constant) are the k input at time t. From this, the total error for the above Type I/II errors can be seen as:

$$\hat{\Sigma}_t(x, y) := \frac{1}{2} \sum_x \sum_y ||\hat{\delta}_t(x, y)||^2 \tag{13}$$

$$\Gamma_t(x, y) := \frac{1}{2} \sum_x \sum_y |\lambda_t(x, y)| ||\gamma_t(x, y)||^2. \tag{14}$$

Theorem 3.2 *Let us assume input has stopped and further assume the above total label errors defined for Type I/II error, then the following flow*

$$\frac{d}{dt}\hat{\mu}_t(x, y) = \left[\hat{\delta}_t(x, y) + \Phi(\lambda_t, \gamma_t)\right]\hat{\mu}_t(x, y) \tag{15}$$

$$\hat{\mu}_0(x, y) = \phi_0(x, y)$$

where $\Phi(\lambda_t, \gamma_t) = -|\lambda_t(x, y)|\gamma_t(x, y)$ *provides an estimator such that the total error* $V_t(x, y) := \hat{\Sigma}_t(x, y) + \Gamma_t(x, y)$ *has a negative semi-definite derivative. In turn, this provides a stable coupled feedback system together with Eq. (9) where the ideal configuration* $\mu^*(x, y)$ *is replaced with an estimator* $\hat{\mu}_t^*(x, y)$.

Proof Computing the total error $V_t(x, y) := \hat{\Sigma}_t(x, y) + \Gamma_t(x, y)$ and dropping the spatial dependency (for reading ease), yields the following:

$$\begin{aligned}\frac{\partial V_t}{\partial t} &= \sum_{x,y} \hat{\delta}_t \cdot \frac{\partial \hat{\delta}_t}{\partial t} + \lambda_t \gamma_t \frac{\partial \hat{\mu}_t}{\partial t} \\ &= \sum_{x,y} \hat{\delta}_t \cdot \left[\frac{\partial \mu_t}{\partial t} - \frac{\partial \hat{\mu}_t}{\partial t}\right] + \lambda_t \gamma_t \frac{\partial \hat{\mu}_t}{\partial t}\end{aligned}$$

$$
\begin{aligned}
&= \sum_{x,y} \hat{\delta}_t \frac{\partial \mu_t}{\partial t} - \hat{\delta}_t \frac{\partial \hat{\mu}_t}{\partial t} + \lambda_t \gamma_t \frac{\partial \hat{\mu}_t}{\partial t} \\
&= \sum_{x,y} \underbrace{\hat{\delta}_t \frac{\partial \mu_t}{\partial t}}_{\leq 0} - \frac{\partial \hat{\mu}_t}{\partial t} \left[\hat{\delta}_t - \lambda_t \gamma_t \right] \\
&\leq - \frac{\partial \hat{\mu}_t}{\partial t} \left[\hat{\delta}_t - \lambda_t \gamma_t \right] \\
&\leq 0
\end{aligned}
$$

As one can see from coupling both the estimator and autonomous model, a useful qualitative behavior emerges. In particular, when the "demon" is satisfied with the agreement between $\mu_t(x, y)$ and their ideal $\mu^*(x, y)$ configuration, it is assumed that the total input $\lambda_t(x, y)$ will then remain constant. That is, either the "demon" never needed to apply a correction or has otherwise stopped providing inputs. Nevertheless, in this case, $\hat{\mu}_t^*(x, y)$ should "follow" $\mu_t(x, y)$. On the other hand, when the total input error $\lambda_t(x, y)$ grows due to persistent input, $\hat{\mu}_t^*(x, y)$ will be increasingly driven towards $\lambda_t(x, y)$ irrespective of the agreement between $\hat{\mu_t}^*(x, y)$ and $\mu_t(x, y)$. Ultimately, the demon has control of the seemingly accurate autonomous flow and can override systems actions.

4 Results

In this section, we present results using graph curvature to indirectly control network entropy. We caution the reader that these results are preliminary and to motivate theory presented. This said, we conduct experiments primarily focused involve scale-free networks as it provides natural topological hubs to test particular inputs can impede (induce fragility) via varying levels of input. For all experiments, we generate networks via the Python NetworkX package and utilize the classic definition of network entropy [9].

The first set of experiments focuses on controlling network entropy via discrete Ollivier-Ricci flow seen in Fig. 2. As there exists an intimate connection that relates that changes in entropy are positively correlated with changes in Ricci curvature, i.e., $\Delta H \times \Delta Ric \geq 0$, we generate scale-free networks with node sizes of $n = \{100, 200, 400, 600\}$ with uniform edge weights. From this, we target the node with the highest degree and begin "injecting" input and allow for our flow to evolve as described by the coupled feedback equations in Eqs. (9) and (15). To be more precise, at time $t = \{30, 75, 120, 175\}$ we make an input of values $p = \{-2, 2, 4, -4\}$, respectively. The resulting changes in network entropy as well as average Olliver-Ricci curvature can be seen as solid colored lines in Fig. 2. Remarkably, we see a very close relationship between network entropy and that of network curvature. Furthermore, to validate our ability to "change direction"

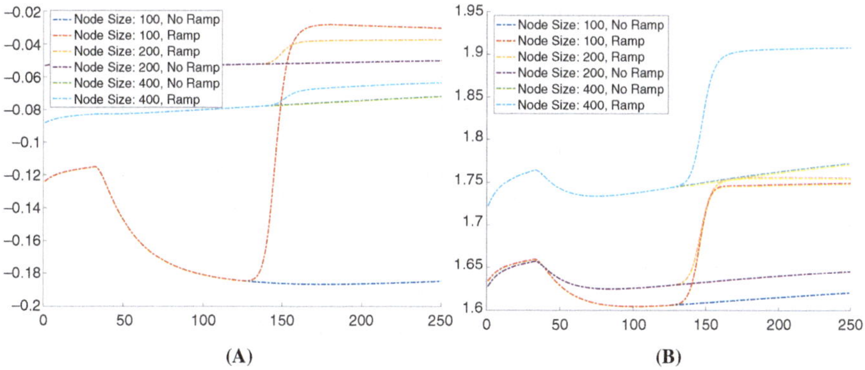

Fig. 2 Results on scale-free networks of varying node sizes and the impact of operator input in "injecting" curvature of a single node associated with the highest topological degree. (**a**) Average Ollivier-Ricci curvature. (**b**) Average network entropy. Note: due to scaling, values for curvature and entropy differ; however, $\Delta H \times \Delta Ric \geq 0$

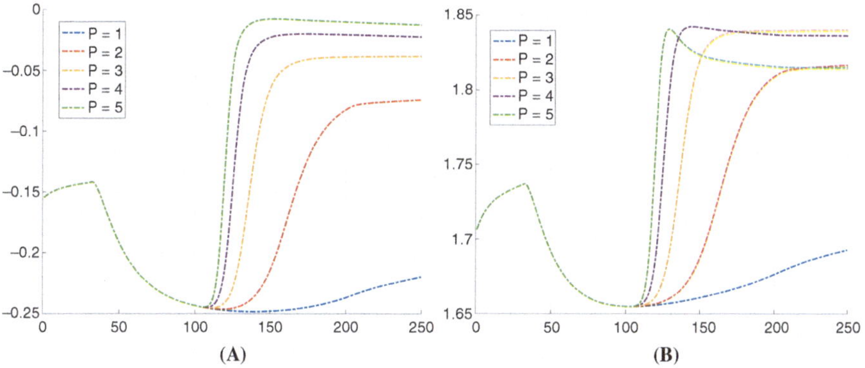

Fig. 3 Results on scale-free networks of node size $n = 200$ and the impact of operator input at varying levels of p from $t = 100$ to $t = 200$ for the node associated with the highest topological degree. (**a**) Average Ollivier-Ricci curvature. (**b**) Average network entropy. Note: the degree of operator input naturally controls (increases) both curvature and entropy in the aforementioned time region

in terms of altering network entropy, we re-run the same experiment with a slight change by "turning off" input at $t = 120$; i.e., for $t = \{30, 75, 120, 175\}$ we make an input of values $p = \{-2, 2, 0, 0\}$, respectively. Once again, we see the natural impact and differences of operator input.

On the other hand, we also want to measure how the degree of input (e.g., choosing the constant p) alters networks entropy as well as the impact of altering more than one hub node in a given network. To this end, we generate scale-free networks of node size $n = 200$. From this, at time $t = \{30, 75, 100, 200\}$ we make an input of values $p = \{-2, 2, \theta, -\theta\}$ where $\theta = \{5, 4, 3, 2, 1\}$. As one can see from Fig. 3, we see exactly this behavior which also correlates to the degree of operator input. Next, we make a slight alteration to this experiment and now at iterations $\{30, 75, 100, 200\}$, we make an operator input of values $p = \{-2, 2, 4, -4\}$ similar to the first experiment for a scale-free network of node size $n = 400$. However, we now plot changes in network entropy and network curvature as a function of altering the top n nodes with the highest degree. Again, we see the behavior that is to be expected in increasing network robustness as seen in Fig. 4. For this experiment, Fig. 5 shows Type I and Type II error for completeness.

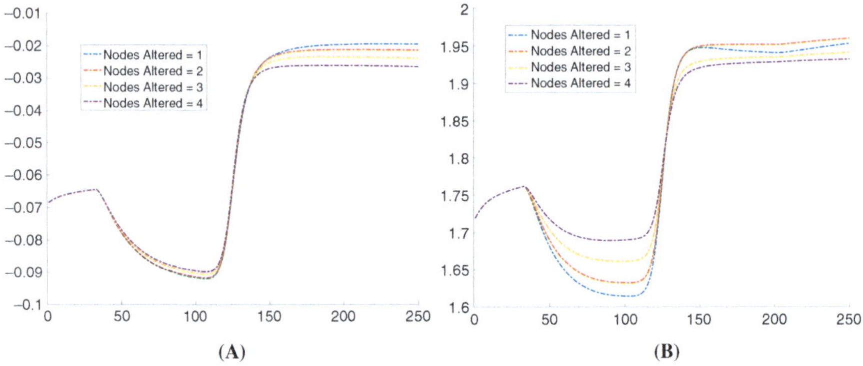

Fig. 4 Results on scale-free networks of node size $n = 400$ and the impact of providing operator input to several nodes associated with the highest topological degree. (**a**) Average Ollivier-Ricci curvature. (**b**) Average network entropy. Note: the number of nodes an operator interacts with naturally controls (increases/decreases) both curvature and entropy

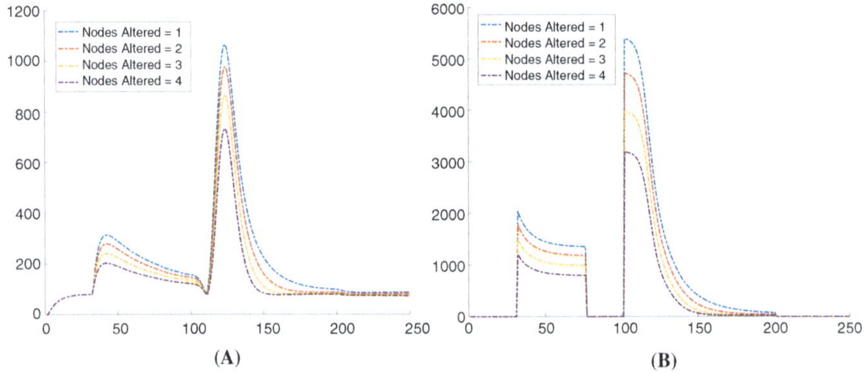

Fig. 5 We present errors results associated with Fig. 4 experiment related to node alteration and operator input. (**a**) Type I error. (**b**) Type II error. Specifically, such errors relate to the geometric discrete Ricci flow (Type I) compared to that of the "demonic" input (Type II) and where the resulting steady state result seeks to minimize both joint errors

5 Conclusions and Future Work

We propose a network control framework that couples the discrete Ollivier-Ricci flow with operator input from a feedback perspective. To this end, we provide the necessary stability conditions in the sense of Lyapunov. This said, there exists several avenues that we are currently pursuing. The above framework has potential biological application towards the real-world setting in which we often seek to understand how can we induce fragility on targets that are deemed "undruggable" [6]. We also aim to extend the above framework for non-constant user input, time-delayed response, and as applied to specific application domains. As such, this work has laid the foundation for which further examination is needed.

Acknowledgements This work was supported by the U.S. Air Force Office of Scientific Research (AFOSR) grant FA9550-18-1-0130 and National Science Foundation (NSF) grant ECCS-1749937. R. Sandhu also discloses he is apart of Backspacer Technologies Incorporated for which no funding or support was used to conduct this study.

References

1. Barabasi, A.: The network takeover. Nat. Phys. **8**, 14–16 (2012)
2. Demetrius, L., Manke, T.: Robustness and network evolution: entropic principle. Phys. A: Stat. Mech. Appl. **346**, 682–696 (2005)
3. Varadhan, S.R.S.: Large Deviations and Applications. SIAM, New York (1984)
4. Ollivier, Y.: Ricci curvature of Markov chains on metric spaces. J. Funct. Anal. **256**, 810–864 (2009)
5. Ollivier, Y.: Ricci curvature of metric spaces. C. R. Math. Acad. Sci. Paris. **345**, 643–646 (2007)
6. Sandhu, R., Georgiou, T., Reznik, E., Zhu, L. , Kolesov, I., Senbabaoglu, Y., Tannenbaum, A.: Graph curvature for differentiating cancer networks. Nat. Sci. Rep. **5**, 12323 (2015)
7. Sandhu, R., Georgiou, T., Tannenbaum, A.: Ricci curvature: an economic indicator for market fragility and systemic risk. Sci. Adv. **2**, e1501495 (2016)
8. Weber, M., Saucan, E., Jost, J.: Characterizing complex networks with Forman-Ricci curvature and associated geometric flows. J. Compl. Netw. **5**, 527–550 (2017)
9. West, J., Bianconi, G., Severini, S., Teschendorff, A.: Differential network entropy reveals cancer system hallmarks. Nat. Sci. Rep. **2**, 802 (2012)
10. Weber, M., Jost, J., Saucan, E.: Forman-Ricci flow for change detection in large dynamic data sets. Axioms **5**, 26 (2016)
11. Villani, C.: Optimal Transport, Old and New. Springer, Berlin (2008)
12. Benamou, J., Brenier, Y.: A computational fluid mechanics solution to the Monge-Kantrovich mass transfer problem. Numer. Math. **84**, 375–393 (2000)
13. Jordan, R., Kinderlehrer, D., Otto, F.: The variational formulation of the Fokker-Planck equation. SIAM J. Math. Anal. **29**, 1–17
14. Chow, B., Lu, P., Ni, L.: Hamilton's Ricci Flow, vol. 77. American Mathematical Society, Providence (2006)
15. Sturm, K.: On the geometry of metric measure spaces. Acta Math. **196**, 65–131 (2006)
16. Saucan, E., Weber, M.: Forman's Ricci curvature-from networks to hypernetworks. In: The International Conference on Complex Networks and Applications, pp. 706–717. Springer, Cham (2018)

17. Pouryahya, M., Mathews, J., Tannenbaum, A.: Comparing three notions of discrete Ricci curvature on biological networks (2017). Preprint. arXiv: 1712.02943
18. Samal, A., Sreejith, R.P., Gu, J., Liu, S., Saucan, E., Jost, J.: Comparative analysis of two discretizations of Ricci curvature for complex networks. Nat. Sci. Rep. **8**, 8650 (2018)
19. Doyle, J., Francis, B., Tannenbaum, A.: Feedback Control Theory. Courier Corporation, New York (2013)
20. Thomson, W.: The sorting demon of Maxwell. R. Soc. Proc. **9**, 113–114 (1879)
21. Zhu, L., Karasev, P., Kolesov, I., Sandhu, R., Tannenbaum, A.: Guiding image segmentation on the fly: interactive segmentation from a feedback control perspective. IEEE Trans. Automat. Control **63**, 3276–3289 (2018)
22. Lott, J., Villani, C.: Ricci curvature for metric-measure spaces via optimal transport. Ann. Math. **169**, 903–991 (2009)

Two Dimensional Opinion Dynamics of Real Opinion and Official Stance

Akira Ishii and Nozomi Okano

Abstract Multi-dimensional opinion dynamics including both trust and distrust between people is presented. Opinion dynamics of two dimensions is applied to real opinion and official stance of people, agents and governments. As examples of the application of the two-dimensional opinion dynamics, we present the case of "Romeo and Juliet" and the conflict of two countries. Analyzing the movement in society, including the difference between the real opinion and the official stance gives us great freedom to the analysis.

1 Introduction

In general, opinion is not simple. For example, in the famous play of Romeo and Juliet by Shakespeare, real opinions of Romeo and Juliet were attracted to each other. However, the families to which Romeo and Juliet belonged were in conflict, so they could not express that they were attracted to each other. This is an example of the conflict between real opinion and official stance.

In international conflicts, if the conflict of interest between the two countries is very serious, the diplomatic efforts may not be made. In that case, the conflict of interest between the two countries correspond to real opinion and the diplomatic attitude corresponds to official stance. At the case that the difference between the official stance and the real opinion is very large, the consensus formation in the official stance would be very difficult because of the large distance between the real opinion of the two countries.

In order to consider such conflicts in opinion dynamics, we should consider both official stance and real opinion. In this paper, we present the two dimensional opinion dynamics of both official stance and real opinion as a case of multi-dimensional opinion dynamics theory.

A. Ishii (✉) · N. Okano
Tottori University, Tottori, Japan
e-mail: ishii@tottori-u.ac.jp; http://www.damp.tottori-u.ac.jp/~lab3/index.html

N. Masuda et al. (eds.), *Proceedings of NetSci-X 2020: Sixth International Winter School and Conference on Network Science*, Springer Proceedings in Complexity,
https://doi.org/10.1007/978-3-030-38965-9_10

In the approaches of opinion dynamics, rather than the binary approach which asks whether someone agree or disagree, we will try to tackle this problem by taking an approach that deals with the transition of opinions continuously. The opinion dynamics theory that handles the transition of opinions continuously is the Bounded Confidence Model [1–5]. Using these theories, we can handle only the trust relationship between persons in the simulation, so that we always obtain consensus formation in the simulation. Recently, Ishii and Kawahata presented a new theory including both trust and distrust between people as an extension of the Bounded Confidence Model [6–12]. Using this new opinion dynamics theory, we can obtain various simulations for society.

The expressed opinion or official stance of an agent may be different from its real opinion. The problem of the difference between expressed opinion or official stance and real opinion is first considered by Asch [13]. Recently, many works introducing official stance (expressed opinion) and real opinion are published [14–17]. However, these works assume the consensus formation and do not consider the distrust between people. The distrust is sometimes very significant for opinion dynamics.

Based on the above new opinion dynamics theory of Ishii and co-workers, we present here a new theory; multi-dimensional opinion dynamics theory. Though the theory can be used as multi-dimensional opinion dynamics theory, we apply this theory to the two-dimensional opinion dynamics.

The two dimensions in this paper means the official stance and the real opinion explained above. The purpose of this paper is to present the possibility of analysis of the conflict between real opinion and official stance in the frame of opinion dynamics theory.

2 Theory

2.1 One Dimensional Opinion Dynamics Theory

Our model is based on the original bounded confidence model of Hegselmann-Krause [3]. For a fixed agent, say i, where $1 \leq i \leq N$, we denote the agents opinion at time t by $I_i(t)$. As shown in Fig. 1, person i can be affected by surrounding people. According to Hegselmann-Krause [3], opinion formation of agent i can be described as follows.

$$I_i(t+1) = \sum_{j=1}^{N} D_{ij} I_j(t) \tag{1}$$

This can be written in the following form.

$$\Delta I_i(t) = \sum_{j=1}^{N} D_{ij} I_j(t) \Delta t \tag{2}$$

where it is assumed that $D_{ij} \geq 0$ for all i, j in the model of Hegselmann-Krause. Based on this definition, $D_{ij} = 0$ means that the opinion of agent i is not affected by the opinion of agent j. In this Bounded Confidence Model, it is expected implicitly that the final goal of the negotiation among people is the formation of consensus.

However, in the real society in the world, the formation of consensus among people is sometimes very difficult. We can find many such examples in the international politics in the world history. Even in domestic problems, the opinions between people pursuing economic development and people claiming nature conservation are not compatible and agreement of the two people is very difficult. Since it is not possible to define the payoff matrix for such serious political conflict, application of game theory may be difficult. Thus, in order to deal with problems that are difficult to form consensus among these people, it is necessary to include the lack of trust between people in our opinion dynamics theory. Here, as a result of exchanging opinions, consider the possibility that the opinions of two people with different opinions change move in different directions. We consider the distribution of opinions in the positive and negative directions of a one-dimensional axis. In this case, the value range of $I_i(t)$ is $-\infty \leq I_i(t) \leq +\infty$. Here, we assume that $I_i(t) > 0$ means positive opinion and $I_i(t) < 0$ means negative opinion. Even in the limitation of Hegselmann-Krause model, one can assume that $1/2 \leq I_i(t) \leq 1$ corresponds to positive opinion and $0 \leq I_i(t) \leq 1/2$ corresponds to negative opinion. However, our definition is intuitive, easy to understand and easy to apply to various examples.

We modify the meaning of the coefficient D_{ij} as the coefficient of trust between the agent i and the agent j. We assume here that $D_{ij} > 0$ if there is a trust relationship between the two persons, and $D_{ij} < 0$ if there is distrust relationship between the two persons.

According to our previous theory [7], we consider here that people disregard the opinion far removed from their opinions without agreeing or repelling. Also, opinions that are very close to himself/herself will not be particularly affected. To include the two effects, we use the following function instead of $D_{ij} I_j(t)$ as follows,

$$D_{ij} \Phi(I_i, I_j)(I_j(t) - I_i(t)) \tag{3}$$

where

$$\Phi(I_i, I_j) = \frac{1}{1 + exp(\beta(|I_i - I_j| - b))} \tag{4}$$

This function is Sigmoid function and it works as a smooth cut-off function at $|I_i - I_j| = b$. The typical graph of this function is shown in Fig. 1. Using this Sigmoid function, we assume that if the opinions of the two are too far apart, they will not be totally influenced by each other's opinion. Moreover, because of the

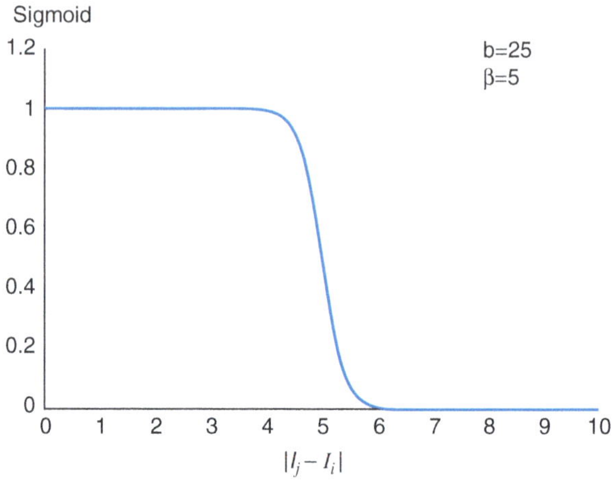

Fig. 1 The typical graph of the Sigmoid function (4) as smooth cut-off function

factor $I_j(t) - I_i(t)$, the opinion $I_i(t)$ is not affected by the opinion $I_j(t)$ if the opinion $I_j(t)$ is almost same as the opinion $I_i(t)$.

For the factor $I_j(t) - I_i(t)$, we consider that $I_j(t) - I_i(t)$ gives the same effect whether $I_i(t)$ and $I_j(t)$ are both positive or both negative or either positive or negative. This is very natural that, for example, even between conservatives, there are intense debates between those with moderate conservatives and those with radical conservatives.

Influences of mass media and government statements can not be ignored in the formation of public opinion. Such mass media effect can also work even for negotiations of small size group. Since formula of our theory above is similar to the model of hit phenomena [18] where the popularity of certain topic is analyzed using the sociophysics model, we introduce here the effects of mass media similar to the way of Ref. [18]. Let $A(t)$ be the pressure at time t from the outside and denote the reaction difference for each agent is denoted by the coefficient c_i. The coefficient c_i can have different values for each person and c_i can be positive or negative. If the coefficient c_i is positive, the person i moves the opinion toward the direction of the mass media. On the contrary, if the coefficient c_i is negative, the opinion of the person change against the mass media direction.

Therefore, including such mass media effects, the change in opinion of the agent can be expressed as follows.

$$\Delta I_i(t) = c_i A(t)\Delta t + \sum_{j=1}^{N} D_{ij}\Phi(I_i(t), I_j(t))(I_j(t) - I_i(t))\Delta t \tag{5}$$

We assume here that D_{ij} is an asymmetric matrix; D_{ij} and D_{ji}, $D_{ij} \neq D_{ji}$ and D_{ij} and D_{ji} can have different signs.

Long-term behavior requires attenuation, which means that topics will be forgotten over time. Here we introduce exponential attenuation. The expression is as follows.

$$\Delta I_i(t) = -\alpha I_i(t)\Delta t + c_i A(t)\Delta t + \sum_{j=1}^{N} D_{ij}\Phi(I_i(t), I_j(t))(I_j(t) - I_i(t))\Delta t \tag{6}$$

This is the equation of the one-component version of our new opinion dynamics theory.

2.2 *Multi-Dimensional Opinion Dynamics Theory*

We extend the above one dimensional opinion dynamics theory to multi-dimensional theory. First, we assume that the multi-dimensional opinion can be write as the following multi-component vector.

$$\Theta_i(t) = (I_i^{(1)}(t), I_i^{(2)}(t), \cdots) \tag{7}$$

The effect of mass media can be extend that the term of mass media effects is $\chi_i A(t)$ where

$$\chi_i = (C_i^{(1)}, C_i^{(2)}, \cdots) \tag{8}$$

The effects from other people's opinion is Ω_{ij} as an extension of D_{ij} to multi-dimension.

Thus, multi-dimension opinion dynamics can be write down as the following equation.

$$\Theta_i(t + \Delta t) - \Theta_i(t) = \left[\chi_i A(t) + \sum_j \Omega_{ij}\Theta_j(t) \right] \Delta t \tag{9}$$

Namely,

$$d\Theta_i(t) = \left[\chi_i A(t) + \sum_j \Omega_{ij}\Theta_j(t) \right] dt \tag{10}$$

2.3 *Two Dimensional Opinion Dynamics Theory*

For two dimension, we can write the two-component opinion as follows,

$$\Theta_i(t) = (I_i^{(1)}(t), I_i^{(2)}(t)) \tag{11}$$

The effect of mass media in two dimension is $\chi_i A(t)$ where

$$\chi_i = (C_i^{(1)}, C_i^{(2)}) \tag{12}$$

The effects from other people's opinion is Ω_{ij} as an extension of D_{ij} to multi-dimension.

$$\Omega_{ij} = \begin{pmatrix} D_{ij}^{(1)} & E_{ij}^{(12)} \\ E_{ij}^{(21)} & D_{ij}^{(2)} \end{pmatrix}$$

Thus, two-dimension opinion dynamics can be write down as the following equation.

$$\Theta_i(t + \Delta t) - \Theta_i(t) = \left[\chi_i A(t) + \sum_j \Omega_{ij} \Theta_j(t) \right] \Delta t \tag{13}$$

Namely,

$$d\Theta_i(t) = \left[\chi_i A(t) + \sum_j \Omega_{ij} \Theta_j(t) \right] dt \tag{14}$$

Actual equations of two dimension opinion dynamics are as follows,

$$dI_i^{(1)} = \left[c_i^{(1)} A(t) + \sum_j D_{ij}^{(1)} I_j^{(1)}(t) + \sum_j E_{ij}^{(12)} I_j^{(2)}(t) \right] dt \tag{15}$$

$$dI_i^{(2)} = \left[c_i^{(2)} A(t) + \sum_j E_{ij}^{(21)} I_j^{(1)}(t) + \sum_j D_{ij}^{(2)} I_j^{(2)}(t) \right] dt \tag{16}$$

Apply similar interaction as (3) to the effect of other persons, actual equations of two dimension opinion dynamics are as follows.

$$dI_i^{(1)} = \left[c_i^{(1)} A(t) + \sum_j D_{ij}^{(1)} \Phi(I_i^{(1)}(t), I_j^{(1)}(t))(I_j^{(1)}(t) - I_i^{(1)}(t)) \right. \\ \left. + \sum_j E_{ij}^{(12)} \Phi(I_i^{(2)}(t), I_j^{(2)}(t))(I_j^{(2)}(t) - I_i^{(2)}(t)) \right] dt \tag{17}$$

$$dI_i^{(2)} = \left[c_i^{(2)} A(t) + \sum_j E_{ij}^{(21)} \Phi(I_i^{(1)}(t), I_j^{(1)}(t))(I_j^{(1)}(t) - I_i^{(1)}(t)) \right. \\ \left. + \sum_j D_{ij}^{(2)} \Phi(I_i^{(2)}(t), I_j^{(2)}(t))(I_j^{(2)}(t) - I_i^{(2)}(t)) \right] dt \tag{18}$$

Similar to one dimension opinion dynamics of Ishii [7], we introduce here the strength of will for both component separately.

$$\mu_i = \begin{pmatrix} m_i^{(1)} & 0 \\ 0 & m_i^{(2)} \end{pmatrix}$$

Using the above strength of will, the two-dimensional opinion dynamics equations is as follows.

$$\mu_i d\Theta_i(t) = \left[\chi_i A(t) + \sum_j \Omega_{ij} \Theta_j(t) \right] dt \tag{19}$$

For each component of the opinion, we can write the equation of the two-dimensional opinion dynamics theory as follows.

$$m_i^{(1)} dI_i^{(1)} = \left[c_i^{(1)} A(t) + \sum_j D_{ij}^{(1)} \Phi(I_i^{(1)}(t), I_j^{(1)}(t))(I_j(t) - I_i(t)) \right. \\ \left. + \sum_j E_{ij}^{(12)} \Phi(I_i^{(2)}(t), I_j^{(2)}(t))(I_j^{(2)}(t) - I_i^{(2)} \right] dt \tag{20}$$

$$m_i^{(2)} dI_i^{(2)} = \left[c_i^{(2)} A(t) + \sum_j E_{ij}^{(21)} \Phi(I_i^{(1)}(t), I_j^{(1)}(t))(I_j^{(1)}(t) - I_i^{(1)}(t)) \right. \\ \left. + \sum_j D_{ij}^{(2)} \Phi(I_i^{(2)}(t), I_j^{(2)}(t))(I_j^{(2)}(t) - I_i^{(2)}(t)) \right] dt \tag{21}$$

2.4 Discussion of Two-Dimensional Opinion Dynamics Theory

In this two-dimensional opinion dynamics equation, we introduce four coefficients $D_{ij}^{(1)}$, $D_{ij}^{(2)}$, $E_{ij}^{(12)}$ and $E_{ij}^{(21)}$. $D_{ij}^{(1)}$ is the strength of the effect from the official stance of the person j to the official stance of the person i. This coefficient is same as the coefficient of one-dimensional opinion dynamics [6, 7].

$D_{ij}^{(2)}$ is the strength of the effect from the real opinion of the person j to the real opinion of the person i. Since the real opinion is not open to public, the effect from the real opinion of the person j to the real opinion of the person i can not be observed in real situation. The people, agents or governments just suggest the real opinion of the other.

$E_{ij}^{(12)}$ is the strength of the effect from the real opinion of the person j to the official stance of the person i. This effect means that, addition to the official stance of person j, person i try to guess the real opinion of person j. It means that persons try to look at the other person's real opinion. The similar situation is also important for negotiation between nations. For example, on the negotiation between US and China, US government try to looks at the real opinion of the Chinese government and Chinese government try to looks at the real opinion of US government during the negotiation. this coefficient $E_{ij}^{(12)}$ is significant for such movements.

$E_{ij}^{(21)}$ is the strength of the effect from the official stance of the person j to the real opinion of the person i. This effect means that real opinion of people can be affected by the official stance of other persons. When the turmoil of the revolution grows, people's real opinions may be influenced and changed by the official stance of the people around them.

As described above, this two-component opinion dynamics theory can be applied to the situation where the real opinion are mixed with the official stance.

3 Results and Discussion

3.1 Real Opinion and Official Stance

First, we show the calculation of Fig. 2. In this calculation, we assume that the coefficients for the official stance $I_A^{(1)}(t)$, $I_B^{(1)}(t)$ and the real opinion $I_A^{(2)}(t)$, $I_B^{(2)}(t)$ are completely same. Thus, the official stance and the real opinion are same.

Figure 3 is the case that the agent A and B have trust each other and they reach the consensus agreement on the official stance. However, the real opinions of them are different. In Fig. 3, we cut off the interaction between the official stance and the real opinion. Figure 4 is, addition to Fig. 3, we set the strength of the effect from the real opinion of the other agent to the official stance of the agent to be positive. Positive means that the agents trust each other on real opinion. Because of this positive effect from the real opinion to the official stance, the consensus formation is rapid.

On the other hand, in Fig. 5, we set that, in addition to Fig. 3, we set the strength of the effect from the real opinion of the other agent to the official stance of the agent to be negative. Negative means that the agents distrust each other. Because of this negative effect from the real opinion to the official stance, the consensus formation is very slow. This no consensus situation on the official stance is caused by the distrust on the real opinion.

3.2 Romeo and Juliet

Next, we consider the situation that the A and B have trust each other on their real opinions but their official stances should be conflicted. This situation is very similar to the famous play "Romeo and Juliet" by William Shakespeare [19]. We calculate

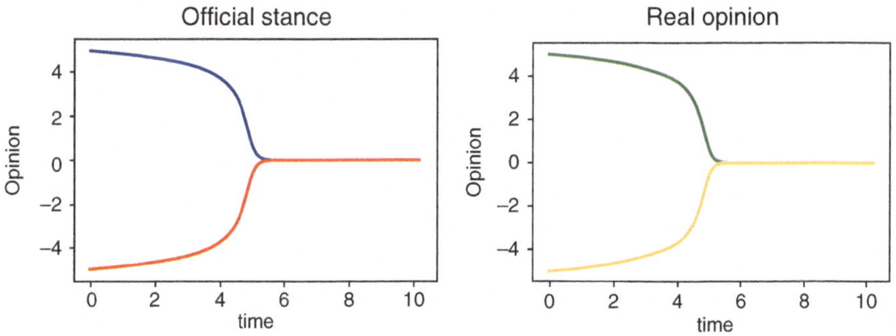

Fig. 2 Calculation of official stance and real opinion for two agents.The left is official distance and the right is real opinion. The vertical axis means opinion and the horizontal axis means time. $D_{AB}^{(1)} = 1$, $D_{BA}^{(1)} = 1$, $D_{AB}^{(2)} = 1$, $D_{BA}^{(2)} = 1$, $E_{AB}^{(12)} = 1$, $E_{BA}^{(12)} = 1$, $E_{AB}^{(21)} = 1$, $E_{BA}^{(21)} = 1$

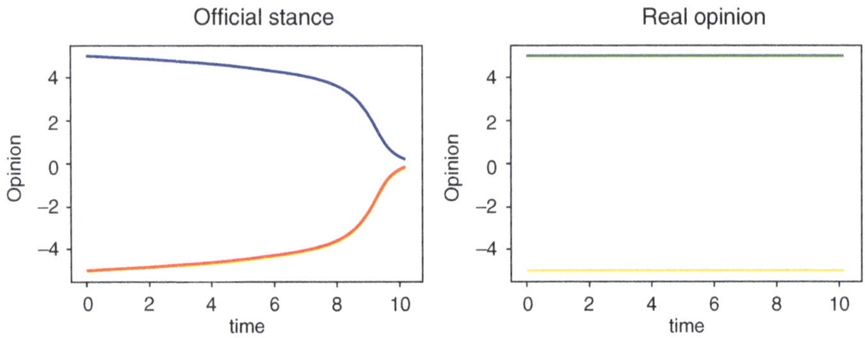

Fig. 3 Calculation of official stance and real opinion for two agents. The left is official distance and the right is real opinion. The vertical axis means opinion and the horizontal axis means time. $D^{(1)}_{AB} = 1$, $D^{(1)}_{BA} = 1$, $D^{(2)}_{AB} = 0$, $D^{(2)}_{BA} = 0$, $E^{(12)}_{AB} = 0$, $E^{(12)}_{BA} = 0$, $E^{(21)}_{AB} = 0$, $E^{(21)}_{BA} = 0$

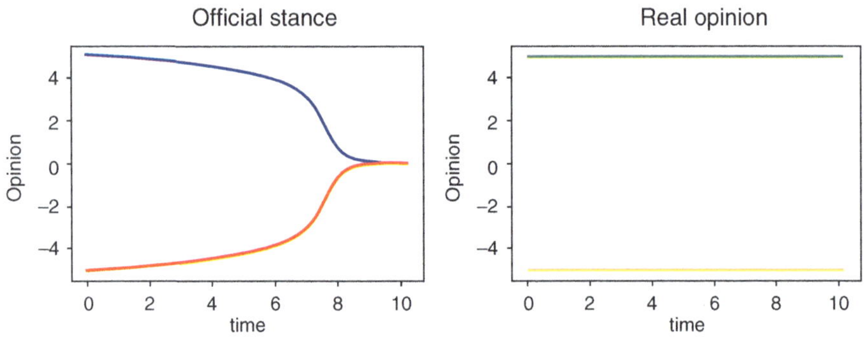

Fig. 4 Calculation of official stance and real opinion for two agents. The left is official distance and the right is real opinion. The vertical axis means opinion and the horizontal axis means time. $D^{(1)}_{AB} = 1$, $D^{(1)}_{BA} = 1$, $D^{(2)}_{AB} = 0$, $D^{(2)}_{BA} = 0$, $E^{(12)}_{AB} = 0.5$, $E^{(12)}_{BA} = 0.5$, $E^{(21)}_{AB} = 0$, $E^{(21)}_{BA} = 0$

the original situation of "Romeo and Juliet" in Fig. 6a, where we set that the official stances of the agent A and B have no trust each other and the real opinions of the agent A and B have trust each other. There are many works for the love affair of Romeo and Juliet using mathematical model [20–22]. However, in such works, calculations were done only for intentions of Romeo and Juliet.

In contrast to these works, we introduce both official stance and real opinion for Romeo and Juliet. In Fig. 6b, we introduce the negative effect of official stance to the real opinions of Romeo and Juliet. In this case, since Romeo and Juliet take care of the official stances, the calculated real opinion of Romeo and Juliet is far from consensus formation. This is the case that Romeo and Juliet do not fall into love because of the constraint of the conflict between their families. In Fig. 6c, we introduce the positive effect of the real opinion to the official stances of Romeo and Juliet. In this case, because of their real opinions, the official stances of Romeo and

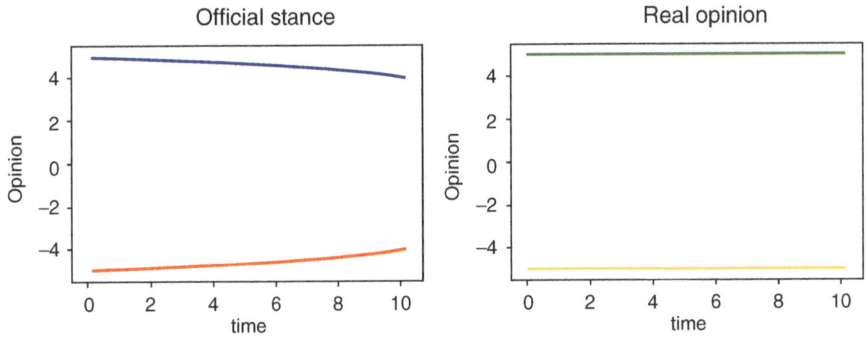

Fig. 5 Calculation of official stance and real opinion for two agents. The left is official distance and the right is real opinion. The vertical axis means opinion and the horizontal axis means time. $D_{AB}^{(1)} = 1$, $D_{BA}^{(1)} = 1$, $D_{AB}^{(2)} = 0$, $D_{BA}^{(2)} = 0$, $E_{AB}^{(12)} = -0.5$, $E_{BA}^{(12)} = -0.5$, $E_{AB}^{(21)} = 0$, $E_{BA}^{(21)} = 0$

Juliet tend to approach. If this situration was realized, the play of Romeo and Juliet would be a happy end story.

3.3 *Conflict of Two Countries*

Here we consider the case that the agent A and B try to obtain consensus formation but their real opinion is far from the consensus formation. Such case can be found in society as the negotiation between two country or between two company. In Fig. 7a we show that A and B get consensus formation in official stance but not in real opinion. In this case, since both A and B trust the other's real opinion on the official stance, A and B can obtain the consensus formation each other on the official stance. In Fig. 7b, we show the case that A and B try to get consensus formation in official stance but not in real opinion. It means that we set to be $E_{AB}^{(12)} = -0.2$ and $E_{BA}^{(12)} = -0.2$. In this case, both A and B distrust the other's real opinion, so that the consensus formation in the official stance is incomplete compared with the case of Fig. 7a because of the effect of distrust of the real opinion of the other. The case of Fig. 7c is that A and B obtain consensus formation in official stance where A and B trust the other's real opinion. For the real opinion, A and B have no trust on the real opinion of the other but the the consensus formation on the official stance affect the real opinion due to $E_{AB}^{(21)} = 0.5$ and $E_{BA}^{(21)} = 0.5$. Thus, we found that the real opinion of A and B come closer in this calculation, Fig. 7c.

On the problem of conflict of countries, the landscape theory [23] and related theory [24] are well known. On these theories, the size, the propensity and the distance between each two country are considered to obtain the energy or the conflict minimization. However, in these theories they do not pay attention to the real opinion we include in the present theory. Usually, the states compromise in

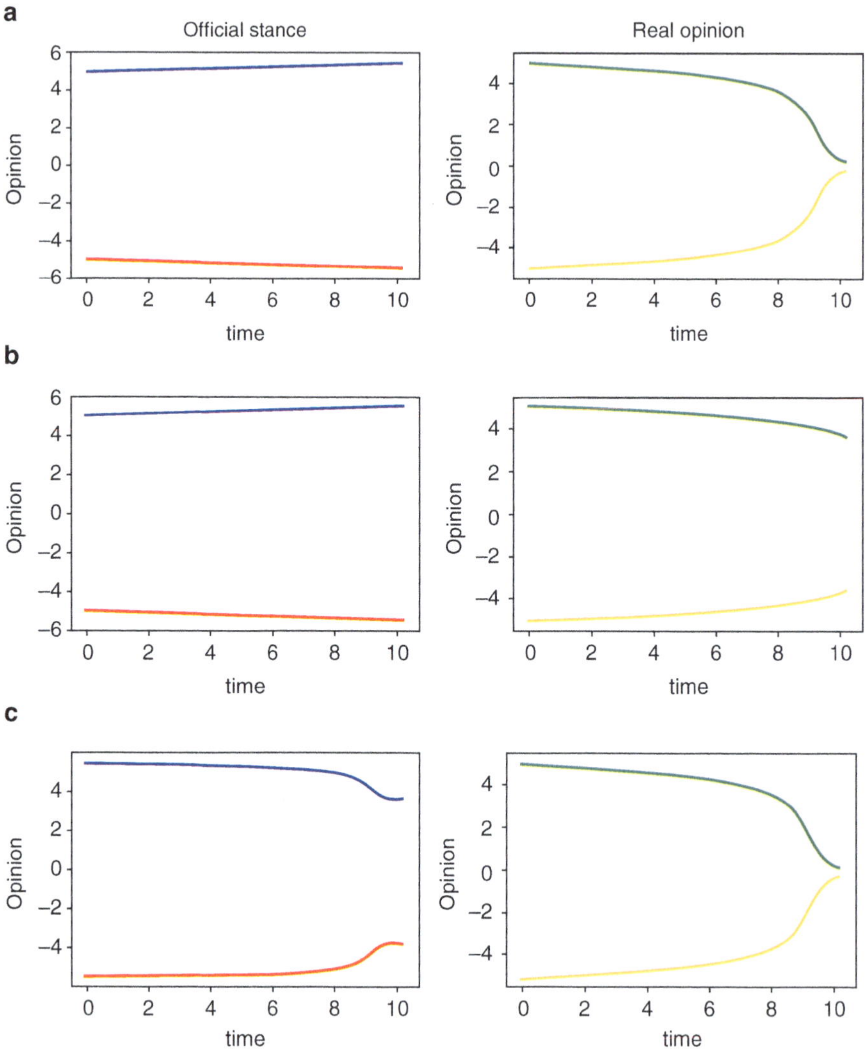

Fig. 6 Calculation of official stance and real opinion for two agents, Romeo and Juliet. The left is official distance and the right is real opinion. The vertical axis means opinion and the horizontal axis means time. The parameters of the calculations are $D_{AB}^{(1)} = -0.5$, $D_{BA}^{(1)} = -0.5$, $D_{AB}^{(2)} = 1$, $D_{BA}^{(2)} = 1$. For three calculations, we set as follows. (**a**) $E_{AB}^{(12)} = 0$, $E_{BA}^{(12)} = 0$, $E_{AB}^{(21)} = 0$, $E_{BA}^{(21)} = 0$. (**b**) $E_{AB}^{(12)} = 0$, $E_{BA}^{(12)} = 0$, $E_{AB}^{(21)} = -0.5$, $E_{BA}^{(21)} = -0.5$. (**c**) $E_{AB}^{(12)} = 0.5$, $E_{BA}^{(12)} = 0.5$, $E_{AB}^{(21)} = 0$, $E_{BA}^{(21)} = 0$

diplomatic negotiations though the real opinion of the states are still apart. Thus, the present theory can be used to analyze such compromise though the real opinion is not compromised.

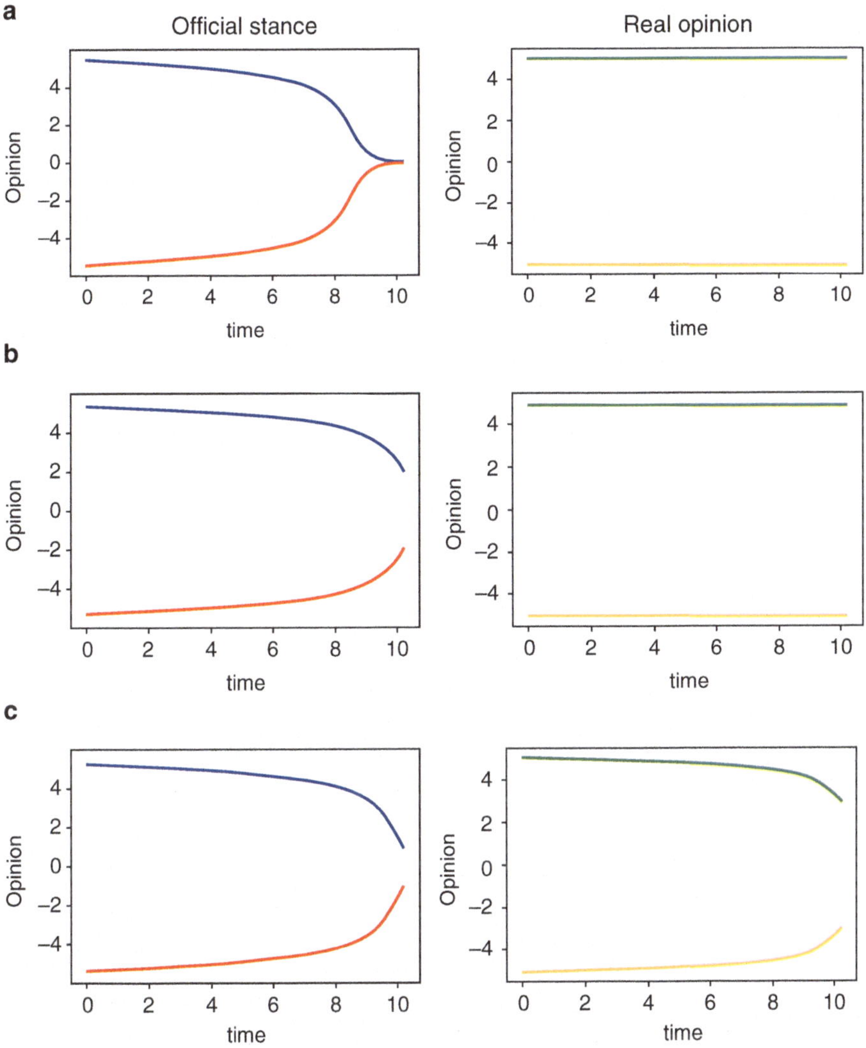

Fig. 7 Calculation of official stance and real opinion for two countries. The left is official distance and the right is real opinion. The vertical axis means opinion and the horizontal axis means time. The parameters of the calculations are $D^{(1)}_{AB} = 1$, $D^{(1)}_{BA} = 1$, $D^{(2)}_{AB} = 0$, $D^{(2)}_{BA} = 0$. For three calculations, we set as follows. (a) $E^{(12)}_{AB} = 0.2$, $E^{(12)}_{BA} = 0.2$, $E^{(21)}_{AB} = 0$, $E^{(21)}_{BA} = 0$. (b) $E^{(12)}_{AB} = -0.2$, $E^{(12)}_{BA} = -0.2$, $E^{(21)}_{AB} = 0$, $E^{(21)}_{BA} = 0$. (c) $E^{(12)}_{AB} = -0.5$, $E^{(12)}_{BA} = -0.5$, $E^{(21)}_{AB} = 0.5$, $E^{(21)}_{BA} = 0.5$.

3.4 Discussion

As we see in the two example calculations of Romeo and Juliet and conflict of two countries, we can do a lot of variation of social simulation using this two component opinion dynamics theory of official stance and real opinion. As we include distrust into calculation, we can even do the simulation of the social devision where effect of real opinion would be very important. For example, in the case of religious conflict, analysis of their official stance would be not enough. The analysis of their real opinion would be very important.

The application of the presented two dimensional opinion dynamics theory is not the analysis of negotiations using real opinions and official stance, because, usually, we cannot obtain the correct information of the real opinion of the negotiation partner. The application of this theory should be the guess of the real opinion of the negotiation partner using this two-dimensional opinion dynamics. Such application is possible even using social media informations.

4 Conclusion

Multi-dimensional opinion dynamics including both trust and distrust between people is presented. Opinion dynamics theory of two dimensions is applied to real opinion and official stance of people, agents and governments. The two dimension, the official stance and the real opinion is very natural to analyze negotiation, discussion and love. Analyzing the negotiations, including the difference between the real opinion and the official stance gives great freedom to the analysis. Therefore, the two-dimensional opinion dynamics theory we present here is very useful and powerful for analyzing social phenomena.

Acknowledgements This work was supported by JSPS KAKENHI Grant Number JP19K04881. The authors thank for the meaningful discussion on the two dimension opinion dynamics of official stance and real opinion with Prof. H.Yamamoto of Rissho University and Prof. H. Takikawa of Tohoku University.

References

1. Deffuant, G., Neau, D., Amblard, F., Weisbuch, G.: Mixing beliefs among interacting agents. Adv. Complex Syst. **3**, 87–98 (2000)
2. Weisbuch, G., Deffuant, G., Amblard, F., Nadal, J.P.: Meet, discuss and segregate! Complexity **7**(3), 55–63 (2002)
3. Hegselmann, R., Krause, U.: Opinion dynamics and bounded confidence models, analysis, and simulation. J. Artif. Soc. Soc. Simul. **5** (2002)
4. Castellano, C., Fortunato, S., Loreto, V: Statistical physics of social dynamics. Rev. Mod. Phys. **81**, 591 (2009)

5. Sîrbu A., Loreto V., Servedio V.D.P., Tria F.: Opinion dynamics: models, extensions and external effects. In: Loreto, V., et al. (eds.) Participatory Sensing, Opinions and Collective Awareness. Understanding Complex Systems, p. 42. Springer, Cham (2017)
6. Ishii, A., Kawahata, Y.: Opinion dynamics theory for analysis of consensus formation and division of opinion on the internet. In: Proceedings of the 22nd Asia Pacific Symposium on Intelligent and Evolutionary Systems (IES2018), pp. 71–76 (2018). arXiv: 1812.11845 [physics.soc-ph]
7. Ishii, A.: Opinion dynamics theory considering trust and suspicion in human relations. In: Morais D., Carreras A., de Almeida A., Vetschera R. (eds.) Group Decision and Negotiation: Behavior, Models, and Support, GDN 2019. Lecture Notes in Business Information Processing, vol. 351. Springer, Cham (2019)
8. Ishii, A., Kawahata, Y.: Opinion dynamics theory considering interpersonal relationship of trust and distrust and media effects. In: The 33rd Annual Conference of the Japanese Society for Artificial Intelligence, vol. 33 (2019)
9. Kawahata, Y., Ishii, A.: Consensus formation online using sociophysics method. In: Proceedings of the 19th International Conference on Group Decision and Negotiation in 2019 a Joint GDN-EWG/BOR Meeting, GDN2019 (2019). arXiv: 1907.07946
10. Ishii, A., Kawahata, Y.: New Opinion dynamics theory considering interpersonal relationship of both trust and distrust. In: IEEE/WIC/ACM International Conference on Web Intelligence Volume (2019)
11. Okano, N., Ishii, A.: Isolated, untrusted people in society and charismatic person using opinion dynamics. In: IEEE/WIC/ACM International Conference on Web Intelligence-Companion Volume (2019)
12. Okano, N., Ishii, A.: Sociophysics approach of simulation of charismatic person and distrusted people in society using opinion dynamics. In: Proceedings of IES2019. Proceedings in Adaptation, Learning and Optimization (2019)
13. Asch, S.E.: Effects of group pressure upon the modification and distortion of judgments. In: Groups, Leadership and Men; Research in Human Relations, pp. 177–190. Carnegie Press, Oxford (1951)
14. Wang, S.W., Huang, C.Y., Sun, C.T.: Modeling self-perception agents in an opinion dynamics propagation society. Simulation **90**(3), 238–248 (2014)
15. Huang, C.-Y., Wen, T.-H.: A novel private attitude and public opinion dynamics model for simulating pluralistic ignorance and minority influence. J. Artif. Soc. Soc. Simul. **17**(3), 8 (2014)
16. Ye, M., Qin, Y., Govaert, A., Anderson, B.D., Cao, M.: An influence network model to study discrepancies in expressed and private opinions. Automatica **107**(7), 371–381 (2019)
17. León-Medina, F.J., Tena-Sànchez, J.M., Miguel, F.J.: Fakers becoming believers: how opinion dynamics are shaped by preference falsification, impression management and coherence heuristics. Qual. Quant. (2019) https://doi.org/10.1007/s11135-019-00909-2
18. Ishii, A., Arakaki, H., Matsuda, N., Umemura, S., Urushidani, T., Yamagata, N., Yoshida, N.: The 'hit' phenomenon: a mathematical model of human dynamics interactions as s stochastic process. New J. Phys. **14**, 063018 (2012)
19. Levenson, J.L. (ed.): Romeo and Juliet. The Oxford Shakespeare. Oxford University Press, Oxford (2000). ISBN: 0-19-281496-6
20. Strogatz, S.H.: Love affairs and differential equations. Math. Mag. **61**, 35 (1988)
21. Strogatz, S.H.: Nonlinear Dynamics and Chaos. With Applications to Physics, Biology, Chemistry, and Engineering. Springer, Berlin (1994)
22. Wauer, J., Schwarzer, D., Cai, G.Q., Lin, Y.K.: Dynamical models of love with time-varying fluctuations. Appl. Math. Comput. **188**, 1535–1548 (2007)
23. Axelrod, R., Bennett, D.S.: A landscape theory of aggregation. Br. J. Polit. Sci. **23**, 211–233 (1993)
24. Galam S.: Spontaneous coalition forming. Why some are stable? In: Bandini S., Chopard B., Tomassini M. (eds.) Cellular Automata. ACRI 2002. Lecture Notes in Computer Science, vol 2493. Springer, Berlin (2002)

On the Fundamental Equation of User Dynamics and the Structure of Online Social Networks

Masaki Aida, Chisa Takano, and Masaki Ogura

Abstract Online social networks suffer from explosive user dynamics such as flaming that can seriously affect social activities in the real world because the dynamics have growth rates that can overwhelm our rational decision making faculties. Therefore, a deeper understanding of user dynamics in online social networks is a fundamental problem in computer and information science. One of the effective user dynamics models is the networked oscillation model; it uses a second-order differential equation with Laplacian matrix. Although our previous study indicates that the oscillation model provides us with a minimal but effective model of user interactions, there still remains the open problem as to the existence of a first-order fundamental differential equation that respects the structure of the original network. This paper fills in this gap and shows that, by doubling the dimension of the state space, we can explicitly but naturally construct a fundamental equation that fully respects the structure of the original network.

1 Introduction

The widespread adoption of information networks, has dramatically activated the exchange of information among individuals, and the dynamics of users in online social networks is beginning to have a major impact beyond online communities; social activities in the real world are being influenced. In particular, explosive user

M. Aida (✉)
Graduate School of Systems Design, Tokyo Metropolitan University, Hino, Tokyo, Japan
e-mail: aida@tmu.ac.jp

C. Takano
Graduate School of Information Sciences, Hiroshima City University, Hiroshima, Japan
e-mail: takano@hiroshima-cu.ac.jp

M. Ogura
Graduate School of Information Science and Technology, Osaka University, Osaka, Japan
e-mail: m-ogura@ist.osaka-u.ac.jp

N. Masuda et al. (eds.), *Proceedings of NetSci-X 2020: Sixth International Winter School and Conference on Network Science*, Springer Proceedings in Complexity,
https://doi.org/10.1007/978-3-030-38965-9_11

dynamics such as the flaming phenomenon that occurs in online social networks can spread far faster than human rational decision making can respond, which can cause major social unrest. Therefore, understanding the dynamics of online social networks in an engineering framework is an urgent and important issue.

Studies into the dynamics of online social networks have examined various models that reflect the diversity of characteristics of user dynamics. User dynamics that describe the adoption and abandonment of a particular social networking service (SNS) have been modeled by the SIR model, which is a traditional epidemiological model, and the irSIR model, which is an extension of that model [1, 2]. The consensus problem including user opinion formation is typical of the dynamics in online social networks [3, 4]. This is modeled by a first-order differential equation with respect to time using a Laplacian matrix that represents the social network structure. The differential equation used in this model is a sort of continuous-time Markov chain on the network. First-order differential equations with respect to time are also used in modeling of the temporal change of social network structure (how to link or to delink the nodes), and there are models that change in a continuous-time Markov chain [5]. In addition to theoretical modeling, user dynamics analysis based on real network observations has also been studied [6, 7].

This paper focuses on explosive user dynamics such as flaming, which is defined as the divergence of the intensity of user dynamics. Since both epidemiological models and continuous-time Markov chains on networks describe the transition to the final state (steady state), they cannot describe the divergence of the intensity. Moreover, it is difficult to clarify the structure of the theoretical model behind user dynamics from just an analysis of actual data. User dynamics in online social networks, including the explosive user dynamics, is generated by interactions between users. It is difficult to fully understand the details of interactions between users, but we can apply the concept of the minimal model; it models the simple interactions exhibited by a wide type of user interactions. Based on the minimal model, it has been proposed to apply the oscillation model on networks to describe user dynamics in online social networks [8]. In the oscillation model approach, network dynamics is described by the wave equation on networks. The oscillation energy of each node calculated from the oscillation model gives a generalization of node centrality and includes the conventional node centrality measures (degree centrality and betweenness centrality) [9–11] commonly used in network analysis [12]. Also, by considering that the occurrence of explosive user dynamics such as flaming in online social networks demonstrate the characteristic that the oscillation energy diverges with time, we can discuss appearance factors of explosive user dynamics in relation to the structure of the online social network [8].

This paper examines a fundamental equation [8] of the oscillation model on networks. The fundamental equation can explicitly describe the causal relation of the influence of the network structure on user dynamics. We give solutions to two major unresolved issues with the fundamental equation. Specifically, we derive all solutions of the wave equation from the fundamental equation and draw a concurrence between the link structure of the networks (represented by the fundamental equation) and that of the wave equation. Surprisingly, they are solved naturally and simultaneously.

2 Oscillation Model on Networks

This section briefly summarizes the oscillation model on networks according to [8].

Let $G(V, E)$ be a simple directed graph (without self-loop and duplicated links) with n nodes representing a social network, $V = \{1, \ldots, n\}$ denote the set of nodes, and E denote the set of links. Hereafter, node IDs are denoted by Roman characters $1 \le i, j \le n$ and the oscillation modes are denoted by Greek characters $0 \le \mu, \nu \le n-1$.

For a pair of adjacent nodes $i, j \in V$, we let the link weight of directed link $(i \to j) \in E$ be denoted by w_{ij}, and define the adjacency matrix $\mathcal{A} = [\mathcal{A}_{ij}]_{1 \le i,j \le n}$ as

$$\mathcal{A}_{ij} := \begin{cases} w_{ij}, & (i \to j) \in E, \\ 0, & (i \to j) \notin E. \end{cases} \tag{1}$$

In addition, if the nodal degree of out-links from node i is given as $d_i := \sum_{j \in \partial i} w_{ij}$, the degree matrix $\mathcal{D} := \mathrm{diag}(d_1, \ldots d_n)$, where ∂i denotes the set of out-neighbors of node i. Finally, Laplacian matrix $\mathcal{L}$ is defined as $\mathcal{L} := \mathcal{D} - \mathcal{A}$.

Let ${}^t\boldsymbol{m} = (m_1, \ldots, m_n)$ denote a left eigenvector associated with the eigenvalue 0 of Laplacian matrix $\mathcal{L}$. We say that the directed graph is *symmetrizable* if and only if $m_i > 0$ for all $i \in V$ and

$$m_i \, w_{ij} = m_j \, w_{ji},$$

for all pairs of adjacent nodes $(i \to j) \in E$. Hereafter, we denote the Laplacian matrix of a symmetrizable directed graph as $\mathcal{L}_0$. The Laplacian matrix $\mathcal{L}_0$ can be transformed into a symmetric matrix $\boldsymbol{S}_0$ by the similarity transformation using $\boldsymbol{M} := \mathrm{diag}(m_1, \ldots, m_n)$, as

$$\boldsymbol{S}_0 := \boldsymbol{M}^{+1/2} \, \mathcal{L}_0 \, \boldsymbol{M}^{-1/2};$$

$\boldsymbol{S}_0$ and $\mathcal{L}_0$ have the same eigenvalues. Furthermore, the eigenvalues are nonnegative and we sort them as

$$0 = \lambda_0 \le \lambda_1 \le \cdots \le \lambda_{n-1}.$$

In addition, we choose the eigenvectors $\boldsymbol{v}_\mu$ of $\boldsymbol{S}_0$ associated with λ_μ in such a way that the eigenvectors form an orthonormal eigenbasis, $\boldsymbol{v}_\mu \cdot \boldsymbol{v}_\nu = \delta_{\mu\nu}$.

Next, we consider a simple and universal interaction model among users through recourse to the concept of the minimal model [8]. We assume that the state of node i

at time t (representing a user in the online social network) can be described by a one-dimensional parameter, $x_i(t)$. Also, it is assumed that the influence exists between adjacent nodes, such that they are influenced by the other's state quantities and tend to harmonize. We specifically assume that the strength of the influence between a pair of adjacent nodes, i and j, is proportional to the absolute value of the difference between their state quantities, $|x_i(t) - x_j(t)|$. Then, the equation of motion (EoM) of state vector $\boldsymbol{x}(t) = {}^t(x_1(t), \ldots, x_n(t))$ can be denoted as

$$\frac{\mathrm{d}^2}{\mathrm{d}t^2}\boldsymbol{x}(t) = -\mathcal{L}\boldsymbol{x}(t). \tag{2}$$

This equation is called the *wave equation* on networks, and the above modeling is called the oscillation model on networks.

We can calculate the oscillation energy of the whole network from the solution, $\boldsymbol{x}(t)$, of EoM (2). In particular, if the social network is a symmetrizable directed graph, we can calculate the oscillation energy of each node from that of the whole network and, furthermore, the oscillation energy of each node gives a generalized notion of node centrality. Node centrality is a quantitative index indicating how important a particular node is in a network, and there are various different node centrality measures depending on the definition of importance used. The representative indices are the degree centrality and the betweenness centrality, but the oscillation model on networks gives a framework that can explain both indices in a unified manner. For example, if the weight of all links is 1, the oscillation energy for each node becomes the degree centrality for the non-biased usage condition of the network. Also, by taking the number of routes passing through a link (or the amount of passing traffic) as the link weight, the oscillation energy for each node gives an value related to the betweenness centrality, again for the non-biased usage condition of the network. In particular, the oscillation energy of each node can generalize the node centrality even in various network usage situations such as having a biased usage condition where a specific node is the information source [12].

If the social network is not symmetrizable, the oscillation energy of the whole network may diverge with time depending on the network structure. This corresponds to the phenomenon where the strength of the user dynamics activity on the network diverges, like flaming on online social networks. Since it is known that such divergence is not generated by symmetrizable directed graphs, the oscillation model on networks gives a model offers a generation mechanism of explosive user dynamics caused by the network structure [12].

3 Fundamental Equation of the Oscillation Model on Networks

This section briefly summarizes the fundamental equation of the oscillation model on networks according to [8].

We specifically consider the situation where the network is not symmetrizable, and can be decomposed into a symmetrizable and one-way link parts. We first discuss the difficulty in expressing the solution of the wave equation (4) as a product of the solutions arising from the decomposition. We then show that, by reducing the second-order differential equation (4) to a first-order equation, we can obtain a product-form solution that reflects the decomposition.

We start with the decomposition of Laplacian matrix $\mathcal{L}$ into the Laplacian matrix of symmetrizable directed graph, $\mathcal{L}_0$, and that of a one-way link graph, $\mathcal{L}_{\mathrm{I}}$, as

$$\mathcal{L} = \mathcal{L}_0 + \mathcal{L}_{\mathrm{I}}, \tag{3}$$

where the one-way link graph is a directed graph that has at most only one-way links between nodes. The decomposition (3) is not unique and any directed graph can be decomposed as shown in (3). Since the non-uniqueness of the decomposition (3) leads to the selection of orthogonal bases in the state space through the choice of $\boldsymbol{S}_0$, we can choose a convenient decomposition that makes the Laplacian matrix of a one-way link graph $\mathcal{L}_{\mathrm{I}}$ simple.

The cause of the divergence in the oscillation energy is the influence of the one-way link graph, since the divergence of the oscillation energy is not inherent in symmetrizable directed graphs. In order to directly express the influence of a one-way link graph on a symmetrizable directed graph, let us rewrite the EoM using the coordinate system obtained by converting $\mathcal{L}_0$ into a diagonal matrix. Let the orthonormal basis determined from $\boldsymbol{S}_0$ based on the decomposition (3) be $\{\boldsymbol{v}_\mu\}_{0\leq\mu\leq n-1}$. By using the orthogonal matrix $\boldsymbol{P} := [\boldsymbol{v}_0, \boldsymbol{v}_1, \ldots, \boldsymbol{v}_{n-1}]$, $\mathcal{L}_0$ can be diagonalized as

$$\boldsymbol{\Lambda}_0 := {}^t\boldsymbol{P}\, \boldsymbol{S}_0\, \boldsymbol{P} = {}^t\boldsymbol{P} \left(\boldsymbol{M}^{+1/2}\, \mathcal{L}_0\, \boldsymbol{M}^{-1/2}\right) \boldsymbol{P},$$

where $\boldsymbol{\Lambda}_0 = \mathrm{diag}(\lambda_0, \ldots, \lambda_{n-1})$. Let us define $\boldsymbol{\psi}(t) := {}^t\boldsymbol{P}\, \boldsymbol{M}^{+1/2}\, \boldsymbol{x}(t)$ and $\boldsymbol{\Lambda}_{\mathrm{I}} := {}^t\boldsymbol{P} \left(\boldsymbol{M}^{+1/2}\, \mathcal{L}_{\mathrm{I}}\, \boldsymbol{M}^{-1/2}\right) \boldsymbol{P}$. Then, the EoM (2) can be transformed into

$$\frac{\mathrm{d}^2\boldsymbol{\psi}(t)}{\mathrm{d}t^2} = -\boldsymbol{\Lambda}\, \boldsymbol{\psi}(t) = -(\boldsymbol{\Lambda}_0 + \boldsymbol{\Lambda}_{\mathrm{I}})\, \boldsymbol{\psi}(t) \tag{4}$$

where $\boldsymbol{\Lambda} := \boldsymbol{\Lambda}_0 + \boldsymbol{\Lambda}_{\mathrm{I}}$.

The solution of the wave equation (4) for $\boldsymbol{\Lambda}_{\mathrm{I}} = \boldsymbol{O}$ (null matrix) is easily obtained. In order to explicitly describe the causal relation of the influence of $\boldsymbol{\Lambda}_{\mathrm{I}}$, it is preferable that the solution of (4) be cast in product-form; it consists of the solutions of the wave equations related to $\mathbf{\Lambda}_0$ and $\mathbf{\Lambda}_\mathrm{I}$. Unfortunately, attempting the product-form solution of the wave equation (4) will not succeed. This is because the wave equation (4) is a second-order differential equation with respect to time, so the equation yields an extra cross term. Hence, to obtain a first-order differential equation with respect to time, we define the following matrix,

$$\mathbf{\Omega}^2 = \mathbf{\Lambda} = \mathbf{\Lambda}_0 + \mathbf{\Lambda}_\mathrm{I}. \tag{5}$$

This means $\mathbf{\Omega}$ is the square root of matrix $\mathbf{\Lambda}$ and it is unique if we choose $\mathbf{\Omega}$ to be semi-positive definite. If we define $\mathbf{\Omega}_0 := \mathbf{\Lambda}_0^{1/2}$, the square root matrix $\mathbf{\Omega}$ is decomposed as

$$\mathbf{\Omega} = \mathbf{\Omega}_0 + \mathbf{\Omega}_\mathrm{I}. \tag{6}$$

By using the diagonal matrix $\boldsymbol{M}$ that symmetrizes $\mathcal{L}_0$ into $\boldsymbol{S}_0$, the square root matrices $\mathcal{H}_0$ of $\mathcal{L}_0$, and $\mathcal{H}$ of $\mathcal{L}$ are defined, respectively, as

$$\mathcal{H}_0 := \boldsymbol{M}^{-1/2}\,(\boldsymbol{P}\,\mathbf{\Omega}_0\,{}^t\boldsymbol{P})\,\boldsymbol{M}^{+1/2},$$
$$\mathcal{H} := \boldsymbol{M}^{-1/2}\,(\boldsymbol{P}\,\mathbf{\Omega}\,{}^t\boldsymbol{P})\,\boldsymbol{M}^{+1/2}.$$

Also, we define $\mathcal{H}_\mathrm{I}$ by using the decomposition

$$\mathcal{H} = \mathcal{H}_0 + \mathcal{H}_\mathrm{I}. \tag{7}$$

Note that $\mathcal{H}_\mathrm{I}$ is not the square root of $\mathcal{L}_\mathrm{I}$.

By using the square root matrix $\mathbf{\Omega}$ of $\mathbf{\Lambda}$, we introduce the following two different wave equations:

$$+\mathrm{i}\,\frac{\mathrm{d}\,\boldsymbol{\psi}^+(t)}{\mathrm{d}t} = \mathbf{\Omega}\,\boldsymbol{\psi}^+(t), \qquad -\mathrm{i}\,\frac{\mathrm{d}\,\boldsymbol{\psi}^-(t)}{\mathrm{d}t} = \mathbf{\Omega}\,\boldsymbol{\psi}^-(t). \tag{8}$$

The solutions of the wave equations (8) satisfy the following equation (double sign correspondence) as

$$\frac{\mathrm{d}^2\boldsymbol{\psi}^\pm(t)}{\mathrm{d}t^2} = \mp\mathrm{i}\Omega\,\frac{\mathrm{d}\,\boldsymbol{\psi}^\pm(t)}{\mathrm{d}t} = -\mathbf{\Omega}^2\,\boldsymbol{\psi}^\pm(t) = -(\mathbf{\Lambda}_0 + \mathbf{\Lambda}_\mathrm{I})\,\boldsymbol{\psi}^\pm(t).$$

This means that the solutions of the wave equations (8) solve the original wave equation (4).

Conversely, let us confirm that the solution of the wave equation (4) does not necessarily solve (8). For constants c^+ and c^-, let us consider a linear combination of the solutions of the two different equations (8), $c^+\,\boldsymbol{\psi}^+(t) + c^-\,\boldsymbol{\psi}^-(t)$. The linear combination solves (4)

$$\frac{\mathrm{d}^2}{\mathrm{d}t^2}(c^+\,\boldsymbol{\psi}^+(t)+c^-\,\boldsymbol{\psi}^-(t)) = \mathrm{i}\,\boldsymbol{\Omega}\,\frac{\mathrm{d}}{\mathrm{d}t}(-c^+\,\boldsymbol{\psi}^+(t)+c^-\,\boldsymbol{\psi}^-(t))$$
$$= -\,\boldsymbol{\Omega}^2\,(c^+\,\boldsymbol{\psi}^+(t)+c^-\,\boldsymbol{\psi}^-(t)) = -\,\boldsymbol{\Lambda}\,(c^+\,\boldsymbol{\psi}^+(t)+c^-\,\boldsymbol{\psi}^-(t)).$$

However, the linear combination satisfies neither of equations in (8). This problem is discussed later.

Next, let us consider the possibility of the product-form solution of $\boldsymbol{\psi}^{\pm}(t)$ for the wave equation (8). The goal here is to write solution $\boldsymbol{\psi}^{\pm}(t)$ in product-form, i.e. $\boldsymbol{\psi}^{\pm}(t) = \boldsymbol{\Psi}_0^{\pm}(t)\,\boldsymbol{\psi}_{\mathrm{I}}^{\pm}(t)$ by using solutions of the two wave equations with respect to $\boldsymbol{\Omega}_0$ and $\boldsymbol{\Omega}_{\mathrm{I}}$. By choosing the initial condition of $\boldsymbol{\Psi}_0^{\pm}(0) = \boldsymbol{I}$ ($n \times n$ unit matrix), that is, $\boldsymbol{\psi}^{\pm}(0) = \boldsymbol{\psi}_{\mathrm{I}}^{\pm}(0)$, and by using the decomposition (6), we introduce the following differential equations:

$$\pm\mathrm{i}\,\frac{\mathrm{d}}{\mathrm{d}t}\,\boldsymbol{\psi}_0^{\pm}(t) = \boldsymbol{\Omega}_0\,\boldsymbol{\psi}_0^{\pm}(t), \tag{9}$$

$$\pm\mathrm{i}\,\frac{\mathrm{d}}{\mathrm{d}t}\,\boldsymbol{\psi}_{\mathrm{I}}^{\pm}(t) = \left(\boldsymbol{\Psi}_0^{\pm}(-t)\,\boldsymbol{\Omega}_{\mathrm{I}}\,\boldsymbol{\Psi}_0^{\pm}(t)\right)\boldsymbol{\psi}_{\mathrm{I}}^{\pm}(t), \tag{10}$$

where $\boldsymbol{\Psi}_0^{\pm}(t)$ is the diagonal matrix with diagonals ${}^t\boldsymbol{\psi}_0^{\pm}(t) = (\psi_0^{\pm}(0;t), \psi_0^{\pm}(1;t), \ldots, \psi_0^{\pm}(n-1;t))$, that is,

$$\boldsymbol{\Psi}_0^{\pm}(t) = \begin{bmatrix} \psi_0^{\pm}(0;t) & 0 & \cdots & 0 \\ 0 & \psi_0^{\pm}(1;t) & \ddots & \vdots \\ \vdots & \ddots & \ddots & 0 \\ 0 & 0 & 0 & \psi_0^{\pm}(n-1;t) \end{bmatrix}.$$

From the initial condition $\boldsymbol{\Psi}_0^{\pm}(0) = \boldsymbol{I}$,

$$\boldsymbol{\Psi}_0^{\pm}(-t) = \boldsymbol{\Psi}_0^{\pm}(t)^{-1} = \boldsymbol{\Psi}_0^{\mp}(t). \tag{11}$$

For ${}^t\boldsymbol{\psi}_{\mathrm{I}}^{\pm}(t) = (\psi_{\mathrm{I}}^{\pm}(0;t), \psi_{\mathrm{I}}^{\pm}(1;t), \ldots, \psi_{\mathrm{I}}^{\pm}(n-1;t))$, the structure of the product-form solution is expressed as

$$\boldsymbol{\psi}^{\pm}(t) = \boldsymbol{\Psi}_0^{\pm}(t)\,\boldsymbol{\psi}_{\mathrm{I}}^{\pm}(t) = \begin{pmatrix} \psi_0^{\pm}(0;t)\,\psi_{\mathrm{I}}^{\pm}(0;t) \\ \psi_0^{\pm}(1;t)\,\psi_{\mathrm{I}}^{\pm}(1;t) \\ \vdots \\ \psi_0^{\pm}(n-1;t)\,\psi_{\mathrm{I}}^{\pm}(n-1;t) \end{pmatrix}.$$

By substituting $\boldsymbol{\psi}^{\pm}(t) = \boldsymbol{\Psi}_0^{\pm}(t)\,\boldsymbol{\psi}_{\mathrm{I}}^{\pm}(t)$ into the wave equations (8), and using the differential equations (9) and (10), and the relation (11), we obtain

$$\begin{aligned}
\pm\mathrm{i}\,\frac{\mathrm{d}\,\boldsymbol{\psi}^{\pm}(t)}{\mathrm{d}t} &= \pm\mathrm{i}\,\frac{\mathrm{d}}{\mathrm{d}t}\,(\boldsymbol{\Psi}_0^{\pm}(t)\,\boldsymbol{\psi}_{\mathrm{I}}^{\pm}(t)) \\
&= \boldsymbol{\Omega}_0\,\boldsymbol{\Psi}_0^{\pm}(t)\,\boldsymbol{\psi}_{\mathrm{I}}^{\pm}(t) + \boldsymbol{\Psi}_0^{\pm}(t)\left(\boldsymbol{\Psi}_0^{\pm}(-t)\,\boldsymbol{\Omega}_{\mathrm{I}}\,\boldsymbol{\Psi}_0^{\pm}(t)\right)\boldsymbol{\psi}_{\mathrm{I}}^{\pm}(t) \\
&= (\boldsymbol{\Omega}_0 + \boldsymbol{\Omega}_{\mathrm{I}})\,\boldsymbol{\psi}^{\pm}(t) = \boldsymbol{\Omega}\,\boldsymbol{\psi}^{\pm}(t).
\end{aligned}$$

This implies that the attempt to derive the product-form solution has succeeded.

Summarizing the above, the solution of the wave equation (8) is also the solution of the original wave equation (4) and can be expressed as the product-form solution with respect to $\boldsymbol{\psi}_0^{\pm}(t)$ and $\boldsymbol{\psi}_{\mathrm{I}}^{\pm}(t)$. Therefore, the causal relation of the influence of the one-way link graph can be explicitly described.

From the above examination, the wave equations (8) that describe the causality of the oscillation dynamics can be considered as more fundamental than the original wave equation (4) (or the original EoM (2)). For this reason, we call the wave equations (8) the fundamental equations of oscillation dynamics on directed graphs. Similarly, the first-order differential equations with respect to time that the original EoM (2) can be rewritten into are

$$+\mathrm{i}\,\frac{\mathrm{d}\,\boldsymbol{x}^{+}(t)}{\mathrm{d}t} = \mathcal{H}\boldsymbol{x}^{+}(t), \qquad -\mathrm{i}\,\frac{\mathrm{d}\,\boldsymbol{x}^{-}(t)}{\mathrm{d}t} = \mathcal{H}\boldsymbol{x}^{-}(t), \tag{12}$$

they are the fundamental equations that are mathematically equivalent to (8).

4 Fundamental Equation and Quantum Theory

Let us rewrite the fundamental equations (12) into a single equation.

First, we set the components of the n-dimensional vectors $\boldsymbol{x}^{+}(t)$ and $\boldsymbol{x}^{-}(t)$ as

$$\begin{aligned}
\boldsymbol{x}^{+}(t) &= {}^{t}(x_1^{+}(t),\ x_2^{+}(t),\ \ldots,\ \psi_n^{+}(t)), \\
\boldsymbol{x}^{-}(t) &= {}^{t}(x_1^{-}(t),\ x_2^{-}(t),\ \ldots,\ \psi_n^{-}(t)).
\end{aligned}$$

By combining them, we define the new $2n$-dimensional vector $\widehat{\boldsymbol{x}}(t)$ as

$$\widehat{\boldsymbol{x}}(t) := {}^{t}(x_1^{+}(t),\ x_1^{-}(t),\ x_2^{+}(t),\ x_2^{-}(t),\ \ldots,\ x_n^{+}(t),\ x_n^{-}(t)).$$

Also, for Laplacian matrix $\mathcal{L}$, we define the following $2n \times 2n$ square matrix

$$\widehat{\mathcal{L}} := \mathcal{L} \otimes \boldsymbol{E}, \tag{13}$$

where $\boldsymbol{E}$ denotes the 2×2 unit matrix and $\otimes$ denotes Kronecker product [13], which is, for $\mathcal{L} = [\mathcal{L}_{ij}]_{1 \le i,j \le n}$,

$$\widehat{\mathcal{L}} = \begin{bmatrix} \mathcal{L}_{11}\,\boldsymbol{E} & \mathcal{L}_{12}\,\boldsymbol{E} & \cdots & \mathcal{L}_{1n}\,\boldsymbol{E} \\ \mathcal{L}_{21}\,\boldsymbol{E} & \mathcal{L}_{22}\,\boldsymbol{E} & \cdots & \mathcal{L}_{2n}\,\boldsymbol{E} \\ \mathcal{L}_{31}\,\boldsymbol{E} & \mathcal{L}_{32}\,\boldsymbol{E} & \cdots & \mathcal{L}_{3n}\,\boldsymbol{E} \\ \vdots & \vdots & \ddots & \vdots \\ \mathcal{L}_{n1}\,\boldsymbol{E} & \mathcal{L}_{n2}\,\boldsymbol{E} & \cdots & \mathcal{L}_{nn}\,\boldsymbol{E} \end{bmatrix} = \begin{bmatrix} \mathcal{L}_{11} & 0 & \mathcal{L}_{12} & 0 & \cdots & \mathcal{L}_{1n} & 0 \\ 0 & \mathcal{L}_{11} & 0 & \mathcal{L}_{12} & \cdots & 0 & \mathcal{L}_{1n} \\ \mathcal{L}_{21} & 0 & \mathcal{L}_{22} & 0 & \cdots & \mathcal{L}_{2n} & 0 \\ 0 & \mathcal{L}_{21} & 0 & \mathcal{L}_{22} & \cdots & 0 & \mathcal{L}_{2n} \\ \mathcal{L}_{31} & 0 & \mathcal{L}_{32} & 0 & \cdots & \mathcal{L}_{3n} & 0 \\ 0 & \mathcal{L}_{31} & 0 & \mathcal{L}_{32} & \cdots & 0 & \mathcal{L}_{3n} \\ \vdots & \vdots & \vdots & \vdots & \ddots & \vdots & \vdots \\ \mathcal{L}_{n-1,1} & 0 & \mathcal{L}_{n-1,2} & 0 & \cdots & \mathcal{L}_{n-1,n} & 0 \\ 0 & \mathcal{L}_{n-11} & 0 & \mathcal{L}_{n-1,2} & \cdots & 0 & \mathcal{L}_{n-1,n} \\ \mathcal{L}_{n1} & 0 & \mathcal{L}_{n2} & 0 & \cdots & \mathcal{L}_{nn} & 0 \\ 0 & \mathcal{L}_{n1} & 0 & \mathcal{L}_{n2} & \cdots & 0 & \mathcal{L}_{nn} \end{bmatrix}.$$

In order to express both $\mathcal{L} = \mathcal{H}^2$ and $\mathcal{L} = (-\mathcal{H})^2$ simultaneously, the square root $\widehat{\mathcal{H}}$ of $\widehat{\mathcal{L}}$ is defined as

$$\widehat{\mathcal{H}} = \mathcal{H} \otimes \begin{bmatrix} 1 & 0 \\ 0 & -1 \end{bmatrix}.; \tag{14}$$

$\widehat{\mathcal{H}}$ satisfies $\widehat{\mathcal{L}} = \widehat{\mathcal{H}}^2$.

By using the above $2n$-dimensional notations, the components of the fundamental equations (12) can be expressed as the one equation of

$$\mathrm{i}\,\frac{\mathrm{d}\widehat{\boldsymbol{x}}(t)}{\mathrm{d}t} = \widehat{\mathcal{H}}\widehat{\boldsymbol{x}}(t). \tag{15}$$

It is worth to note that this equation has essentially the same structure as the Dirac equation found in relativistic quantum theory [8, 14].

5 Problems with the Fundamental Equation of the Oscillation Model

The two expressions of the fundamental equations (8) and (12) are mathematically equivalent and they can be transformed into each other by using a simple linear transformation. On the other hand, there are two crucial problems with the fundamental equation (12) as listed below.

- The solutions of the fundamental equation (12) are also solutions of the original wave equation (2). Unfortunately, the converse is not true, as shown in Sect. 3. If we are to claim that the fundamental equation (12) is really *fundamental*, it should be possible to derive all solutions of the original wave equation (2) from the fundamental equation (12).
- The square root matrix $\mathcal{H}$ of the Laplacian matrix that is appeared in the fundamental equation (12) does not reflect the structure of social networks

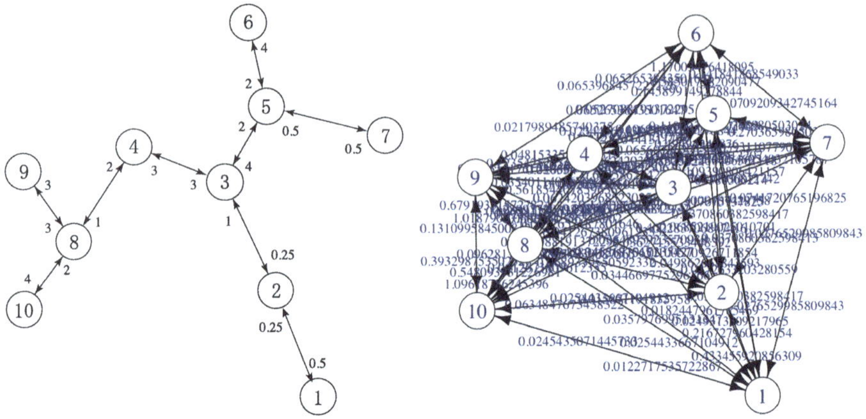

link structure of $\mathcal{L}$ link structure of $\mathcal{H}$

Fig. 1 Link structures and link weights described by Laplacian matrix $\mathcal{L}$ and its square root matrix $\mathcal{H}$

described by the Laplacian matrix $\mathcal{L}$. Since it is unacceptable in practice to hypothesize some direct relationships between nodes where links do not exist in the social network structure described by the Laplacian matrix, the Laplacian matrix $\mathcal{L}$ and its square root matrix $\mathcal{H}$ should have completely identical link structures.

Here, we describe the latter problem via an example. The figure on the left of Fig. 1 shows an example of a social network structure. For the Laplacian matrix $\mathcal{L}$ describing the left figure, the figure on the right describes link structure of the square root matrix $\mathcal{H}$ of $\mathcal{L}$. Even if the structure of a social network is sparse, the link structure of its square root matrix is a complete graph, in general. This means that some direct relationships exist between all users yielding an unacceptable situation. Conversely, if we give $\mathcal{H}$ as a sparse matrix, $\mathcal{L}$ is also sparse. However, their link structures are not identical, in general, which is also an unacceptable situation.

To solve these problems at the same time, we discuss the following proposition in the next section.

Proposition 1 *By utilizing one advantage of the structure of the* 2*n-dimensional wave equation (15), the following two properties hold simultaneously:*

- *The fundamental equation can generate all solutions of the original wave equation (2), and*
- *the matrix* $\widehat{\mathcal{H}}$ *appearing in the fundamental equation can be chosen so that its link structure completely matches the link structure represented by the Laplacian matrix.*

6 Fundamental Equation Reflecting Social Network Structure

As shown in (14), the square root matrix $\widehat{\mathcal{H}}$ of $\widehat{\mathcal{L}} := \mathcal{L} \otimes \boldsymbol{E}$ is chosen not as a semi-positive definite matrix, so the choice of $\widehat{\mathcal{H}}$ is not unique. The first attempt utilizes this degree of freedom of the choice to yield matching link structures, $\widehat{\mathcal{H}}$ and $\widehat{\mathcal{L}}$.

First, we decompose $\mathcal{H}$ into diagonal matrix $\mathcal{H}^{(\mathrm{d})}$ and the other matrix $-\mathcal{H}^{(\mathrm{a})}$, which has only non-diagonal components, as

$$\mathcal{H} = \mathcal{H}^{(\mathrm{d})} - \mathcal{H}^{(\mathrm{a})}.$$

Since $\mathcal{H}^2 = \mathcal{L}$ and

$$\mathcal{H}^2 = (\mathcal{H}^{(\mathrm{d})} - \mathcal{H}^{(\mathrm{a})})^2 = (\mathcal{H}^{(\mathrm{d})})^2 - \mathcal{H}^{(\mathrm{d})}\mathcal{H}^{(\mathrm{a})} - \mathcal{H}^{(\mathrm{a})}\mathcal{H}^{(\mathrm{d})} + (\mathcal{H}^{(\mathrm{a})})^2,$$

the link structures of both $\mathcal{H}$ and $\mathcal{L}$ are identical, if $(\mathcal{H}^{(\mathrm{a})})^2 = \mathbf{O}$. In order to realize this relation, we consider the following $2n \times 2n$ matrix

$$\widehat{\mathcal{H}}^{(\mathrm{a})} = \mathcal{H}^{(\mathrm{a})} \otimes \frac{1}{2}\begin{bmatrix} 1 & 1 \\ -1 & -1 \end{bmatrix}. \tag{16}$$

The 2×2 matrix used here exhibits nilpotency and so has the following property

$$\begin{bmatrix} 1 & 1 \\ -1 & -1 \end{bmatrix}^2 = \begin{bmatrix} 0 & 0 \\ 0 & 0 \end{bmatrix},$$

so $(\widehat{\mathcal{H}}^{(\mathrm{a})})^2 = \mathbf{O}$.

Let this nilpotent 2×2 matrix be $\boldsymbol{X}$; by choosing a certain 2×2 matrix $\boldsymbol{Y}$, we introduce

$$\widehat{\mathcal{H}} = \mathcal{H}^{(\mathrm{d})} \otimes \boldsymbol{Y} - \mathcal{H}^{(\mathrm{a})} \otimes \boldsymbol{X}.$$

Here, we consider the possibility of whether or not the following relation is realized:

$$\widehat{\mathcal{H}}^2 = \mathcal{L} \otimes \boldsymbol{E}.$$

From the expansion of $\widehat{\mathcal{H}}^2$, we obtain

$$\begin{aligned}\widehat{\mathcal{H}}^2 &= (\mathcal{H}^{(\mathrm{d})} \otimes \boldsymbol{Y} - \mathcal{H}^{(\mathrm{a})} \otimes \boldsymbol{X})^2 \\ &= (\mathcal{H}^{(\mathrm{d})})^2 \otimes \boldsymbol{Y}^2 - (\mathcal{H}^{(\mathrm{d})}\mathcal{H}^{(\mathrm{a})}) \otimes (\boldsymbol{Y}\boldsymbol{X}) \\ &\quad - (\mathcal{H}^{(\mathrm{a})}\mathcal{H}^{(\mathrm{d})}) \otimes (\boldsymbol{X}\boldsymbol{Y}) + (\mathcal{H}^{(\mathrm{a})})^2 \otimes \boldsymbol{X}^2\end{aligned}$$

$$= (\mathcal{H}^{(\mathrm{d})})^2 \otimes \boldsymbol{Y}^2 - (\mathcal{H}^{(\mathrm{d})}\mathcal{H}^{(\mathrm{a})}) \otimes (\boldsymbol{Y}\boldsymbol{X}) - (\mathcal{H}^{(\mathrm{a})}\mathcal{H}^{(\mathrm{d})}) \otimes (\boldsymbol{X}\boldsymbol{Y}).$$

Therefore, the sufficient condition for $\widehat{\mathcal{H}}^2 = \mathcal{L} \otimes \boldsymbol{E}$ can be written as

$$\boldsymbol{Y}^2 = \boldsymbol{E}, \quad \boldsymbol{X}\boldsymbol{Y} = \boldsymbol{Y}\boldsymbol{X} = \boldsymbol{E}, \quad \boldsymbol{X}^2 = \mathbf{O}. \tag{17}$$

Here, the second condition implies $\boldsymbol{Y} = \boldsymbol{X}^{-1}$, but it is known that no nilpotent matrix has an inverse. Thus we cannot choose $\boldsymbol{Y}$ that satisfies the condition (17).

The next step is to relax the condition $\boldsymbol{Y} = \boldsymbol{X}^{-1}$. As one example that satisfies the following relation

$$\boldsymbol{Y}^2 = \boldsymbol{E}, \quad \boldsymbol{X}^2 = \mathbf{O}, \tag{18}$$

let us consider the following matrix [15]

$$\widehat{\mathcal{H}} := \widehat{\mathcal{H}}^{(\mathrm{d})} - \widehat{\mathcal{H}}^{(\mathrm{a})} = \mathcal{H}^{(\mathrm{d})} \otimes \begin{bmatrix} 1 & 0 \\ 0 & -1 \end{bmatrix} - \mathcal{H}^{(\mathrm{a})} \otimes \frac{1}{2}\begin{bmatrix} 1 & 1 \\ -1 & -1 \end{bmatrix}. \tag{19}$$

The corresponding fundamental equation is expressed as

$$\begin{aligned} \mathrm{i}\,\frac{\mathrm{d}\widehat{\boldsymbol{x}}(t)}{\mathrm{d}t} &= \widehat{\mathcal{H}}\widehat{\boldsymbol{x}}(t) \\ &= \left(\mathcal{H}^{(\mathrm{d})} \otimes \begin{bmatrix} 1 & 0 \\ 0 & -1 \end{bmatrix} - \mathcal{H}^{(\mathrm{a})} \otimes \frac{1}{2}\begin{bmatrix} 1 & 1 \\ -1 & -1 \end{bmatrix}\right)\widehat{\boldsymbol{x}}(t). \end{aligned} \tag{20}$$

Here, we obtain

$$\begin{aligned} \widehat{\mathcal{H}}^2 &= (\mathcal{H}^{(\mathrm{d})})^2 \otimes \begin{bmatrix} 1 & 0 \\ 0 & 1 \end{bmatrix} - (\mathcal{H}^{(\mathrm{d})}\mathcal{H}^{(\mathrm{a})}) \otimes \frac{1}{2}\begin{bmatrix} 1 & 1 \\ 1 & 1 \end{bmatrix} - (\mathcal{H}^{(\mathrm{a})}\mathcal{H}^{(\mathrm{d})}) \otimes \frac{1}{2}\begin{bmatrix} 1 & -1 \\ -1 & 1 \end{bmatrix} \\ &= (\mathcal{H}^{(\mathrm{d})})^2 \otimes \boldsymbol{E} - (\mathcal{H}^{(\mathrm{d})}\mathcal{H}^{(\mathrm{a})} + \mathcal{H}^{(\mathrm{a})}\mathcal{H}^{(\mathrm{d})}) \otimes \frac{1}{2}\,\boldsymbol{E} \\ &\quad - (\mathcal{H}^{(\mathrm{d})}\mathcal{H}^{(\mathrm{a})} - \mathcal{H}^{(\mathrm{a})}\mathcal{H}^{(\mathrm{d})}) \otimes \frac{1}{2}\begin{bmatrix} 0 & 1 \\ 1 & 0 \end{bmatrix}. \end{aligned} \tag{21}$$

So, if $\mathcal{H}^{(\mathrm{d})}\mathcal{H}^{(\mathrm{a})} = \mathcal{H}^{(\mathrm{a})}\mathcal{H}^{(\mathrm{d})}$, that is, $\mathcal{H}^{(\mathrm{d})}$ and $\mathcal{H}^{(\mathrm{a})}$ are commutable (the order of product of matrices is commutable), we obtain the relation $\widehat{\mathcal{H}}^2 = \mathcal{L} \otimes \boldsymbol{E}$. However, the commutation relation does not hold in general, and it is limited to the case when $\mathcal{H}^{(\mathrm{d})}$ is proportional to the unit matrix $\boldsymbol{I}$.

From the above discussion, it can be seen that for a general Laplacian matrix $\widehat{\mathcal{H}}^2 = \mathcal{L} \otimes \boldsymbol{E}$ cannot be satisfied while matching the links actually present. Conversely, in order to match the link structures of $2n \times 2n$ matrices $\widehat{\mathcal{H}}$ and $\widehat{\mathcal{L}}$, we can recognize that two solutions of the fundamental equations (12) should be mixed by the influence of the third term of the right-hand side of (21). Expressing such a mixture of solutions yields the benefit of expressing the fundamental equation (15) by using a $2n$-dimensional vector and a $2n \times 2n$ square matrix.

Halting the attempt to realize $\widehat{\mathcal{H}}^2 = \mathcal{L} \otimes \boldsymbol{E}$, We aim to reproduce the original equation of motion by pursuing the benefit of expressing the fundamental equation as a $2n$-dimensional vector. Here, as discussed in Sect. 3, remember that the solutions of the wave equations (8) and that of the original equation of motion (4) are not the same. The solutions of the wave equations (8) are always the solution of the original equation of motion (4), but the linear combination of the solutions of the two different wave equations (8) is also a solution of (4). Therefore, no problem is created if $\widehat{\mathcal{H}}^2$ mixes the solutions of the two fundamental equations (8) even if they have different signs; on the contrary, it is a desirable situation.

From the fundamental equation (20), the second derivative of $\widehat{\boldsymbol{x}}(t)$ is written as

$$\begin{aligned}\frac{\mathrm{d}^2 \widehat{\boldsymbol{x}}(t)}{\mathrm{d}t^2} &= -\mathrm{i}\, \frac{\mathrm{d}\widehat{\boldsymbol{x}}(t)}{\mathrm{d}t} \widehat{\mathcal{H}}\widehat{\boldsymbol{x}}(t) = -\widehat{\mathcal{H}}^2 \widehat{\boldsymbol{x}}(t) \\ &= -\Bigg((\mathcal{H}^{(\mathrm{d})})^2 \otimes \begin{bmatrix} 1 & 0 \\ 0 & 1 \end{bmatrix} - (\mathcal{H}^{(\mathrm{d})} \mathcal{H}^{(\mathrm{a})} + \mathcal{H}^{(\mathrm{a})} \mathcal{H}^{(\mathrm{d})}) \otimes \frac{1}{2} \begin{bmatrix} 1 & 0 \\ 0 & 1 \end{bmatrix} \\ &\qquad - (\mathcal{H}^{(\mathrm{d})} \mathcal{H}^{(\mathrm{a})} - \mathcal{H}^{(\mathrm{a})} \mathcal{H}^{(\mathrm{d})}) \otimes \frac{1}{2} \begin{bmatrix} 0 & 1 \\ 1 & 0 \end{bmatrix} \Bigg) \widehat{\boldsymbol{x}}(t). \end{aligned} \tag{22}$$

By extracting the differential equations for $\boldsymbol{x}^+(t)$ and $\boldsymbol{x}^-(t)$, we obtain

$$\begin{aligned}\frac{\mathrm{d}^2 \boldsymbol{x}^+(t)}{\mathrm{d}t^2} &= -\left((\mathcal{H}^{(\mathrm{d})})^2 - \frac{1}{2} (\mathcal{H}^{(\mathrm{d})} \mathcal{H}^{(\mathrm{a})} + \mathcal{H}^{(\mathrm{a})} \mathcal{H}^{(\mathrm{d})}) \right) \boldsymbol{x}^+(t) \\ &\quad - \left(-\frac{1}{2} (\mathcal{H}^{(\mathrm{d})} \mathcal{H}^{(\mathrm{a})} - \mathcal{H}^{(\mathrm{a})} \mathcal{H}^{(\mathrm{d})}) \right) \boldsymbol{x}^-(t), \end{aligned} \tag{23}$$

$$\begin{aligned}\frac{\mathrm{d}^2 \boldsymbol{x}^-(t)}{\mathrm{d}t^2} &= -\left((\mathcal{H}^{(\mathrm{d})})^2 - \frac{1}{2} (\mathcal{H}^{(\mathrm{d})} \mathcal{H}^{(\mathrm{a})} + \mathcal{H}^{(\mathrm{a})} \mathcal{H}^{(\mathrm{d})}) \right) \boldsymbol{x}^-(t) \\ &\quad - \left(-\frac{1}{2} (\mathcal{H}^{(\mathrm{d})} \mathcal{H}^{(\mathrm{a})} - \mathcal{H}^{(\mathrm{a})} \mathcal{H}^{(\mathrm{d})}) \right) \boldsymbol{x}^+(t). \end{aligned} \tag{24}$$

Here, by adding both sides of the differential equations (23) and (24), we obtain

$$\frac{\mathrm{d}^2}{\mathrm{d}t^2} (\boldsymbol{x}^+(t) + \boldsymbol{x}^-(t)) = -\left((\mathcal{H}^{(\mathrm{d})})^2 - \mathcal{H}^{(\mathrm{d})} \mathcal{H}^{(\mathrm{a})} \right) (\boldsymbol{x}^+(t) + \boldsymbol{x}^-(t)). \tag{25}$$

This equation corresponds to the original equation of motion (2). In addition, the solutions $\boldsymbol{x}^+(t)$ and $\boldsymbol{x}^-(t)$ of the fundamental equations satisfy, respectively, the same fundamental equation even if multiplied by a constant, so Eq. (25) shows that the linear combination of the solutions of the fundamental equations (20) is also the solution of the original equation of motion (2).

From the above, by setting the fundamental equation to (20), it is possible to not only perfectly match the link structures between nodes represented by $\widehat{\mathcal{H}}$ and $\mathcal{L}$, but also generate all solutions of the original equation of motion (2).

The correspondences of $\mathcal{H}$ to the Laplacian matrix $\mathcal{L}$, the adjacency matrix $\mathcal{A}$, and the degree matrix $\mathcal{D}$ are obtained as

$$\mathcal{L} = (\mathcal{H}^{(\mathrm{d})})^2 - \mathcal{H}^{(\mathrm{d})}\,\mathcal{H}^{(\mathrm{a})}, \quad \mathcal{A} = \mathcal{H}^{(\mathrm{d})}\,\mathcal{H}^{(\mathrm{a})}, \quad \mathcal{D} = (\mathcal{H}^{(\mathrm{d})})^2. \tag{26}$$

More specifically, we obtain the simple relations of

$$\mathcal{H}^{(\mathrm{d})} = \mathrm{diag}\left(\sqrt{d_1},\ \dots,\ \sqrt{d_n}\right), \tag{27}$$

and, since $\mathcal{H}^{(\mathrm{a})} = (\mathcal{H}^{(\mathrm{d})})^{-1}\,\mathcal{A}$, $\mathcal{H}^{(\mathrm{a})} = [\mathcal{H}^{(\mathrm{a})}_{ij}]_{1\le i,j\le n}$ is obtained by

$$\mathcal{H}^{(\mathrm{a})}_{ij} := \begin{cases} w_{ij}/\sqrt{d_i}, & (i \to j) \in E, \\ 0, & (i \to j) \notin E. \end{cases} \tag{28}$$

Incidentally, the existence of simple relations (27) and (28) is due to the selection of the nilpotent matrix $\begin{bmatrix} 1 & 1 \\ -1 & -1 \end{bmatrix}$. If we choose the nilpotent matrix of $\begin{bmatrix} 1 & -1 \\ 1 & -1 \end{bmatrix}$, the adjacent matrix is obtained as $\mathcal{A} = \mathcal{H}^{(\mathrm{a})}\,\mathcal{H}^{(\mathrm{d})}$, and the relations are complicated. This is because it complicates the use of the property of the Laplacian matrix that the row sum is zero.

Figure 2 shows an example of social network structures of $\widehat{\mathcal{L}}$ and $\widehat{\mathcal{H}}$.

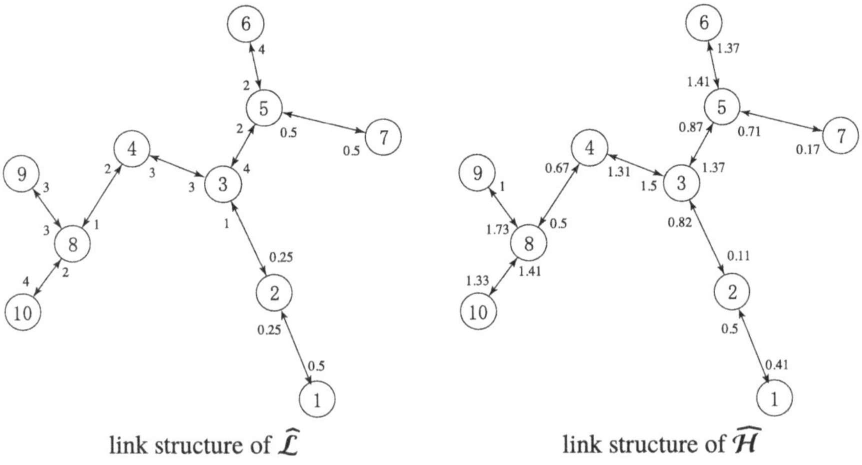

Fig. 2 Link structures and link weights described by the Laplacian matrix $\widehat{\mathcal{L}}$ and the proposed matrix $\widehat{\mathcal{H}}$

7 Conclusions

The two key remaining problems with the fundamental equation (12) of the oscillation model on networks have been solved. One is the problem that the solutions of (12)–(20) do not represent all the solutions of the original wave equation (2), and the other is that the link structures expressed by $\mathcal{H}$ and those expressed by $\mathcal{L}$ do not coincide.

This paper examined solutions to the latter problem and clarified that the two problems can be solved naturally and simultaneously. The constraints of the matching the link structures of $\mathcal{H}$ and $\mathcal{L}$ while keeping the characteristic of modeling that clearly describes the causality of the fundamental equation (8) is retained by considering the wave equation (15) as a $2n$-dimensional vector. By utilizing one advantage of the structure of $2n$-dimensional wave equation (15), the solutions of the fundamental equations (12) mix naturally and generate the solutions of the original equation of motion (2) of the n-dimensional vector, so solving the fundamental equation (20) gives all solutions of (2). While $\widehat{\mathcal{H}}^2 \neq \widehat{\mathcal{L}} := \mathcal{L} \otimes \boldsymbol{E}$, $\widehat{\mathcal{H}}$ is not the square root matrix of $\widehat{\mathcal{L}}$, the following $n \times 2n$ matrix

$$\boldsymbol{I} \otimes (1, 1) = \begin{bmatrix} 1\ 1\ 0\ 0 \cdots 0\ 0 \\ 0\ 0\ 1\ 1 \cdots 0\ 0 \\ \vdots\ \vdots\ \vdots\ \vdots\ \ddots\ \vdots\ \vdots \\ 0\ 0\ 0\ 0 \cdots 1\ 1 \end{bmatrix},$$

can be used to obtain

$$(\boldsymbol{I} \otimes (1, 1))\widehat{\mathcal{H}}^2 \widehat{\boldsymbol{x}} = \mathcal{L}\boldsymbol{x}. \tag{29}$$

where $(\boldsymbol{I} \otimes (1, 1))\widehat{\boldsymbol{x}} = \boldsymbol{x}$.

Acknowledgements This research was supported by Grant-in-Aid for Scientific Research (B) No. 17H01737 (2017–2019) and No. 19H04096 (2019–2021), and Grant-in-Aid for Scientific Research (C) No. 18K11271 (2018–2020) and No. 18K13777 (2018–2020) from the Japan Society for the Promotion of Science (JSPS).

References

1. Nekovee, M., Moreno, Y., Bianconi, G., Marsili, M.: Theory of rumour spreading in complex social networks. Phys. A Stat. Mech. Appl. **374**(1), 457–470 (2007)
2. Cannarella, J., Spechler, J.A.: Epidemiological modeling of online social network dynamics (2014). arXiv: 1401.4208
3. Olfati-Saber, R., Murray, R.M.: Consensus problems in networks of agents with switching topology and time-delays. IEEE Trans. Autom. Control **49**(9), 1520–1533 (2004)

4. Wang, L., Xiao, F.: Finite-time consensus problems for networks of dynamic agents. IEEE Trans. Autom. Control **55**(4), 950–955 (2010)
5. Snijders, T.A.B., Koskinen, J., Schweinberger, M.: Maximum likelihood estimation for social network dynamics. IEEE Trans. Autom. Control **4**(2), 567–588 (2010)
6. Cha, M., Mislove, A., Gummadi, K.P.: A measurement-driven analysis of information propagation in the Flickr social network. In: Proceedings of the 18th International Conference on World Wide Web (WWW'09), pp. 721–730 (2009)
7. Zhao, X., Sala, A., Wilson, C., Wang, X., Gaito, S., Zheng, H., Zhao, B.Y.: Multi-scale dynamics in a massive online social network. In: Proceedings of the 2012 Internet Measurement Conference (IMC'12), pp. 171–184 (2012)
8. Aida, M., Takano, C., Murata, M.: Oscillation model for describing network dynamics caused by asymmetric node interaction. IEICE Trans. Commun. **E101-B**(1), 123–136 (2018)
9. Wasserman, S., Faust, K.: Social Network Analysis: Methods and Applications. Cambridge University Press, Cambridge (1994)
10. Carrington, P.J., Scott, J., Wasserman, S.: Models and Methods in Social Network Analysis. Cambridge University Press, Cambridge (2005)
11. Mislove, A., Marcon, M., Gummadi, K.P., Druschel, P., Bhattacharjee, B.: Measurement and analysis of online social networks. In: Proceedings of the ACM SIGCOMM Conference on Internet Measurement, pp. 29–42 (2007)
12. Takano, C., Aida, M.: Revealing of the underlying mechanism of different node centralities based on oscillation dynamics on networks. IEICE Trans. Commun. **E101-B**(8), 1820–1832 (2018)
13. Brewer, J.: Kronecker products and matrix calculus in system theory. IEEE Trans. Circuits Syst. **25**(9), 772–781 (1978)
14. Bjorken, J.D., Drell, S.D.: Relativistic Quantum Mechanics. McGraw-Hill, New York (1965)
15. Aida, M.: Introduction to Network Dynamics. Morikita Publishing, Tokyo (2020)

Beyond Social Fragmentation: Coexistence of Cultural Diversity and Structural Connectivity Is Possible with Social Constituent Diversity

Hiroki Sayama and Junichi Yamanoi

Abstract Social fragmentation caused by widening differences among constituents has recently become a highly relevant issue to our modern society. Theoretical models of social fragmentation using the adaptive network framework have been proposed and studied in earlier literature, which are known to either converge to a homogeneous, well-connected network or fragment into many disconnected subnetworks with distinct states. Here we introduced the diversities of behavioral attributes among social constituents and studied their effects on social network evolution. We investigated, using a networked agent-based simulation model, how the resulting network states and topologies would be affected when individual constituents' cultural tolerance, cultural state change rate, and edge weight change rate were systematically diversified. The results showed that the diversity of cultural tolerance had the most direct effect to keep the cultural diversity within the society high and simultaneously reduce the average shortest path length of the social network, which was not previously reported in the earlier literature. Diversities of other behavioral attributes also had effects on final states of the social network, with some nonlinear interactions. Our results suggest that having a broad distribution of cultural tolerance levels within society can help promote the coexistence of cultural diversity and structural connectivity.

H. Sayama (✉)
Waseda Innovation Lab, Waseda University, Tokyo, Japan

Center for Collective Dynamics of Complex Systems, Binghamton University, State University of New York, Binghamton, NY, USA
e-mail: sayama@binghamton.edu

J. Yamanoi
School of Commerce, Waseda University, Tokyo, Japan

N. Masuda et al. (eds.), *Proceedings of NetSci-X 2020: Sixth International Winter School and Conference on Network Science*, Springer Proceedings in Complexity,
https://doi.org/10.1007/978-3-030-38965-9_12

1 Introduction

Social fragmentation caused by widening differences among constituents has recently become a highly relevant issue to our modern society, as various forms of gaps and conflicts are emerging from cultural, political, economic, ethnic, religious, linguistic, and other driving factors. Researchers have developed theoretical models of social fragmentation using the adaptive network framework [1–5], where the topologies of social ties between constituents and their states co-evolve simultaneously through homophily, social contagion, and/or other social processes. Such adaptive social network models are known to either converge to a homogeneous, well-connected network (=loss of cultural diversity), or fragment into many disconnected subnetworks with distinct states (=loss of structural connectivity). From a viewpoint of social capital and innovation, however, neither of these two social states would be desirable, because the former would mean the loss of information and the latter the loss of communication. In order to keep our society active and innovative, we should maintain cultural/informational diversity within our society high while also maintain information exchange and communication actively ongoing. This can be conceptualized as a structurally well-connected network with diverse node states. Earlier theoretical models of adaptive social network dynamics did not succeed in demonstrating how such outcomes could occur.

We note that those earlier models typically used stylized assumptions that behavioral attributes of social constituents were spatially homogeneous, and therefore, they may not have fully reproduced richer macroscopic outcomes, such as potential coexistence of diverse cultures within a connected network structure. To overcome this limitation, here we introduced the diversities of behavioral attributes among social constituents into an adaptive social network model, and computationally investigated their effects on social network evolution.

The rest of the paper is organized as follows. Section 2 describes our networked agent-based simulation model. Section 3 describes the design of our computational experiments and outcome measures. Section 4 summarizes the results. Section 5 concludes the paper with a brief discussion on the implications of the results for relevant research fields and real-world socio-cultural dynamics.

2 Model

For the purpose of this study, we developed a computational adaptive social network model of cultural diffusion dynamics by using our previous work on cultural integration in corporate merger [6, 7] as the basis and implementing some revisions to it to allow representation of social constituent diversity.

In this model, we simulate the dynamics of an adaptive social network made of two initially distant cultural groups, each consisting of 50 individual constituents (nodes). Individuals are connected to each other through directed weighted edges, which represent the direction and intensity of cultural information flow. Constituents within each group and across the two groups are initially connected randomly with 20% and 2% edge densities, respectively, to represent initially modularized social structure. Edge weights are initially random, with weights sampled from a uniform distribution between 0 and 1. This initial network structure captures the state of two groups that is distant from each other both structurally and culturally.

Each individual constituent has its cultural state as a vector in a 10-dimensional continuous cultural space, based on previous empirical studies on measuring organizational cultural dimensions [8, 9]. The distance between two cultures is characterized by the Euclidean distance between their two vectors in the cultural space. The cultural distributions among individual constituents are initialized as follows: First, two cultural "center" vectors are created for the two groups, separated by 3.0 (in an arbitrary unit) in the cultural space. Then individual cultural vectors are created for individuals in each group by adding a random number drawn from a normal distribution with a mean of 0 and a standard deviation of 0.1 (in the same unit used above) to each component of the cultural center vector of that group. This creates an initial condition in which the average between-group cultural difference is approximately seven times larger than the average within-group cultural difference.

Such an initial condition made of two distant, distinct cultural clusters may not be a popular choice for models studied in complex systems, network science, and statistical physics, where more randomized, homogeneous initial conditions are typically preferred. However, such random homogeneous conditions are extremely rare and unrealistic in real society, even for initial conditions. Rather, large-scale social systems emerge and evolve through numerous encounters and interactions between multiple smaller communities that are often culturally distinct from each other at the beginning. The heterogeneous, clustered initial conditions adopted in our study were intended to capture such social encounter situations, with the aim to increase the realism and applicability of our model and results in view of actual social self-organization and evolution.

Each iteration in simulation consists of simulating actions for all individual constituents in a sequential order. In its turn to take actions, an individual first selects an information source from its local in-neighbors with 99% probability (in this case the selection probabilities are proportional to edge weights), or with 1% probability, from anyone in the connected component to which the focal individual belongs. If there is no edge in the latter case, a new directed edge is created from the source to the focal individual with a minimal edge weight 0.01. Then, the individual decides to either accept or reject the source's cultural vector based on the distance between the received cultural vector and its own. The probability of cultural acceptance P_A is an exponentially decreasing function of the cultural distance, defined as

$$P_A = \left(\frac{1}{2}\right)^{\frac{|v_i - v_j|}{d}}, \tag{1}$$

where v_i and v_j are the cultural vectors of the individual's own and of the selected source, respectively, and d is the cultural tolerance, or the characteristic cultural distance at which P_A becomes 50%. If the received culture is accepted, the individual's cultural vector is updated as

$$v_i \rightarrow (1 - r_s)\, v_i + r_s v_j, \tag{2}$$

where r_s is the rate of cultural state change, and the edge weight from the source to the focal individual w_{ij} is updated as

$$w_{ij} \rightarrow \text{logistic}\left(\text{logit}\left(w_{ij}\right) + r_w\right), \tag{3}$$

where r_w is the rate of edge weight change. Or, if the received culture is rejected, no change occurs to the focal individual's culture, but the edge weight is updated in an opposite direction as

$$w_{ij} \rightarrow \text{logistic}\left(\text{logit}\left(w_{ij}\right) - r_w\right). \tag{4}$$

The above formula that combines logit and logistic functions guarantees that the updated edge weight is always constrained between 0 and 1. When the edge weight falls below 0.01, the edge is considered insignificant and is removed from the network. Additional details of these model assumptions, parameter settings, and their rationale can be found in our earlier work [6, 7].

In the present study, we use $d = 0.5$, $r_s = 0.5$ and $r_w = 0.5$ as their mean values within the social network, and we systematically vary their variances among social constituents as the key experimental parameters. More details are given in the following section.

3 Experiments

We computationally investigated how the resulting social network states and topologies would change as social constituents' behavioral attributes were systematically diversified within the simulated society. The standard deviations of d, r_s, and r_w were varied from 0 to 0.5 at interval 0.1, which makes the total number of parameter value combinations $|\{0, 0.1, 0.2, 0.3, 0.4, 0.5\}|^3 = 6^3 = 216$. We ran 100 independent simulation runs for each specific combination of parameter values (and therefore, the total number of simulations $=$ 21,600 runs). Each run was simulated for 500 iterations.

After each simulation run was completed, the following two quantities were measured as outcome variables on the final network configuration:

- <*CD*>: average cultural distance between constituents in the initially distant two groups
- <*SPL*>: average shortest path length within the whole network

In these two measures, social fragmentation transitions can be captured as a positively correlated increase or decrease of both quantities. Namely, (<*CD*>, <*SPL*>) = (high, high) implies social fragmentation, while (<*CD*>, <*SPL*>) = (low, low) implies social assimilation with loss of cultural diversity.

4 Results

Figure 1 shows the baseline behaviors of the proposed model in the low behavioral diversity parameter region ($d \leq 0.1$, $r_s \leq 0.1$, $r_w \leq 0.1$), in which previously reported social fragmentation transitions are clearly observed as transitions between (<*CD*>, <*SPL*>) = (high, high) and (<*CD*>, <*SPL*>) = (low, low) behaviors (Fig. 1, black trend curve). Meanwhile, none of the simulation results showed (<*CD*>, <*SPL*>) = (high, low) behaviors (Fig. 1, red dashed circle) when social constituents were behaviorally homogeneous.

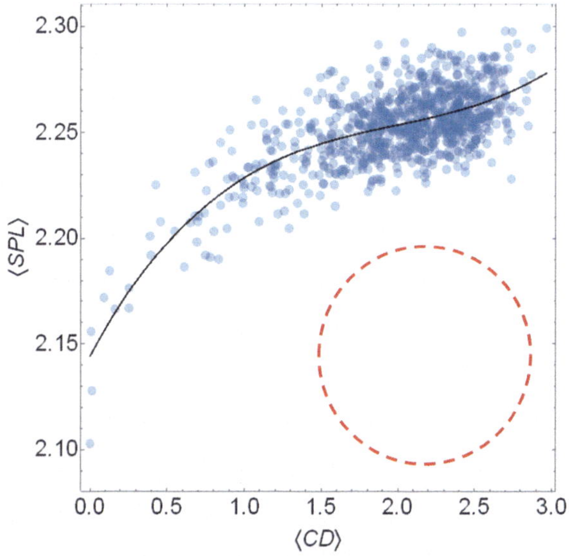

Fig. 1 Scatter plot showing two outcome measures <*CD*> and <*SPL*> of the final network configuration for baseline results obtained in the low behavioral diversity parameter region ($d \leq 0.1$, $r_s \leq 0.1$, $r_w \leq 0.1$). Each dot represents a result of one simulation run. A black cubic trend curve is drawn to illustrate the social fragmentation transition between (<*CD*>, <*SPL*>) = (high, high) and (<*CD*>, <*SPL*>) = (low, low) behaviors. Meanwhile, the dashed circle indicates the missing (<*CD*>, <*SPL*>) = (high, low) behavior that does not occur when social constituents are behaviorally homogeneous

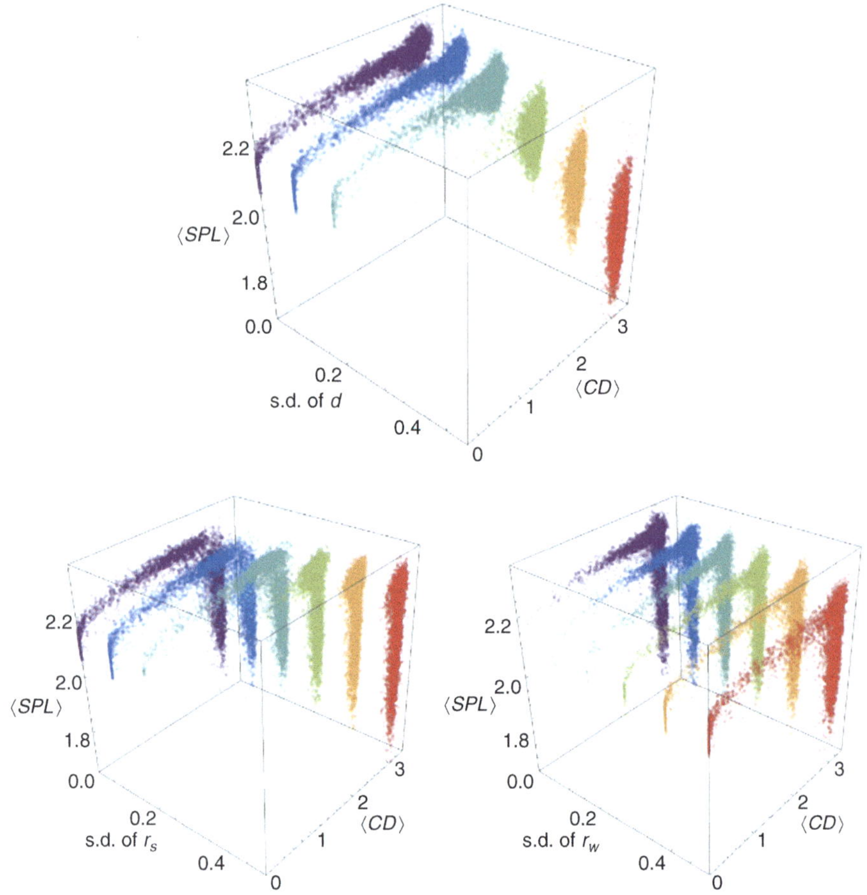

Fig. 2 3D scatter plots showing the effect of each experimental parameter, i.e., standard deviation of d (top), r_s (bottom left) and r_w (bottom right), on two outcome measures <*CD*> and <*SPL*> of the final network configuration. Each dot represents a result of one simulation run, colored according to the parameter value

Figure 2 shows the same (<*CD*>, <*SPL*>) plots for different standard deviations of d (top), r_s (bottom left) and r_w (bottom right). It is seen that greater diversity of d and r_s helps maintain <*CD*> at higher levels. In addition, greater diversity of d also helps lower <*SPL*> more (Fig. 2 top; orange/red dots), which corresponds to the (<*CD*>, <*SPL*>) = (high, low) behavior that was not previously recognized in the literature. Meanwhile, the effect of diversity of r_w is not as clearly seen in this visualization compared to the other two parameters. These results strongly imply that having a broad distribution of cultural tolerance levels within society can help promote the coexistence of cultural diversity and structural connectivity.

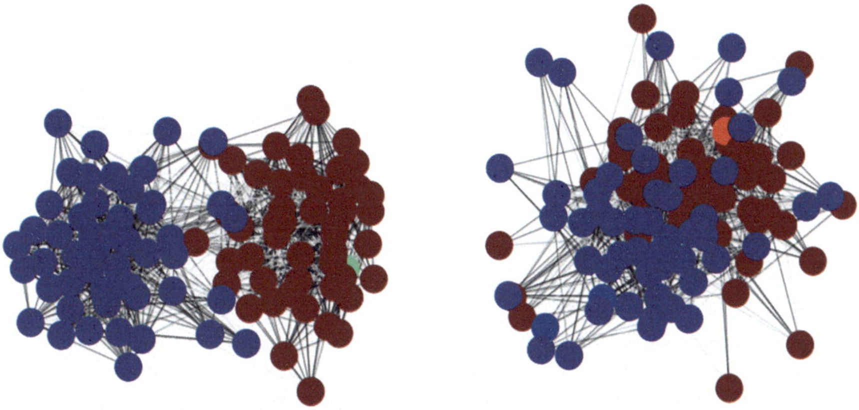

Fig. 3 Two examples of final network configuration. Left: Standard deviations of $(d, r_s, r_w) = (0.0, 0.2, 0.2)$. Right: Standard deviations of $(d, r_s, r_w) = (0.5, 0.2, 0.2)$. Node colors represent the individuals' cultural values (multidimensional vectors projected to a linear color scale)

Table 1 ANOVA table of linear regression of <*CD*> shown in Eq. (5)

	Sum of squares	df	Mean square	F	Sig.
σd	1782.60	1	1782.60	9797.94	$p < 0.0001$
σr_s	1847.26	1	1847.26	10,153.30	$p < 0.0001$
σr_w	445.64	1	445.64	2449.42	$p < 0.0001$
$\sigma d\, \sigma r_s$	1474.26	1	1474.26	8103.16	$p < 0.0001$
$\sigma d\, \sigma r_w$	239.38	1	239.38	1315.75	$p < 0.0001$
$\sigma r_s \sigma r_w$	280.83	1	280.83	1543.57	$p < 0.0001$
Error	3928.56	21,593	0.18		
Total	9998.54	21,599			

All terms were statistically extremely significant

Figure 3 shows typical final network configurations for two experimental settings. In a situation where culturally heterogeneous constituents remain connected (Fig. 3 right), constituents with different levels of cultural tolerance typically occupy different positions in the network. For example, less tolerating constituents tend to form clusters of their own, acting as cultural memory, while more tolerating ones tend to act as a "glue" to connect such culturally distinct clusters, serving as bridges.

We also conducted linear regression analysis to regress each of the two outcome variables onto the three experimental parameters (i.e., standard deviations of the three behavioral attributes; denoted as σd, σr_s, and σr_w below) and their interactions. Results are given in Eqs. (1) and (2), and their ANOVA tables are shown in Tables 1 and 2.

Table 2 ANOVA table of linear regression of <*SPL*> shown in Eq. (6)

	Sum of squares	df	Mean square	F	Sig.
σd	253.02	1	253.02	58, 391.20	$p < 0.0001$
σr_s	0.64	1	0.64	148.78	$p < 0.0001$
σr_w	2.26	1	2.26	521.21	$p < 0.0001$
$\sigma d\, \sigma r_s$	2.80	1	2.80	646.91	$p < 0.0001$
$\sigma d\, \sigma r_w$	2.28	1	2.28	526.55	$p < 0.0001$
$\sigma r_s \sigma r_w$	0.28	1	0.28	63.83	$p < 0.0001$
Error	93.47	21,571	0.00		
Total	354.75	21,577			

All terms were statistically extremely significant. There were several simulation runs in which some individual nodes became disconnected, and such cases were excluded from the calculation (which is the reason that the dfs of Error and Total have smaller values than in Table 1)

$$< CD > \sim 1.87262 + 3.01908\, \sigma d + 2.97431\, \sigma r_s – 2.72074\, \sigma r_w - 8.95723\, \sigma d\, \sigma r_s + 3.60938\, \sigma d\, \sigma r_w + 3.90939\, \sigma r_s\, \sigma r_w \quad (5)$$

$$< SPL > \sim 2.31216 – 0.624629\, \sigma d + 0.0989771\, \sigma r_s – 0.178676\, \sigma r_w - 0.390949\, \sigma d\, \sigma r_s + 0.35265\, \sigma d\, \sigma r_w + 0.122775\, \sigma r_s\, \sigma r_w \quad (6)$$

The linear terms in Eq. (5) imply that the average cultural distance (<*CD*>) is maintained by having the diversities of d and r_s, while the diversity of r_W has a negative effect on the cultural distance. Meanwhile, the linear terms in Eq. (6) indicate that the average shortest path length (<*SPL*>) is reduced by having the diversities of d and r_W, while the diversity of r_s has only a marginal (positive) effect on the average shortest path length.

The nonlinear interaction terms in Eqs. (5) and (6) imply that the interaction between the diversities of d and r_s has a negative effect on both outcome measures, while other interaction terms generally have positive effects on them. Their interactions were visualized in more detail in the 3D scatter/surface plots shown in Fig. 4. These plots illustrate that the interactions of diversity parameters are much more significant on <*CD*> than on <*SPL*>, and that greater diversity of either d or r_s maintain <*CD*> consistently at a higher level.

Among all the terms included in these regression models, the only term whose coefficients point to the (<*CD*>, <*SPL*>) = (high, low) direction is the diversity of cultural tolerance (σd). This result suggests that enhancing the diversity of cultural tolerance has the single most effective way to achieve the social state that maintain high cultural diversity and high structural connectivity simultaneously.

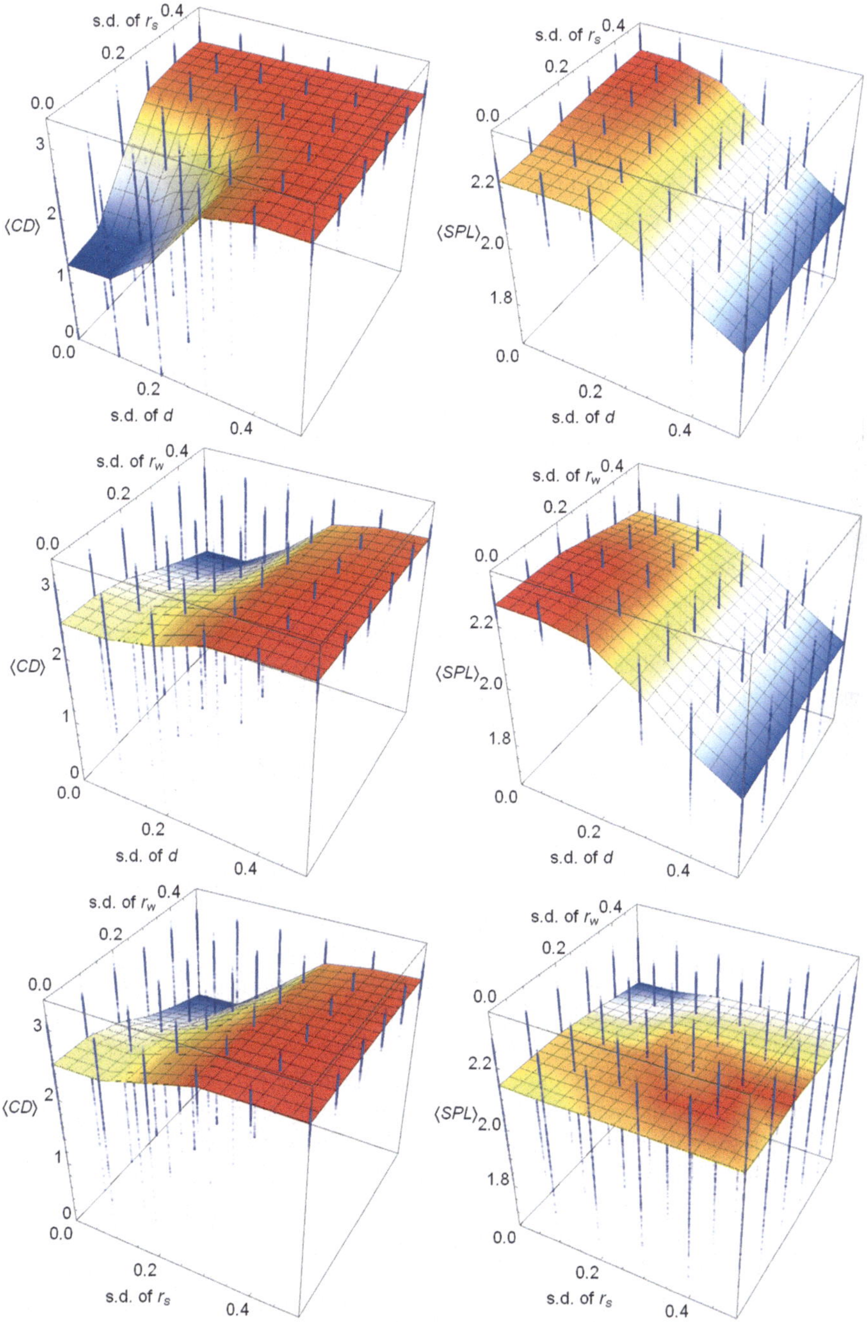

Fig. 4 3D scatter plots showing the interactions between diversities of d and r_s (top), d and r_w (middle), and r_s and r_w (bottom), on outcome measures <CD> (left) and <SPL> (right). Each blue dot represents a result of one simulation run. Average trends are also shown as a surface

5 Conclusions

In this brief paper, we computationally studied the effects of behavioral diversities of social constituents on the resulting cultural diversity and social connectivity using a networked agent-based simulation model. Our results indicated that allowing cultural tolerance levels to differ broadly within society helps promote the coexistence of cultural diversity and structural connectivity, which is a novel macroscopic state of the adaptive social network models that was previously not known in the literature.

Our key finding above is interesting and relevant to network science, complex systems and social/organizational sciences in a couple of distinct ways.

First, it offers a clear demonstration of the risk in assuming that agents in a social system are identical and homogeneous. In view of the complexity of real-world systems, such a simplification is apparently wrong, but it is still widely used in many complex systems/network models of social dynamics. Our results show that inclusion of variations in individual attributes even in the simplest manner may already have huge impacts on the macroscopic outcomes of the system's evolution.

Second, our results point out the importance of *behavioral* diversity, not demographic or other surface-level diversities that are often discussed in the context of social, organizational and political studies. In contrast to demographic properties that cannot be altered easily, behaviors of people are by a large part acquired traits, and therefore, they can be trained and modified through proper intervention. This indicates that our finding may eventually lead to some education/intervention strategies to promote the maintenance of informational diversity and communication, possibly enhancing the creativity and innovation of our society as a whole.

Third, we note that our finding indicates the importance of the *diversity* of cultural tolerance levels, and not the tolerance itself. In today's socio-political climate, cultural tolerance is highly encouraged, but our model does not imply that simply increasing the cultural tolerance levels globally within the social network would lead to beneficial outcomes. Additional experiments with globally enhanced cultural tolerance levels of all the constituents (i.e., simply increasing the value of d for all individuals; results not shown here) did not generate the same outcome as presented in this paper, because such a condition would quickly lead to a loss of cultural diversity. This implies that, at least from the perspective of enhancing both informational diversity and communication, telling people to be just tolerating does not produce the desired outcomes. These findings and implications collectively illustrate the highly non-trivial nature of the cultural dynamics in our society.

This study is still far from completion, and there are several future tasks to conduct. One obvious limitation of our present model is that it is fairly complicated and is not suitable for mathematical analysis. We plan to develop a much more simplified model of the same adaptive social network dynamics so that its behavior can be analyzed and explained mathematically. The other important direction of future research is to compare the dynamics of these adaptive social network models with real-world data of information exchange in order to validate and revise the model assumptions. For this purpose, we are currently working on collecting empirical data of cultural dynamics from social media and other online/offline sources. Recent machine learning tools for content analysis [10, 11] allow researchers to quantify similarities and differences between contents posted by users, and this information can be examined with regard to its potential correlation with temporal changes of future contents posted by the same users as well as their social relationships [12]. Our future goal is to use such empirical data to determine the role and importance of individual behavioral heterogeneity in real-world adaptive social networks, as predicted in the theoretical model presented in this paper.

Acknowledgements This work was supported by JSPS KAKENHI Grant Number 19H04220.

References

1. Holme, P., Newman, M.E.: Nonequilibrium phase transition in the coevolution of networks and opinions. Phys. Rev. E. **74**(5), 056108 (2006)
2. Zanette, D.H., Gil, S.: Opinion spreading and agent segregation on evolving networks. Phys. D: Nonlinear Phenom. **224**(1-2), 156–165 (2006)
3. Centola, D., Gonzalez-Avella, J.C., Eguiluz, V.M., San Miguel, M.: Homophily, cultural drift, and the co-evolution of cultural groups. J. Confl. Resolut. **51**(6), 905–929 (2007)
4. Kozma, B., Barrat, A.: Consensus formation on adaptive networks. Phys. Rev. E. **77**(1), 016102 (2008)
5. Böhme, G.A., Gross, T.: Analytical calculation of fragmentation transitions in adaptive networks. Phys. Rev. E. **83**(3), 035101 (2011)
6. Yamanoi, J., Sayama, H.: Post-merger cultural integration from a social network perspective: a computational modeling approach. Comput. Math. Organ. Theory. **19**(4), 516–537 (2013)
7. Sayama, H., Pestov, I., Schmidt, J., Bush, B.J., Wong, C., Yamanoi, J., Gross, T.: Modeling complex systems with adaptive networks. Comput. Math. Appl. **65**(10), 1645–1664 (2013)
8. O'Reilly, C.A., Chatman, J., Caldwell, D.F.: People and organizational culture: a profile comparison approach to assessing person-organization fit. Acad. Manag. J..**34**, 487–516 (1991)
9. Chatterjee, L., Lubatkin, M., Schweiger, D., Weber, Y.: Cultural differences and shareholder value in related mergers: linking equity and human capital. Strateg. Manag. J. **13**, 319–334 (1992)
10. Mikolov, T., Sutskever, I., Chen, K., Corrado, G.S., Dean, J.: Distributed representations of words and phrases and their compositionality advances. In: Advances in Neural Information Processing Systems (NIPS 2013), vol. 26, pp. 3111–3119 (2013)
11. Le, Q., Mikolov, T.: Distributed representations of sentences and documents. In: International Conference on Machine Learning, pp. 1188–1196 (2014)
12. Cao, Y., Dong, Y., Kim, M., MacLaren, N., Kulkarni, A., Dionne, S., Yammarino, F., Sayama, H.: Capturing the production of innovative ideas: An online social network experiment and "Idea Geography" visualization. In: Proceedings of CSS 2019: 10th Anniversary International Conference on Computational Social Science (2019)

Part III
Resilience and Robustness

Complex Networks Antifragility under Sustained Edge Attack-Repair Mechanisms

Alexandru Topîrceanu, Mihai Udrescu, and Radu Mărculescu

Abstract Resilience is an important property of real-world complex networks with many applications in technological, biological, and social systems. While many natural systems are particularly resilient, some are antifragile, namely, they become stronger when being subjected to attacks, volatility, or errors. In this paper, we consider an edge-attack and local edge-repair response mechanism over several synthetic and real-world datasets, on which we quantify both antifragility (as the dynamics of the largest connected component) and the cost incurred by edge repairs. Our findings show that (1) random repairs generate a stronger antifragile response, thus confirming that antifragility manifests itself in the context of random, rather than deterministic events; and (2) antifragile behavior is fostered by strongly clustered topologies (e.g., real-world networks and the synthetic Watts–Strogatz model with degree distribution). Our results represent a first step towards designing highly resilient networks and developing new methods for thwarting the antifragile response of harmful and hostile systems.

1 Introduction

The term *antifragility* describes a system that grows stronger (up to a point) when being exposed to attacks, volatility or randomness [1]. Thus, antifragility goes

A. Topîrceanu · M. Udrescu (✉)
Department of Computer and Information Technology, Politehnica University of Timişoara, Timişoara, Romania
e-mail: alext@cs.upt.ro; mudrescu@cs.upt.ro

R. Mărculescu
Department of Electrical and Computer Engineering, Carnegie Mellon University, Pittsburgh, PA, USA

Department of Electrical and Computer Engineering, The University of Texas at Austin, Austin, TX, USA
e-mail: radum@utexas.edu

N. Masuda et al. (eds.), *Proceedings of NetSci-X 2020: Sixth International Winter School and Conference on Network Science*, Springer Proceedings in Complexity,
https://doi.org/10.1007/978-3-030-38965-9_13

beyond the conventional concepts of robustness (i.e., mere resistance to attacks) and fragility [2].

Antifragility is the defining characteristic of many complex (organic) systems that were shaped by evolution in adverse environments, as opposed to top-down designed synthetic systems. As such, the first practical explorations of antifragile behavior were performed in biological and social systems [3]. To this end, the authors of [4] propose a non-causal method of preparing organizations and social systems for future unexpected events. Another useful application is the prevention of antifragile behavior when trying to dismantle potentially hostile systems [5] such as criminal networks [6, 7]. Also, the authors of [8] investigate how to calibrate chemotherapy such that cancer cells' behavior does not become antifragile. There are also various applications of antifragility in engineering, in areas as diverse as wireless communications [9], cloud computing [10], power grids [11], intelligent autonomous systems [12], and cyber-physical systems [13]. For instance, in April 2011, a large crash in Amazon cloud put out of service some of their main clients (e.g., Quora, Reddit), while Netflix remained functional; it turned out that Netflix uses a failure injection tool for SDN (software-defined networking) called "Chaos Monkey" [14], in order to make their systems antifragile.

The exploration of antifragility is new to network science, having been only recently addressed in [15]. Similar to [6, 7, 9–11], we interpret a *robust* network as being characterized by a high connectivity between nodes, whereas a *fragile* network as having a low connectivity. Accordingly, *antifragility* increases network connectivity when subjected to attacks. Quantitatively, we measure the robust, fragile, and antifragile behaviors using the largest connected component size (LCS) and the number of connected components (NCC), as these parameters are directly related to network's communication capacity [16–18]. Thus, the objectives of our analysis are as follows:

- Analyze the response to attacks at micro-scale (i.e., node-to-node local interaction through links) that triggers an antifragile behavior at macro-scale (i.e., network-level); such an antifragile behavior indicates a highly resilient network.
- Identify the topological properties that foster an antifragile behavior in complex networks.
- Quantify the costs entailed by the response mechanisms that induce an antifragile network behavior.

Our approach is to consider attacks as continuous removal of links at micro-scale. Without a proper response to such aggression, the network evolves towards structural and functional collapse. Consequently, to induce antifragility, the network must benefit from some link repairs. To this end, we design a decentralized repair mechanism where the nodes affected by the removal of incident links are themselves creating new links in response to the attacks.

To uncover the topological features that foster antifragile behavior in complex networks, we simulate multiple attack-repair scenarios on some generic synthetic topologies (random, mesh, small-world and scale-free), as well as real-world network topologies. Accordingly, the *antifragile* network behavior is detected at macro-scale if, as simulation unfolds, the connectivity of the network (measured via LCS and NCC variations) does increase under sustained attacks. The *fragile* network behavior under attack corresponds to the network's reduced tolerance to incurred faults (i.e., destroyed links), leading to degraded LCS and NCC [19–21].

As the state-of-the-art applications of antifragility aim at improving the communication capacity of the system, we focus only on edge attacks (i.e., a node with no edges is considered to be removed from the network). As such, the contributions related to our generalized edge attack-repair mechanism that aims at inducing antifragility are:

- We consider sustained attacks, where network responses to current attacks must happen in real time, in order to meet the deadline before the next wave of attacks.
- We propose a decentralized (node-level) repair mechanism that requires only local information processing.
- We provide an analytical framework based on Markov chains, that connects the network topological features to the time-to-dismantle a network, given an attack-repair mechanism.
- We analyze the network topological features that foster antifragility at macro-scale, by factoring in the costs entailed by the decentralized edge repairs.

The remainder of this paper is organized as follows: In Sect. 2, we present the proposed attack-repair simulation mechanism, along with the formal models for network analysis. In Sect. 3, we show our simulation results. Finally, we discuss the implications of our results and draw some conclusions in Sect. 4.

2 Methods

A network $G = (V, E)$ is the pair of node (vertex) and link (edge) sets, V and E respectively. The number of nodes $|V|$ is fixed, whereas the number of links $|E|$ is variable during simulation ($|E| \leq \frac{|V|(|V|-1)}{2}$); $|E|$ may decrease until the network is considered destroyed (or dysfunctional). Similar to [22], the LCS and NCC metrics quantify the evolution of structural fragility of network G. More precisely, a network is considered fragile if its LCS decreases rapidly during simulation, and antifragile if its LCS increases during the attack-repair process up to a specific stress point.

To consider this scenario, we start our simulations with disconnected networks ($NCC > 1$); depending on the attack-repair ratios, the LCS may counterintuitively increase, even though the network is losing edges overall (e.g., the attacks mostly remove edges in smaller components, while edge repairs connect the new nodes to the largest connected component). This is the *antifragile* effect that we intend to quantify.

2.1 Edge Attacks

The simulation that we employ in all scenarios finishes after 100 iterations; from our experience, this is sufficient to highlight both the fragile and antifragile behaviors. At each iteration, a fixed ratio α of edges (which we call the attack rate) is removed; therefore, the number of removed edges is $r = \alpha \cdot |E|$. In [15], four α values are considered ($\alpha \in \{1\%, 2\%, 5\%, 10\%\}$) to conclude that $\alpha = 0.05$ (5%) is the optimal trade-off between speed and amplitude of network destruction.

From literature, we find that while most attack strategies focus primarily on node centralities [23, 24], but random attacks are also used; thus, we consider the following types of attacks:

- Random attacks: Any edge $e_{ij} \in E$ in the network can be removed with an equal probability.
- Targeted attacks: Any edge $e_{ij} \in E$ can be removed with probability p_{ij} proportional with the average centrality of adjacent vertices v_i and v_j. For any vertex v_i, its centrality $C(v_i)$ can be any of the following: degree (Deg), betweenness (Btw), or eigenvector (Eig). Thus, the probability of removing edge e_{ij} (connecting nodes v_i and v_j) is proportional to $C(e_{ij}) = \frac{1}{2} \cdot (C(v_i) + C(v_j))$: $p_{ij} = \frac{C(e_{ij})}{\sum_{e \in E} C(e)}$.

2.2 Edge Repairs

The response after each attack is a set of edge repairs which happen with rate β. We discuss two different simulation settings: one in which $\beta < 1$ (i.e., fewer edges are repaired than destroyed), and another in which $\beta \geq 1$ (i.e., more edges are repaired than destroyed).

In our decentralized repair decision approach, a key aspect is the selection of vertices that will add new incident edges in response to attacks; these vertices are taken from the set of *affected nodes* V_a, i.e., vertices that lost at least one edge during the current iteration. By removing r edges during the current iteration, a maximum of $\max\{|V_a|\} = 2r$ nodes may become affected (some edges may be incident to the same nodes); the theoretical minimum number of affected nodes is $\min\{|V_a|\} = r + 1$, which corresponds to a star topology. However, in real-world situations, $|V_a|$ is closer to $2r$. The repair strategy determines which of the maximum $2r$ affected nodes will add new edges:

- Random selection: A number of $\beta r < |V_a|$ random vertices from the list of affected nodes V_a.

- Selection paired with the attack strategy: βr vertices from V_a are selected; each affected node in V_a has a probability p_i of being selected that is proportional with the node's centrality (the same centrality considered for attacking).

To perform the repairs, each selected (affected) node $v_i \in V_a$ will add an edge to a target node that is designated according to the following strategies:

- Random repair: The selected node v_i adds a new edge e_{ij} that connects to a randomly designated node v_j.
- Centrality-driven repair: High degree first (HDF), high betweenness first (HBF), and high eigenvector first (HEF). For example, HDF defines probability p_j of designating vertex v_j to connect to the selected affected node v_i via e_{ij} as $p_j = \frac{d_j}{\sum_{\forall \iota \in V}(d_\iota)}$, where d_ι is the degree of node v_ι.

In this paper, we consider attacking edges with a specific targeted strategy and then either repairing them at random or with a paired strategy (e.g., betweenness targeted attack and repair—HBF).

2.3 *Network Topologies*

We first use the following synthetic topologies [25]: random ER (Erdős-Rényi), mesh Me, Watts–Strogatz small-world SW, and scale-free SF. We also append the topology with the Watts–Strogatz model developed by including degree distribution [26], denoted as WD, to serve as a more complex synthetic topology.

We consider several real-world networks which are relevant to the applications of antifragility [6, 10–12]: a US power-grid network[1] (UP) [27], a network representing cloud routers[2] (Rt) [27], an email collaboration network [3] (Em) [28], a criminal network[4] (Mo) [27], and a set of disconnected Twitter ego-networks[5] (Tw) [29] to explore the antifragility of social networks in general. Our aim is to compare the responses to attacks and repairs of these real networks against those observed on synthetic networks.

As explained, all the networks that we use in our simulations are initially disconnected graphs. This enables us to measure a possible increase in LCS, respectively a decrease in NCC, from the very beginning of each simulation. Thus, using a big network like Twitter [29] allows us to make better observations regarding the studied phenomenon. We also add the Molloy-Reed (*M-R*) criterion

[1] Available at http://networkrepository.com/power.php.

[2] Available at http://networkrepository.com/tech-routers-rf.php.

[3] Available at https://www.cs.cmu.edu/~enron/.

[4] Available at http://networkrepository.com/ia-crime-moreno.php.

[5] Available at https://snap.stanford.edu/data/ego-Twitter.html.

Table 1 Graph theoretical measurements for synthetic and real-world networks

Data	N	$\|E\|$	$\langle cc \rangle$	Dmt	Mod	NCC	LCS
ER	5000	4198	0	45	0.963	1067	3336
Me	5000	4276	0.018	82	0.953	1083	3339
SW	5000	4730	0.121	80	0.974	983	2447
SF	5000	6224	0	26	0.871	945	3696
WD	1007	3535	0.472	9	0.960	229	52
UP	4941	5685	0.093	58	0.946	289	4402
Rt	2113	4871	0.274	15	0.702	220	1876
Em	12,625	19,358	0.519	9	0.690	503	12,123
Mo	829	1026	0.006	15	0.692	101	720
Tw	6735	15,388	0.554	16	0.994	467	80

All metrics reflect the initial state of the networks, before simulating multiple iterations of attacks and repairs

[30] to complement the observations of the LCS evolution. *M-R* states that a largest connected component exists if, on average, each node in the network has at least two links, or $M-R = \frac{\langle d^2 \rangle}{\langle d \rangle} > 2$, where $\langle d \rangle$ is the average node degree.

The synthetic datasets we use consist of $V = 5000$ vertices divided into roughly 1000 distinct components. To obtain these customized datasets, we generate the networks according to the conventional algorithms in the literature [25, 30], then remove some of their edges at random until the NCC rises to $\approx$1000. Table 1 details the following measurements on our networks: network size ($N = |V|$ and $|E|$), average clustering coefficient ($\langle cc \rangle$), diameter (Dmt), modularity (Mod), number of connected components (NCC), and largest connected component size (LCS).

2.4 Quantitative Characterization of Antifragility

For each simulation, we obtain two time series, namely the evolution of $LCS(t)$ and $NCC(t)$ over 100 iterations, $t \in \{1, \ldots, 100\}$. To quantify the antifragile response, we use two intuitive measures:

- The *maximum improvement* (I) of LCS for repair rate β, based on the maximum of ensemble averages $I_e(t = k)$, is defined as

$$I_e(t = k) = \frac{\text{average}\{LCS(0 < t \leq k)\}}{LCS(t = 0)}, \quad I_\beta = \max\{I_e(t)\}. \tag{1}$$

- For a repair rate β, when $I > 1$ (i.e., antifragility is present), the *duration of antifragility* D_β is the time interval when $LCS(t) \geq LCS(t = 0)$,

$$D_\beta = \{t_2 - t_1 \mid LCS(t) \geq LCS(t = 0),\ t_1 \leq t \leq t_2\}. \tag{2}$$

If the simulation exhibits any antifragile behavior, then $I \geq 1$ and $0 < D \leq 100$. If the edge repair rate is higher than the attack rate, we obtain a high duration, $D \approx 100$. However, to maintain a robust topology with minimal repair costs and limited resources for response, we are most interested in scenarios where $D > 0$ for a repair rate of $\beta < 1$.

2.5 Analytical Estimation of the Time to Network Dismantling

The time-to-dismantle a network, given the attack-repair mechanisms with rates α and β, is related to the topological features of the network. We link this time-to-dismantle a network to α and β using a Markov model. The rate of disconnection, which means jumping from state S_0 (connected network) to state S_1 (disconnected network with a largest connected component) in one iteration, is the attack rate α (Fig. 1). The transition from state S_1 to state S_2 (disconnected network without a large connected component) is proportional to the attack rate α, but is considerably slower, i.e., $k\alpha$ with $k < 1$. Parameter k depends on the network topological features (i.e., link density, degree distribution) and the attack and repair strategy. For instance, if the number of iterations until the LCS decays under 10% of its initial size is n_d, then $k = \dfrac{1}{n_d}$.

A dismantled network cannot be repaired for $\beta < 1$, therefore S_2 becomes an absorbing state. However, a transition from S_1 to S_0 is possible, given that repairs can initially be effective for graph reconnection; therefore we consider a repair rate of $\beta\alpha$ from S_1 to S_0.

According to the Markov diagram in Fig. 1, the probabilities of being in states S_0, S_1, or S_2 are given by

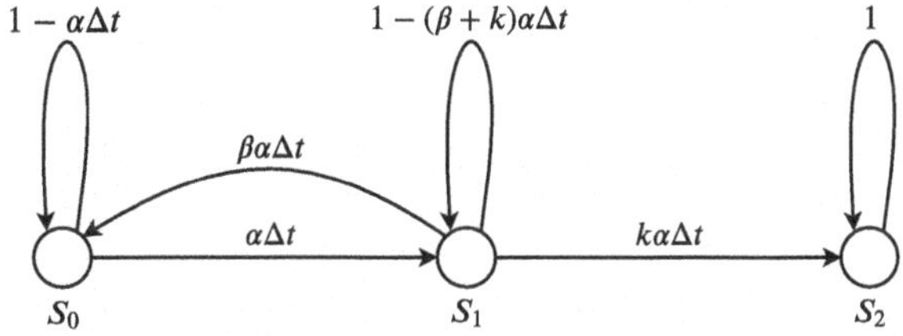

Fig. 1 Markov model representing the process of dismantling a complex network, where S_0 represents a connected network, S_1 represents a network with a large connected component and many (smaller) disconnected components, and state S_2 represents a disconnected network without a large connected component (Molloy-Reed parameter $M-R \leq 2$). The rates of disconnecting and dismantling, α and $k\alpha$, depend on the attack rate α, the attack and repair mechanisms, and the network topology; the connection repair rate $\beta\alpha$ is proportional with edge repair rate β

$$\begin{aligned}\frac{dP_{S_0}(t)}{dt} &= -\alpha P_{S_0}(t) + \beta\alpha P_{S_1}(t),\\ \frac{dP_{S_1}(t)}{dt} &= \alpha P_{S_0}(t) - (\beta + k)\,\alpha P_{S_1}(t),\\ \frac{dP_{S_2}(t)}{dt} &= k\alpha P_{S_1}(t)\,.\end{aligned} \tag{3}$$

To solve the system of equations, we apply the Laplace transform $\mathcal{L}\{\}$ from t to s, then solve in s, finding that the probability of having a network with a large connected component (if the system starts in state S_0) is

$$R(s) = P_{S_0}(s) + P_{S_1}(s) = \frac{s + \alpha(1 + \beta + k)}{s^2 + [\alpha(1 + \beta + k)]\,s + k\alpha^2}. \tag{4}$$

If the system starts in state S_1, then we have

$$R(s) = P_{S_0}(s) + P_{S_1}(s) = \frac{s + \alpha(1 + \beta)}{s^2 + [\alpha(1 + \beta a + k)]\,s + k\alpha^2}. \tag{5}$$

The Mean Time To Failure $MTTF$ (mean time-to-dismantle the network) is

$$\mathcal{L}\left\{\lim_{t\to\infty}\int_0^t \left(P_{S_0}(\tau) + P_{S_1}(\tau)\right) d\tau\right\} = \lim_{s\to 0} R(s)\,. \tag{6}$$

In our case, if the system starts in state S_0, then $MTTF = \dfrac{1 + k(1+\beta)}{k\alpha}$; if it starts in S_1, then $MTTF = \dfrac{1+\beta}{k\alpha}$. Accordingly, simulations can be used to estimate k for various combinations of topologies and attack-repair mechanisms, then compute the expected time-to-dismantle the network (a situation indicated by $M-R \le 2$).

2.6 Cost-Optimality Considerations

Our attack-repair mechanism implies edge repairs at every iteration. In the real world, these edge repairs would incur corresponding costs. For instance, either we consider adding or repairing power lines, creating new physical links between routers, or establishing new social links, we need to minimize the cost of repairs.

We define the absolute cost at iteration t as the sum of degrees for all target nodes receiving new links in that iteration $costAbs(t) = \sum_j d_j(t)$ (where $d_j(t)$ is the degree of node v_j which receives a new link at iteration t). Further, for each iteration, we define the absolute repair efficiency as the gain/cost ratio $LCS(t)/costAbs(t)$.

3 Results

For each of the ten networks, we run four paired centrality simulations, three simulations where each targeted attack is repaired by a random strategy, and three where each random attack is repaired by a targeted strategy; this leads to a total of $10 \cdot (4+3+3) = 100$ distinct scenarios. For each attack-repair scenario, we use an $\alpha = 0.05$ attack rate, but test 6 different repair rates $\beta \in \{0.5, 0.7, 0.9, 1, 1.3, 1.5\}$. The total number of individual simulations is $100 \cdot 6 = 600$, therefore we only present the results for the most relevant attack-repair scenarios. Computing betweenness after an attack round entails a big overhead. However, we get a convenient tradeoff between simulation accuracy and runtime by running the betweenness computation algorithm [31] once every 10 iterations.

In Table 2, we provide the improvements I and durations D (see Sect. 2.4) measured on all datasets in the context of random ($Rand$), degree (Deg), betweenness (Btw), and eigenvector (Eig) paired attack-repairs with repair rates of $\beta = 0.7$ (i.e., reduced repair rate) and $\beta = 1$ (i.e., balanced repair rate).

By analyzing the data in Table 2, we find that antifragility does occur in our simulations, as LCS increases for a limited period although the network loses more edges than it regains ($\beta < 1$).

Of all synthetic networks, we find that SW networks exhibit an improvement $I > 1$ under a reduced repair rate $\beta = 0.7$; this improvement is obtained under random and degree repair-attacks, but not for betweenness and eigenvector repair strategies. When moving to the more complex synthetic topology (WD), both I and D increase noticeably. Here we observe the same pattern, i.e., random attack-repairs offer the highest improvement, followed by degree, eigenvector, and betweenness.

The emergence of antifragile responses in synthetic and real-world networks seems to follow a correlation with the complexity of the underlying topology. Namely, the real-world networks (especially the natural ones) show the highest antifragile improvement of $I \approx 1.0 - 78.8$, followed by WD ($I = 17.08$), then SW ($I = 1.24$), SF ($I = 1.03$), and finally ER and Me ($I < 1$).

Overall, we conclude that the paired *random* repair-attacks are the best combination for triggering an antifragile behavior in both synthetic and real-world networks. Second, the betweenness attacks consistently rank as the most destructive strategy overall, regardless of the repair strategy. Third, we find that the random strategy offers the highest improvements I, on average, with the degree strategy providing slightly longer antifragile durations D.

We structure our graphical results into three categories: (A) impact of paired versus non-paired repairs on synthetic networks; (B) the antifragile response of a more complex synthetic topology (i.e., Watts–Strogatz with degree distribution) and real-world networks; (C) the evolution of $LCS(t)/costAbs(t)$ for networks with antifragile behavior.

Table 2 Topological improvement I and antifragile duration D (in parentheses) for paired centrality attack-repairs on each network, with $\beta = 0.7$ (upper half) and $\beta = 1$ (lower half)

	Network	*Rand*	*Deg*	*Btw*	*Eig*
$\beta = 0.7$	*ER*	0.98 (–)	0.97 (–)	0.94 (–)	0.94 (–)
	Me	0.99 (–)	0.98 (–)	0.95 (–)	0.96 (–)
	SW	**1.19 (17)**	**1.24 (21)**	0.73 (–)	**1.07 (9)**
	SF	**1.03 (16)**	0.99 (–)	0.97 (–)	0.99 (–)
	WD	**17.08 (85)**	**14.76 (100)**	**12.88 (55)**	**11.83 (56)**
	UP	0.98 (–)	0.96 (–)	0.95 (–)	0.97 (–)
	Rt	**1.0 (10)**	0.99 (–)	0.99 (–)	1.0 (–)
	Em	**1.03 (1)**	**1.02 (1)**	**1.03 (1)**	**1.03 (2)**
	Mo	0.99 (–)	0.98 (–)	0.98 (–)	0.99 (–)
	Tw	**78.82 (68)**	**74.49 (96)**	**49.06 (20)**	**67.31 (55)**
$\beta = 1$	*ER*	**1.03 (100)**	**1 (1)**	0.94 (–)	0.96 (–)
	Me	**1.02 (100)**	0.99 (–)	0.95 (–)	0.95 (–)
	SW	**1.49 (100)**	**1.35 (100)**	0.85 (–)	**1.17 (100)**
	SF	**1.21 (100)**	**1.0 (5)**	0.98 (–)	0.99 (–)
	WD	**18.02 (100)**	**14.97 (100)**	**13.84 (100)**	**13.04 (100)**
	UP	**1.0 (6)**	0.98 (–)	0.98 (–)	0.98 (–)
	Rt	**1.10 (68)**	0.99 (–)	0.99 (–)	0.99 (–)
	Em	**1.00 (3)**	**1.0 (7)**	**1.0 (1)**	**1.0 (8)**
	Mo	**1.02 (91)**	0.99 (–)	0.98 (–)	0.98 (–)
	Tw	**81.73 (100)**	**75.22 (100)**	**57.03 (100)**	**70.19 (100)**

A higher I denotes a stronger antifragility, $I < 1$ means no antifragility. A higher D value indicates a longer response measured as the number of attack-repair rounds, a dash (–) means no antifragile response. The antifragile behaviors are shown in bold

3.1 Paired Versus Non-paired Repairs on Synthetic Networks

We explore the impact of different repair strategies (paired—*Deg*, versus non-paired—*Rand*) applied after a *Deg* attack. Specifically, we find two different responses. In Fig. 2a, we showcase the scenario for *ER* networks where the *Deg* repairs are better than the *Rand* repairs for a reduced repair rate $\beta < 0.9$ (similar results are noticed for *SW* and *SF* networks). Then, in Fig. 2b, we show that on meshes (*Me*) the random repairs (*Rand*) are more efficient.

Further, we explore the response in the same context, but with higher repair rates of $\beta > 1$. Again, we find that the *Me* topology responds differently than the other three topologies. We display the representative response for *SF* in Fig. 2c, namely *Deg* repairs are weaker than random repairs (*ER* and *SW* have similar responses). Conversely, for *Me*, *Deg* repairs are more efficient than random ones (Fig. 2d).

From the results in Fig. 2, we draw two important insights:

- In terms of efficiency of random repairs, the mesh (*Me*) topology has a different response than the other three topologies (*ER*, *SW*, *SF*).

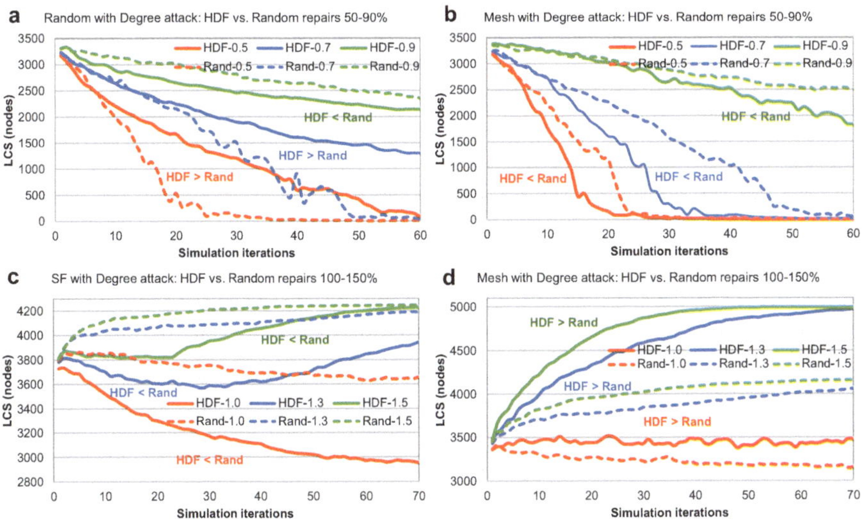

Fig. 2 Evolution of the LCS for networks under degree-driven HDF attack (Deg) with an attack rate of $\alpha = 0.05$; the results highlight the impact of a paired Deg repair versus a non-paired random $Rand$ repair. (**a**) For a random ER network and repair rate $\beta \in \{0.5, 0.7, 0.9\}$ the HDF repair is more efficient than the random repair ($Deg > Rand$)—the same results are obtained on SW and SF; (**b**) for a mesh Me with $\beta \in \{0.5, 0.7, 0.9\}$ the HDF repair is less efficient than the random repair ($Deg < Rand$); (**c**) for a scale-free SF network with $\beta \in \{1, 1.3, 1.5\}$ the HDF repair is less efficient than the random repair ($Deg < Rand$)—we obtain the same results for ER and SW; (**d**) for Me with $\beta \in \{1, 1.3, 1.5\}$, the HDF repair is more efficient than the random repair ($Deg > Rand$)

- There is a transition around $\beta = 1$ between the efficiency of Deg (paired) versus $Rand$ (non-paired) repairs. On meshes, Deg is more efficient than the other repair strategies for $\beta > 1$ and less efficient than $Rand$ for $\beta < 1$. The opposite is true for the other topologies.

3.2 *Antifragility in Complex Synthetic and Real-World Networks*

In Fig. 3a, b, we illustrate the antifragile responses (i.e., high I and D values) for all four paired attack-repair strategies on the WD topology [26].

The edge attack-repair methodology applied to real-world datasets further confirms the propensity for antifragile behavior in complex networks. Indeed, in Fig. 3c, d, we show the responses for all four paired attack-repair strategies on the Tw topology (clustering similar to SW and high modularity), which clearly exhibits an antifragile response for both LCS and $M-R$. The performance of targeting strategies remain the same as the one on the synthetic networks: $Rand >$

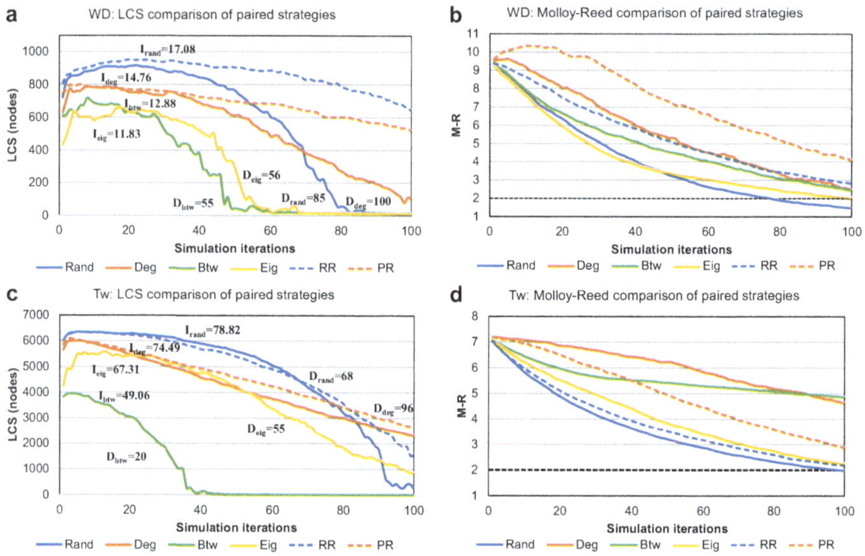

Fig. 3 Evolution of the LCS and $M-R$ parameters for complex networks under paired centrality attacks, with attack rate $\alpha = 0.05$, and repair rate $\beta = 0.7$ (we highlight the improvement I and antifragility duration D for each attack-repair strategy): (**a**) LCS for the WD network; (**b**) Molloy-Reed parameter ($M-R$) for WD; (**c**) LCS for the Twitter network Tw; (**d**) $M-R$ for Tw

$Deg > Eig > Btw$. The duration of response are slightly longer for $Rand$ while Deg, Eig and Btw hardly permit a positive response for $\beta = 0.7$. As a baseline comparison for our $Rand$ and Deg repair mechanisms we use random rewiring (RR) and preferential rewiring (PR) to highlight the difference between local and centralized repairs.

3.3 Antifragility Repair Costs

We analyze the costs, according to the measures in Sect. 2.6, for topologies where antifragility is observed. In Fig. 4 we depict the scaling of the proposed edge repair cost ratio: $LCS(t)/costAbs(t)$ on the SW and SF networks, and on the Mo and Tw networks respectively.

All the cost efficiency plots ($LCS/costAbs$) show that the random strategy is better at first but—as the simulation progresses and a significant number of edges is lost—the network becomes more fragile and, in this context, the degree-driven Deg strategy becomes more efficient. We validate the empirical results for real-world networks using their corresponding rewired versions, which preserve the number of nodes, the number of edges, and the degree distribution (see the dashed RR and PR lines in Fig. 4) according to [32, 33].

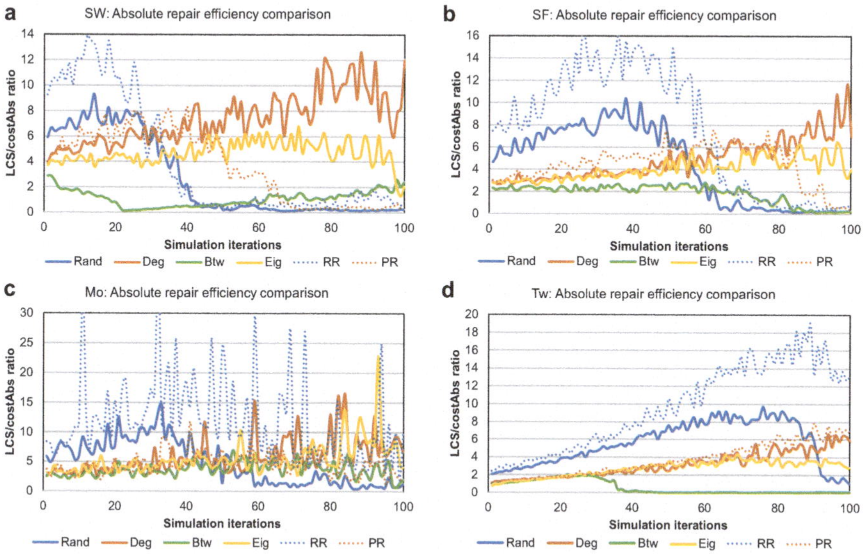

Fig. 4 Scaling of $LCS(t)/costAbs(t)$ on the SW (**a**), SF (**b**), Mo criminal (**c**), and Tw Twitter (**d**) networks using four different paired attack-repair strategies. All plots indicate that the random *Rand* strategy is initially better, but as the network loses links, the centrality-driven strategies (especially *Deg*) become more efficient

4 Conclusion

In this paper, we have introduced a new *edge attack-repair* mechanism meant to explore whether an antifragile behavior is exhibited in complex networks. To the best of our knowledge, the synergy between sustained edge attacks balanced by decentralized edge repairs represents a scientific novelty.

We have first shown that antifragility is more pronounced on more complex synthetic topologies such as WD and real-world networks. We then have found that the random targeting repair strategy provides the highest improvements at first, thus confirming the theory stating that antifragility appears in the context of random solution searches, rather than deterministic ones [1]. We have also found that betweenness-driven attacks are the most destructive on all tested datasets. Another important observation is that natural real-world topologies have a stronger drive towards antifragility than their technological counterparts. Finally, the efficiency analysis based on costs shows that the random strategy is initially better but, as the network becomes more damaged, the degree-driven HDF strategy becomes more cost-effective. Taken together, these results suggest that, for network systems that require a high resilience, the evolutionary strategy of trying solutions at random and then letting the environment perform the selection is more efficient when the system is not too damaged and has enough time to react; otherwise, preferential attachment works best.

We hope that our findings will stimulate new research towards developing dynamic edge reconfiguration models based on the principle of antifragility. Further research will need to consider more sophisticated repair strategies based on other node centralities, i.e., intra-community-first, inter-community-first, etc.

References

1. Taleb, N.N.: Antifragile: How to Live in a World We Don't Understand. Allen Lane, London (2012)
2. Taleb, N.N., Douady, R.: Mathematical definition, mapping, and detection of (anti) fragility. Quant. Financ. **13**(11), 1677–1689 (2013)
3. Danchin, A., Binder, P.M., Noria, S.: Antifragility and tinkering in biology (and in business) flexibility provides an efficient epigenetic way to manage risk. Genes **2**(4), 998–1016 (2011)
4. Derbyshire, J., Wright, G.: Preparing for the future: development of an 'antifragile' methodology that complements scenario planning by omitting causation. Technol. Forecast. Soc. Chang. **82**, 215–225 (2014)
5. Ren, X.-L., Gleinig, N., Helbing, D., Antulov-Fantulin, N.: Generalized network dismantling. Proc. Natl. Acad. Sci. U. S. A. **116**(14), 6554–6559 (2019)
6. Bright, D., Greenhill, C., Britz, T., Ritter, A., Morselli, C.: Criminal network vulnerabilities and adaptations. Global Crime **18**(4), 424–441 (2017)
7. Duxbury, S.W., Haynie D.L.: Criminal network security: an agent-based approach to evaluating network resilience. Criminology **57**(2), 314–342 (2019)
8. Gatenby, R.A., Silva, A.S., Gillies, R.J., Frieden, B.R.: Adaptive therapy. Cancer Res. **69**(11), 4894–4903 (2009)
9. Lichtman, M., Vondal, M.T., Clancy, T.C., Reed, J.H.: Antifragile communications. IEEE Syst. J. **12**(1), 659–670 (2018)
10. Abid, A., Khemakhem, M.T., Marzouk, S., Jemaa, M.B., Monteil, T., Drira, K.: Toward antifragile cloud computing infrastructures. Prog. Comput. Sci. **32**, 850–855 (2014)
11. Fang, Y., Sansavini, G.: Emergence of antifragility by optimum postdisruption restoration planning of infrastructure networks. J. Inf. Syst. **23**(4), 04017024 (2017)
12. Mujumdar, A., Mohalik, S.K., Badrinath R.: Antifragility for intelligent autonomous systems (2018). arXiv: 1802.09159
13. Jones, K.H.: Engineering antifragile systems: a change in design philosophy. Proc. Comput. Sci. **32**, 870–875 (2014)
14. Basiri, A., Behnam, N., De Rooij, R., Hochstein, L., Kosewski, L., Reynolds, J., Rosenthal, C.: Chaos engineering. IEEE Softw. **33**(3), 35–41 (2016)
15. Topirceanu, A., Udrescu M.: Topological fragility versus antifragility: understanding the impact of real-time repairs in networks under targeted attacks. In: 2018 IEEE/ACM International Conference on Advances in Social Networks Analysis and Mining (ASONAM), pp. 1215–1222. IEEE, Barcelona (2018)
16. He, S., Li, S., Ma, H.: Effect of edge removal on topological and functional robustness of complex networks. Phys. A: Stat. Mech. Appl. **388**(11), 2243–2253 (2009)
17. Piraveenan, M., Thedchanamoorthy, G., Uddin, S., Chung, K.S.K.: Quantifying topological robustness of networks under sustained targeted attacks. Soc. Netw. Anal. Min. **3**, 939–952 (2013)
18. Sun, W., Zeng, A.,: Target recovery in complex networks. Eur. Phys. J. B **90**(1), 10 (2017)
19. Iyer, S., Killingback, T., Sundaram, B., Wang, Z.: Attack robustness and centrality of complex networks. PLoS One **8**, e59613 (2013)
20. Albert, R., Jeong, H., Barabási, A.-L.: Error and attack tolerance of complex networks. Nature **406**(6794), 378 (2000)

21. Crucitti, P., Latora, V., Marchiori, M., Rapisarda, A.: Error and attack tolerance of complex networks. Phys. A Stat. Mech. Appl. **340**(1–3), 388–394 (2004)
22. Callaway, D.S., Newman, M.E.J., Strogatz, S.H., Watts, D.J.: Network robustness and fragility: percolation on random graphs. Phys. Rev. Lett. **85**(25), 5468–5471 (2000)
23. Wang, X.F., Chen, G.: Synchronization in scale-free dynamical networks: robustness and fragility. IEEE Trans. Circ. Syst. Fund. Theory Appl. **49**(1), 54–62 (2002)
24. Vespignani, A.: Complex networks: the fragility of interdependency. Nature **464**(7291), 984 (2010)
25. Wang, X.F., Chen, G.: Complex networks: small-world, scale-free and beyond. IEEE Circ. Syst. Mag. **3**(1), 6–20 (2003)
26. Chen, Y.W., Zhang, L.F., Huang, J.P.: The Watts–Strogatz network model developed by including degree distribution: theory and computer simulation. J. Phys. A Math. Theor. **40**(29), 8237 (2007)
27. Rossi, R., Ahmed, N.: The network data repository with interactive graph analytics and visualization (2015). http://networkrepository.com
28. Makse, H.: Software and data. http://www-levich.engr.ccny.cuny.edu/webpage/hmakse/software-and-data/
29. Leskovec, J., Mcauley, J.J.: Learning to discover social circles in ego networks. In: *Advances in Neural Information Processing Systems*, pp. 539–547 (2012)
30. Barabási, A.L.: Network Science. Cambridge University Press, Cambridge (2016)
31. Brandes, U.: A faster algorithm for betweenness centrality. J. Math. Sociol. **25**(2), 163–177 (2001)
32. Dekker, A.H.: Realistic social networks for simulation using network rewiring. In: International Congress on Modelling and Simulation, pp. 677–683 (2007)
33. Pósfai, M., Liu, Y.Y., Slotine, J.J., Barabási, A.L.: Effect of correlations on network controllability. Sci. Rep. **3**, 1067 (2013)

How to Collect Private Signals in Information Cascade: An Empirical Study

Kota Takeda, Masato Hisakado, and Shintaro Mori

Abstract In the information cascade experiment, several subjects sequentially answer a two-choice question, after referring to previous subjects' choices. Information cascade is defined as a tendency to follow the majority choice, even if, one's private signal suggests the minority choice. When information cascade occurs, the private signal is lost, and the collective intelligence mechanism does not work. If the majority's choice is wrong at the onset of the information cascade, it continues to be wrong forever. How can we find the correct choice even when the majority choice is wrong? In this study, we investigate a Bayesian Inference method, which collects private signals in the information cascade, based on the choice behavior of the subjects. Using the empirical data of an experiment, we estimate the probabilistic rule of the choice behavior. We demonstrate that the Bayesian algorithm works and one can know the correct choice even if the majority's choice is wrong.

1 Introduction

Nowadays, people's choices are highly influenced by others' evaluations and choices. When shopping at online stores such as Amazon, we often refer to the product ratings, which are based on other people's ratings and their review comments, and based on this, we decide whether to buy or what to buy [1]. Imitating others' behaviors and choices is called social learning [2]. When we are forced to make decisions in uncertain situations, social learning is known to be a rational strategy, as it often suggests optimal options. On the contrary, social learning is also

K. Takeda (✉) · S. Mori
Graduate School of Science and Technology, Hirosaki University, Hirosaki, Aomori, Japan
e-mail: h19ms103@hirosaki-u.ac.jp

M. Hisakado
Nomura Holdings, Inc., Chiyoda-ku, Tokyo, Japan

N. Masuda et al. (eds.), *Proceedings of NetSci-X 2020: Sixth International Winter School and Conference on Network Science*, Springer Proceedings in Complexity,
https://doi.org/10.1007/978-3-030-38965-9_14

known to be error prone, as the obtained information might be outdated or wrong. Even so, inadvertent filtering of social learning makes it an efficient strategy for decision making [2, 3].

Information cascade is a type of social learning in the sequential choices of several persons, which are taken after referring to previous persons' choices [4, 5]. When one picks the majority choice from the previous persons' choices that is different from the choice suggested by one's private signal, it is called information cascade. Information cascade is also called rational herding, as the majority's choice is more optimal than that suggested by one's private signal, in numerous cases. As the essential mechanism of collective intelligence is to aggregate each person's private signals [6, 7], information cascade acts as an impediment when people stop showing their private signals by following the majority choice.

It is important to remove the "fog" caused by information cascade and collect private signals. Wang et al. proposed an algorithm to infer the true ratings of the products on Amazon [1]. When we shop at Amazon's online store, we observe the ratings and reviews of the products. We expect that ratings, to some extent, would reflect the true values of the products. Wang et al. assume the products' true values and herding effect are two fundamental factors that drive the dynamics of the ratings. Based on the probabilistic rule of the dynamics of the ratings, they proposed an algorithm to remove rating distortions, caused by herding. However, the verification of the true rating is impossible. Nobody knows the true ratings of the products on Amazon.

Instead, we study a Bayesian estimation method to infer private signals using information cascade experiment data [8, 9]. By collecting the private signals, we infer the correct choice. Contrary to Amazon's ratings, there is an answer to each question and based on that we can check the validity of the inference method. In the information cascade experiment, subjects answer two choice questions, one by one, sequentially, after referring previous subjects' choices. We adopt the canonical one of two-choice urn quiz [10]: There are two urns, urn A and urn B. Urn A contains m red balls and $n-m$ blue balls. Urn B contains $n-m$ red balls and m blue balls. $m > n-m$ and urn A contains more red balls than urn B. At the beginning of an experiment, one of the two urns is chosen as the answer, which is called urn X. The quiz is whether X is A or B. The subjects need to choose urn A or urn B. By the choice of the correct urn, subjects receive payoffs.

Subjects decide based on two information,private signal and social information. As the private signal, each subject draws a ball from urn X. If the ball is red, urn X is more likely to be urn A, and vice versa. In addition, subjects answer one by one, after referring choices by previous subjects', which we call social information. In the simplest case, the social information is provided as summary statistic: how many subjects chose urn A or urn B [8, 11]. The history of the choices: Say the first subject chooses urn A, the second subject chooses urn B, etc. is provided as social information in previous experiments [10, 12]. If the subjects refer the history of the choices and choose based on a Bayesian inference, the majority choice coincides with the answer after many subjects' choices. The reversed cascade, where the majority choice is wrong, is fragile [12]. However, if the social information is provided as summary statistics, the reverse cascade is robust [4]. When the subjects choose based on a Bayesian inference and the difference in the choices of each urn exceeds two, information cascade starts. The subjects discard their private signals after that and the majority choice does not change forever. Two experiments studied the possibility and showed that the reversed cascade is robust [8, 11].

In this paper, we reexamine the choice behavior of the subjects in an information cascade experiment. We consider the amount of social information and propose a refined model of the choice behavior of subjects t, where t indicates the order in the subjects' sequence. In particular, we show that Weber's law, which states that the probability of choosing an option depends on the proportion of previous answers choosing that option, and not on the total number of those answers, is applicable for $t \geq 20$ [13]. Next, we study a Bayesian method to infer the private signals [14]. By collecting those, we can know the configuration of the balls in urn X, even in the reverse cascade situation.

We organize the paper as follows: In Sect. 2, we explain the experimental design and introduce a model for the choice behavior. A Bayesian algorithm, to infer the private signals of the subjects is introduced in Sect. 3. In Sect. 4, we reexamine the experimental data and propose a refined model for the choice behavior of the subjects. We show that the algorithm succeeds in gathering private signals from the subjects' choice sequence, in Sect. 5. Section 6 presents the conclusions.

2 Information Cascade Experiment and Bayesian Inference Formula

The experimental implementation of information cascade is based on the "Basic" model [4] where, a two-choice question is the choice of one of the two urns. There are two urns, urn A and urn B. Urn A contains m red and $n-m$ blue balls. Urn B contains $n-m$ red and m blue balls. We adopt $(m,n) \in \{(5,9),(2,3),(7,9),(8,9)\}$. At the beginning, urn X is chosen at random from the two urns A, B and the task is to guess which urn is X. As the private signal, one draws a ball from X. If the ball is red, the posterior probability that urn X is A is m/n. Additionally, if the ball is blue, the posterior probability that urn X is B is m/n. (5, 9) and (8, 9) are the most difficult and the easiest quizzes, respectively. As public information, one knows that C_A subjects choose urn A and C_B subjects choose urn B. We denote the length of the subjects' sequences as T, and the order of a subject as $t \in \{1,\cdots,T\}$. Without loss of generality, we express the correct and wrong signals (choice) by 1 and 0, respectively. If urn X is A and the private signal of subject t is a red ball, we denote it as $S(t)=1$. Additionally, if the private signal of subject t is a blue ball, we denote it as $S(t)=0$. If subject t chooses urn A, we denote the choice as $X(t)=1$. If subject t chooses urn B, we denote the choice as $X(t)=0$. We denote the number of correct and wrong choices up to subject t as $C_1(t)$ and $C_0(t)$. $C_1+C_0=t$ holds. We denote the ratio of correct choices up to subject t as $Z(t) \equiv \sum_{s=1}^{t} X(s)/t = C_1/t$.

Subject $t+1$ refers $S(t+1)$, $C_1(t)$ and $C_0(t) = t - C_1(t)$. We write the probability that subject $t+1$ chooses $X(t+1) = 1$ under the condition $S(t+1) = s$, $Z(t) = z$ as $f(s, z)$.

$$f(s, z) \equiv P(X(t+1) = 1 | S(t+1) = s, Z(t) = z).$$

The choice sequence becomes a nonlinear Pólya urn, that is a binary stochastic process $X(t+1) \in \{0, 1\}, t = 1, \cdots$, and the probability for $X(t) = 1$ is a nonlinear function $f(z)$ [15].

$$f(z) \equiv P(X(t+1) = 1 | Z(t) = z) = f(z).$$

$f(z)$ is given by the average of $f(s, z)$ as $f(z) = \sum_s f(s, z) P(S = s)$. In the information cascade experiment, $P(S = s)$ is given as

$$q \equiv P(S = 1) = \frac{m}{n}.$$

From the Z_2 symmetry of the system $(X, Z, S) \to (1 - X, 1 - Z, 1 - S)$, $f(s, z)$ should obey the next relation.

$$f(1 - s, 1 - z) = 1 - f(s, z).$$

For the first subject, that is $t = 1$, there is no social information, and therefore, we set $Z(0) = 1/2$. The probability for $X(1) = 1$ is then given as

$$P(X(1) = 1 | S(1) = s, Z(0) = \frac{1}{2}) = f(s, \frac{1}{2}).$$

In addition, we usually assume that $f(s, z)$ is monotonic non-decreasing function of z for $s \in \{0, 1\}$.

The number of stable states in nonlinear Pólya urn is the number of downcrossing fixed points $\{z | f(z) = z\}$ [15]. Here, the stable state is an equilibrium where the probability that $Z(t)$ converges to it in the limit $t \to \infty$ is positive. If there is only one downcrossing fixed point z_+, $Z(t)$ converges to z_+ and $\lim_{t \to \infty} Z(t) = z_+$ (Fig. 1a). If there are two downcrossing fixed points $z_- < z_+$, the probabilities that $Z(t)$ converges to z_+ and to z_- are both positive (Fig. 1b). Furthermore, the probabilities depend on the history of the choices $X(t), t = 1, \ldots$ [16].

We adopt a logistic regression model for $f(s, z)$ which preserves the Z_2-symmetry as

$$f(s, z) = \frac{1}{1 + \exp(-\beta_1(s - 1/2) - \beta_2(z - 1/2))}. \tag{1}$$

$\beta_1 (\geq 0)$ represents the strength to rely on one's private signal. $\beta_2 (\geq 0)$ shows the strength of the tendency to follow the majority's choice. In the limit $\beta_1 \to \infty$ and

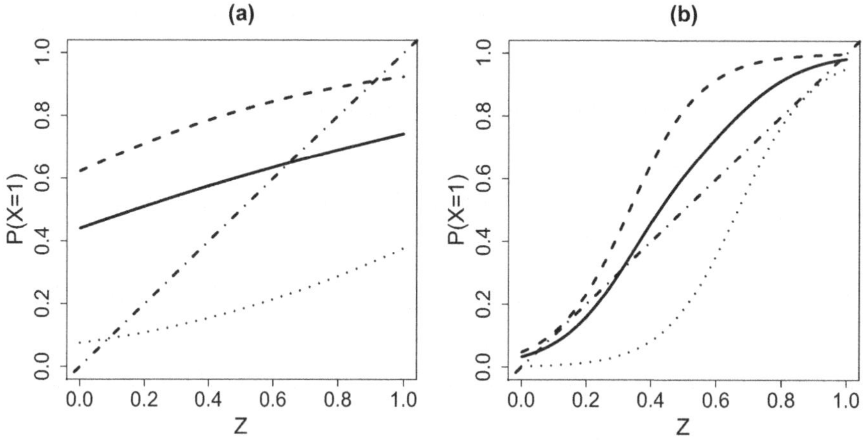

Fig. 1 Plot of $f(1,z)$ (*broken line*), $f(0,z)$ (*dotted line*) and $f(z)$ (*solid line*) of the logistic model (1). We set $q = 2/3$. (**a**) is for $\beta_1 = 3, \beta_2 = 2$ and (**b**) is for $\beta_1 = 3, \beta_2 = 9$, respectively. The downcrossing point of $f(z)$ with the diagonal (*chain line*) is the stable state of the nonlinear Pólya urn. There is one stable state z_+ in (**a**). There are two stable states $z_- < z_+$ in (**b**). The upcrossing point z_u between z_- and z_+ is unstable state of the process

$\beta_2 = 0$, one chooses the same option with one's private signal. If one draws a red ball, one chooses urn A, and vice versa. There is only one stable state z_+. When $\beta_2 > \beta_1$, the subject tends to follow the majority's choice. In the limit $\beta_2 \to \infty$, one discards one's private signal and follows the majority's choice. There are two stable states $z_- < z_+$ and it is probabilistic whether the majority's choice converges to z_+ or to z_-. It crucially depends on the first subject's choice $X(1)$ [16, 17].

In general, the regression coefficients $\beta_i, i = 1, 2$ should depend on the order t of the subjects. If t is small, the amount of social information in $z(t-1)$ is little, one might trust one's private signal and $\beta_1 > \beta_2$. If t is large, there are many subjects who have answered the question previously and subject t might tend to follow the majority's choice. Then, $\beta_2 > \beta_1$ holds. If t becomes too large and one's decision does not depends on t, $\beta_i, i = 1, 2$ converges to some constants. Weber's law states that human's decision depends on $C_1(t), C_0(t)$ through the form of the ratio $Z(t) = C_1(t)/t$ instead of the difference $C_1(t) - C_0(t)$ [13]. In the regression model context, Weber's law suggests that the regression coefficients don't depend on t.

3 Bayesian Estimation of Private Signal

We explain a Bayesian algorithm, which tells the probability $q = P(S = 1)$ using the information $Z(t-1) = z$, $X(t) = x$ and prior q_{prior}. We denote the posterior probability as $q_t(x, z, q_{\text{prior}})$.

$$q_t(x, z, q_{\text{prior}}) \equiv \mathrm{P}(S(t) = 1 | X(t) = x, Z(t-1) = z, q_{\text{prior}}).$$

As there are total n red and blue balls in the two urns, the prior q_{prior} is $q_{\text{prior}} \equiv P(S(t) = 1) = 1/2$. The Bayes' theorem for the posterior $q_t(x, z, q_{\text{prior}})$ is

$$q_t(x, z, q_{\text{prior}}) \propto q_{\text{prior}} \times P(X(t) = x | S(t) = 1, Z(t-1) = z).$$

More explicitly, we can write $q_t(x, z, q_{\text{prior}})$ for $x = 1$ using $f(s, z)$ as

$$q_t(1, z, q_{\text{prior}}) = \frac{q_{\text{prior}} f(1, z)}{q_{\text{prior}} f(1, z) + (1 - q_{\text{prior}}) f(0, z)}.$$

Likewise, for $x = 0$, we have

$$q_t(0, z, q_{\text{prior}}) = \frac{q_{\text{prior}}(1 - f(1, z))}{q_{\text{prior}}(1 - f(1, z)) + (1 - q_{\text{prior}})(1 - f(0, z))}.$$

These equations show the amount of information encoded in $X(t)$. We consider the case $\beta_2 \gg \beta_1$. When $z \sim 1$, almost all preceding subjects choose $X(s) = 1, s < t$. If subject t chooses $X(t) = 1$, as $f(0, z) \simeq 1$ and $f(1, z) \simeq 1$, and $q_t(1, z, q_{\text{prior}}) \simeq q_{\text{prior}}$. We do not receive much information from the choice $X(t) = 1$. If subject t chooses $X(t) = 0$, the relative difference between $1 - f(0, z) \simeq 0$ and $1 - f(1, z) \simeq 0$ is large and $q_t(0, z, q_{\text{prior}}) \gg q_{\text{prior}}$. We receive much information in the choice $X(t) = 0$. Likewise, if $z \sim 0$, we receive much information from the choice $X(t) = 1$. Using the posterior, we can collect private signals and estimate q.

When $\beta_1 \gg 1$ and $\beta_2 = 0$, $f(1, z) = 1$ and $f(0, z) = 0$. We have $q_t(1, z, q_{\text{prior}}) = 1$ and $q_t(0, z, q_{\text{prior}}) = 0$. If $X(t)$ is urn A, subject t must draw a red ball. There is a significant amount of information in $X(t)$.

In order to collect the private signals in the choice sequence $X(t), t = 1, \cdots$, we estimate the expected value of the posterior probability as $\mathrm{E}(q_t | z, q_{\text{prior}})$.

$$\mathrm{E}(q_t | z, q_{\text{prior}}) \equiv \sum_{s=0,1} \sum_{x=0,1} q_t(x, z, q_{\text{prior}}) P(x | s, z) P(S(t) = s). \tag{2}$$

Using Eq. (2), we obtain

$$\mathrm{E}(q(t) | z, q_{\text{prior}}) - q_{\text{prior}}$$
$$= \frac{(q - q_{\text{prior}}) q_{\text{prior}} (1 - q_{\text{prior}}) (f(1, z) - f(0, z))^2}{(q_{\text{prior}} f(1, z) + (1 - q_{\text{prior}}) f(0, z))(q_{\text{prior}}(1 - f(1, z)) + (1 - q_{\text{prior}})(1 - f(0, z)))}.$$

The curve of $\mathrm{E}(q_t | z, q_{\text{prior}})$ vs. q_{prior} passes through the diagonal at $q_{\text{prior}} = 0, 1$ and $q_{\text{prior}} = q$. The solid line in Fig. 2 shows the plots of the curves with the same sets of parameters in Fig. 1. The broken line and the dotted line shows $q_t(x, z, q_{\text{prior}})$ with $x = 1$ and $x = 0$, respectively. One sees that $\mathrm{E}(q_t | z, q_{\text{prior}}) > q_{\text{prior}}$ and

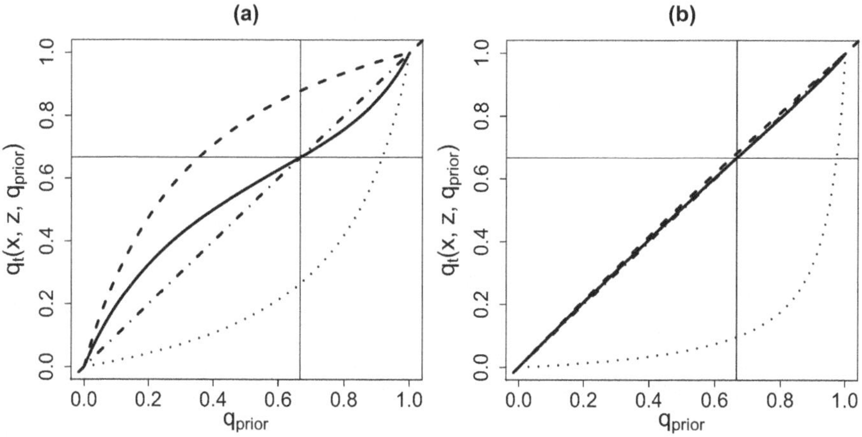

Fig. 2 Plot of $q_t(1, z, q_{\text{prior}})$ (*broken line*), $q_t(0, z, q_{\text{prior}})$ (*dotted line*) and $\text{E}(q_t|z, q_{\text{prior}})$ (*solid line*) vs. q_{prior}. We adopt the same values for the parameters in Fig. 1. We also set $z = 2/3$ in (**a**) and $z = 0.95$ in (**b**)

$\text{E}(q_t|z, q_{\text{prior}}) < q_{\text{prior}}$ hold for $q_{\text{prior}} < q$ and $q_{\text{prior}} > q$, respectively, for $\forall z \in [0, 1]$. If there is one stable state z_+, case of (a), $|q_t(x, z, q_{\text{prior}}) - q_{\text{prior}}|$ are large for both of $x = 1$ and $x = 0$. If there are two stable states and $z \sim z_+$, case of (b), $|q_t(0, z, q_{\text{prior}}) - q_{\text{prior}}| > |q_t(1, z, q_{\text{prior}}) - q_{\text{prior}}|$. The amount of information in $x = 0$ is much larger than that in $x = 1$. As we shall see below, by collecting the difference in the amount of information of $x = 0$ and that of $x = 1$, we can estimate q.

As the curve $(q_{\text{prior}}, \text{E}(q_t|z, q_{\text{prior}}))$ passes through the diagonal at $q_{\text{prior}} = q$, we have the next relation for q as

$$q = \{q_{\text{prior}} \quad \text{s.t.} \quad \text{E}(q_t|z, q_{\text{prior}}) = q_{\text{prior}}\}.$$

By solving the relation $\text{E}(q_t|z, q_{\text{prior}}) - q_{\text{prior}} = 0$, we can estimate q. In the study of experimental data, we replace the expectation value with the average over the choice history $X(s), Z(s-1), s = 1, \cdots, t$. We denote the average over the history $(X(s), Z(s-1), s = 1, \cdots, t$ as $\overline{q_t}(q_{\text{prior}})$.

$$\overline{q_t}(q_{\text{prior}}) \equiv \frac{1}{t} \sum_{s=1}^{t} q_t(X(s), Z(s-1), q_{\text{prior}}). \tag{3}$$

We propose an estimator of q as

$$\hat{q} = \{q_{\text{prior}} | \overline{q_t}(q_{\text{prior}}) = q_{\text{prior}}\}. \tag{4}$$

$\hat{q}$ shows the rate of the majority ball in X. If $\hat{q} > 1/2$, we can correctly guess whether X is A or B. In the derivation of the estimator $\hat{q}$, we replace the expectation value with the time average over a particular sample path. If the system is stationary, one can justify the replacement. However, the nonlinear Pólya urn is not a stationary stochastic process. Later, we demonstrate the validity of the estimator empirically, whereas we do not understand why it works.

4 Choice Behaviors: Result I

We estimate $\beta_i, i = 1, 2$ empirically for $q = m/n \in Q = \{5/9, 2/3, 7/9, 8/9\}$ using the maximum-likelihood method. The detailed information of the experiment can be found in [8] and the experimental data is open [9]. We prepared $I = 200$ questions for $q \in \{5/9, 2/3, 7/9\}$ and $I = 400$ questions for $q = 8/9$. We label the questions as $i = 1, 2, \cdots, I$. We obtained I sequences of answers of length $T = 63$ for $q \in \{5/9, 2/3\}$. The average length T is 54.0 for $q = 7/9$ and 60.5 for $q = 8/9$.

As explained in Sect. 2, subject t answers the question with the private signal $S(t)$ and the ratio of correct choices up to the previous subject $Z(t-1)$. In the previous work, we assume Weber's law and estimate $\beta_1 = 2.92$ and $\beta_2 = 7.64$ using the data for $t \geq 20$. Here, we estimate β_i for subject t using all sample sequence for each $q \in Q$ and verify Weber's law.

Figure 3 depicts the plot of β_1 and β_2 vs. t for each $q \in Q$. The error-bars show 95% confidence intervals. In either case, both β_1 and β_2 fluctuate around a constant value for $t \geq 20$, which suggests that Weber's law holds for $t \geq 20$.

We fit the t-dependence of $\beta_i, i = 1, 2$ assuming the next functional form, which is the sum of the logistic function with three parameters $c_i, i = 1, 2, 3$ and a constant c_4.

$$\beta_i(t) = \frac{c_1}{1 + c_2 \exp(-c_3 t)} + c_4. \tag{5}$$

By the least squares method, we estimate the parameters of $\beta_i(t), i = 1, 2$ for each $q \in Q$. The results are shown in Table 1. The solid and broken lines in Fig. 3 are the results of the fitting. From the solid lines, β_1 converge from 2.0 to 4.0 at $t \geq 10$ for each $q \in Q$. From the broken lines, β_2 converge to approximately 8.0 at $t \geq 20$ for each $q \in Q$. In the next section, we estimate q using the above functional forms for $\beta_i, i = 1, 2$.

5 Estimation of q: Results II

We estimate q by solving $\overline{q_t}(q_{\text{prior}}) = q_{\text{prior}}$ for the sample sequences $(X(s), Z(s-1)), s = 1, \cdots, T$ using Eq. (3). We write the solution as $\hat{q}$. Figure 4 shows the violin plot of $\hat{q}$ and $Z(T)$ for each q. In each figure, the black boxes show the

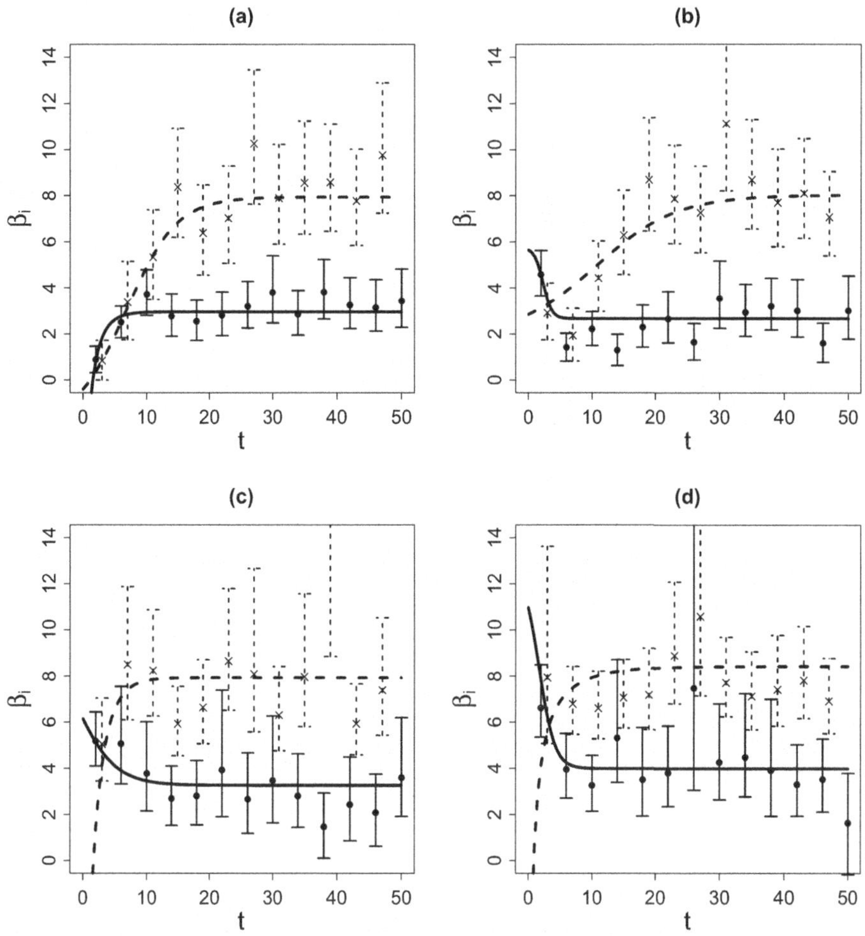

Fig. 3 Plots of β_1 and β_2 vs. t for (**a**) $q = 5/9$, (**b**) $q = 2/3$, (**c**) $q = 7/9$, and (**d**) $q = 8/9$. The error-bars are 95% confidence intervals. The solid and broken lines are the results of the fitting with (5) for β_i, $i = 1, 2$, respectively

Table 1 Estimated values of the parameters in Eq. (5) for β_1(*left table*) and β_2(*right table*)

q	c_1	c_2	c_3	c_4
5/9	−9.07	−0.58	0.55	0.556
2/3	−3.09	31.32	1.48	5.75
7/9	−7.36	0.64	0.32	10.62
8/9	−8.73	4.03	0.81	12.71

q	c_1	c_2	c_3	c_4
5/9	−9.87	0.18	−0.25	7.94
2/3	−6.14	0.19	−0.16	8.02
7/9	−38.94	−0.37	0.48	46.87
8/9	−2.27	−0.91	0.16	10.70

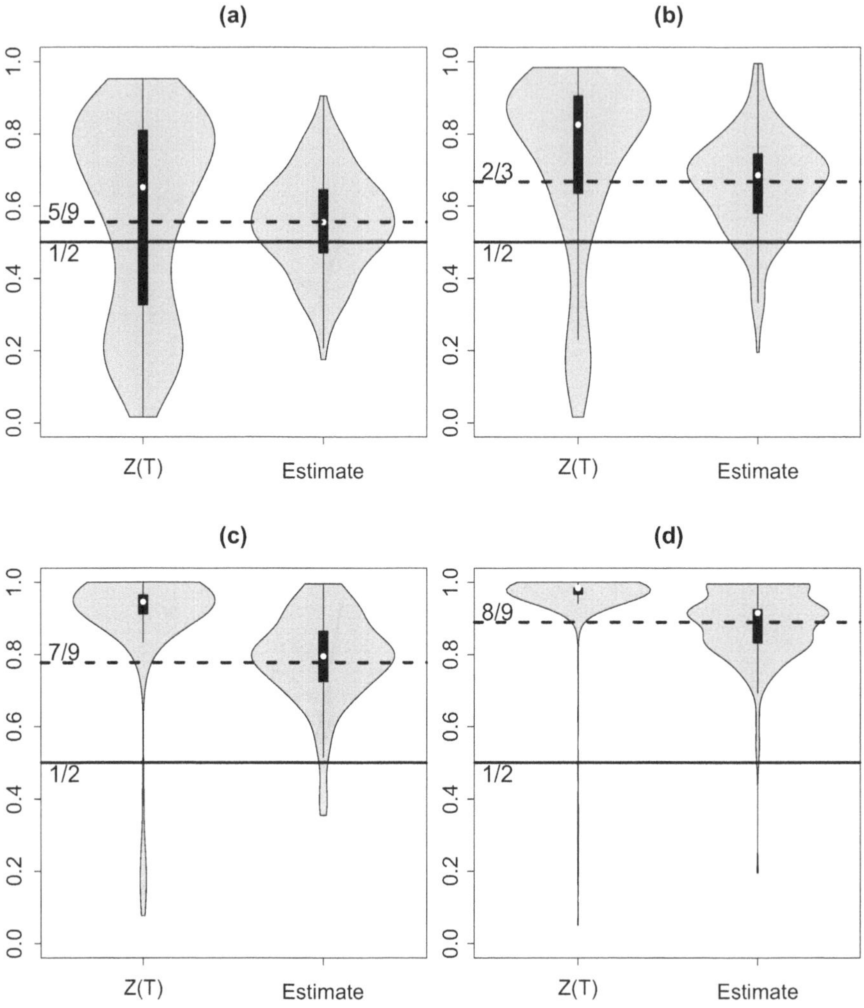

Fig. 4 Violin plot of $\hat{q}$ and $Z(T)$ for (**a**) $q = 5/9$, (**b**) $q = 2/3$, (**c**) $q = 7/9$, and (**d**) $q = 8/9$. The solid line shows $1/2$ and broken lines show $5/9$, $2/3$, $7/9$ and $8/9$

boxplots. In addition, the white dots indicate the median value of each data and the gray regions indicate the density of each data distribution. The statistics of $\hat{q}$ are shown in Table 2.

It can be seen that the average values of $\hat{q}$ almost coincide with q. As q increases from $5/9$ to $8/9$, the accuracy of the classification of correct urn by $\hat{q} > 1/2$ improves. For the most difficult case of $q = 5/9$, the classification fails for 31.5% of the scenarios. For $q = 2/3$, the success rate is 89.5%. On the other hand, if we classify by $Z(T) > 1/2$, the success rates are 64.5 and 84.5% for $q = 5/9$ and $2/3$, respectively. If we infer that X is A by observing that the majority subjects

Table 2 Statistics of $\hat{q}$ and the ratio of $Z(T) > 1/2$

q	$\hat{q}_{25\%}$	$\hat{q}_{50\%}$	$\hat{q}_{75\%}$	$\hat{q}_{mean}$	$\hat{q} > 1/2$	$Z(T) > 1/2$
5/9	0.470	0.555	0.645	0.556	0.685	0.645
2/3	0.580	0.685	0.745	0.661	0.895	0.845
7/9	0.725	0.795	0.865	0.790	0.970	0.940
8/9	0.832	0.915	0.925	0.885	0.995	0.980

have chosen A and vice versa, the accuracy of the naive method is lower than that obtained by $q_t(x, z, q_{\text{prior}})$. As one can clearly see in Fig. 4, the median value of $Z(T)$ is far from q. The subjects' choices are greatly influenced by the social information. From the violin plots, one can also observe two peaks in the distribution of $Z(T)$. It is difficult to estimate q from $Z(T)$. In the case of the distribution of $\hat{q}$, the mode of the distribution almost coincides with the true value. These results indicate that the proposed Bayesian method is extremely effective owing to the success of the collection of the private signals. Notably, we use $q_t(x, z, q_{\text{prior}})$ instead of $X(t)$.

6 Conclusions

In this paper, we proposed a Bayesian method to collect private signals in case of information cascade. The tendency to follow the majority choice is so strong that it becomes difficult to collect private signals, from the choice sequences of the subjects. The Bayesian method to infer the private signal $S(t)$, is based on the probabilistic rule for $X(t)$ under the condition $S(t), Z(t-1)$. We reexamine the choice behavior of subjects empirically, and verify that Weber's law holds for $t \geq 20$. For $t < 20$, we describe the t dependencies of the regression coefficients. We estimate q by solving $\overline{q_t}(q_{\text{prior}}) = q_{\text{prior}}$ in Eq. (4). The mean value of $\hat{q}$, almost coincides with q, which proves the validity of the Bayesian estimation and the logistic regression model for the choice behavior.

References

1. Wang, T., Wang, D.: Why Amason's ratings might mislead you. Big Data **2**, 196 (2014)
2. Rendell, L., Boyd, R., Cownden, D., Enquist, M., Eriksson, K., Feldman, M.W., Fogarty, L., Ghirlanda, S., Lillicrap, T., Laland, K.N.: Why copy others? Insights from the social learning strategies tournament. Science **328**, 208 (2010)
3. Nakayama, K., Hisakado, M., Mori., S.: Nash equilibrium of social-learning agents in a restless multiarmed bandit game. Sci. Rep. **7** (2017). Article number: 1937
4. Bikhchandani, S., Hirshleifer, D., Welch, I.: A theory of fads, fashion, custom, and cultural change as information cascades. J. Polit. Econ. **100**, 992–1026 (1992)
5. Devenow, A., Welch, I.: Eur. Econ. Rev. **40**, 603–615 (1996)

6. Surowiecki, J.: The Wisdom of Crowds. Doubleday, New York (2004)
7. Page, S.E.: The Difference. Princeton University Press, Princeton (2007)
8. Hino, M., Irie, Y., Hisakado, M., Takahashi, T., Mori, S.: Detection of phase transition in generalized Póla urn in information cascade experiment. J. Phys. Soc. Jpn. **85**(3), 034002–034013 (2016)
9. Mori, S., Hisakado, M.: Information cascade experiment: Urn Quiz. In: Sato, A.H. (ed.) Applications of Data-Centric Science to Social Design. Agent-Based Social Systems, vol. 14, pp. 181–191. Springer, Singapore (2016)
10. Anderson, L., Holt, C.: Information cascades in the laboratory. Am. Econ. Rev. **87**(5), 847–862 (1997)
11. Mori, S., Hisakado, M., Takahashi, T.: Phase transition to two-peaks phase in an information cascade voting experiment. Phys. Rev. E **86**, 026109 (2012)
12. Goeree, J.K., Palfrey, T.R., Rogers, B.W., McKelvey, R.D.: Self-correcting information Cascades. Rev. Econ. Stud.**74**, 733–762 (2007)
13. Eguíluz V.M., Masuda, N., Fernández-Gracia, J.: Bayesian decision making in human collectives with binary choices. PLoS One **10**(4), e0121332 (2015). https://doi.org/10.1371/journal.pone.0121332
14. Hisakado M., Mori S.: Information cascade and bayes formula. In: Sato, A.H. (ed.) Applications of Data-Centric Science to Social Design. Agent-Based Social Systems, Chapter 12, vol. 14, pp. 193–202. Springer, Singapore (2019)
15. Hill, B., Lane, D., Sudderth, W.: A strong law for some gener-alized urn processes. Ann. Prob. **8**, 214–226 (1980)
16. Mori, S., Hisakado, M.: Correlation function for generalized Polya urns: finite-size scaling analysis. Phys. Rev. **E92**, 052112 (2015)
17. Mori, S., Hisakado, M.: Finite-size scaling analysis of binary stochastic processes and universality classes of information cascade phase transition. J. Phys. Soc. Jpn. **84**, 054001 (2015)

Space Geometry Effect over the Internet as a Physical-Logical Interdependent Network

Ivana Bachmann, Francisco Sanhueza, and Javier Bustos-Jiménez

Abstract In this article we study the Internet's robustness under physical node failures, given that the physical layer is built over spaces with geometry/shape restrictions. This is of special interest for countries prone to natural catastrophes, and long and narrow geographies such as Chile and Japan. We model the Internet as an interdependent network composed of the Internet's physical layer (Internet backbone) and he Internet's logical layer (Autonomous System level network) coupled. Here, the robustness is tested by measuring the amount of functional nodes on the logical network after randomly removing physical nodes. In this work, we tested six different spatially constrained network models to generate the Internet's physical layer (Yao graphs, geometric preferential attachment, Erdős-Rényi, n-nearest neighbours, Gabriel graphs, and Modified relative neighbourhood model), and three different geometries with width to lengths ratios going from a square geometry to a Chile-like space geometry. Additionally, we study the relation between the amount of physical edges and the Internet's robustness. Our findings suggest that both: the edge addition strategy (i.e. the physical network model used) and the amount of physical edges play an important role on the Internet's robustness. Our results also suggest that Internet based interdependent systems whose robustness is affected by the space geometry (geometry-sensitive) can become more robust by randomly adding few edges. Furthermore, these interdependent systems can become geometry-insensitive after the edge addition, meaning that the robustness of the interdependent system is no longer affected by the space geometry.

I. Bachmann (✉) · F. Sanhueza · J. Bustos-Jiménez
NIC Labs, University of Chile, Santiago, Chile
e-mail: ivana@niclabs.cl; francisco@niclabs.cl; jbustos@niclabs.cl

N. Masuda et al. (eds.), *Proceedings of NetSci-X 2020: Sixth International Winter School and Conference on Network Science*, Springer Proceedings in Complexity,
https://doi.org/10.1007/978-3-030-38965-9_15

1 Introduction

The Internet has been mostly studied as the Autonomous System level network (Border Gateway Protocol network or Internet's logical layer). However, many more layers interact and affect the Internet's logical layer. Each of these layers corresponds to a network with its own set of nodes and edges. The interactions among these layers generate dependencies that affect one another in ways that can lead to cascading failures on the system. Even more, dependencies between networks are known to affect the robustness of the whole system in ways that can not be understood by studying each network in isolation [6]. Cases of massive power grid blackouts in the past due to interdependencies between the power grid and its communication network [11, 32] have motivated the study and analysis of complex systems as interdependent networks. However, for the Internet case few interdependent network models have been proposed [3, 10].

It has been stated before that in order to understand the Internet's behaviour we must first understand the underlying structures that compose it, and how they affect one another [36]. On the one hand there is the *physical Internet network* comprising cables, antennas, routers, etc. On the other hand, there is the *logical Internet network* composed of autonomous systems (AS) [2] which are connected according to the Border Gateway Protocol (BGP) protocol [5]. These networks interact with each other and damage on one network may affect the other.

In particular physical node failures can affect the functionality of the logical Internet network by damaging the information flow between Internet consumers and Internet Service Providers (ISPs). Damaging this flow can leave users without Internet access. Countries such as Chile and Japan are particularly prone to physical failures due to natural catastrophes such as earthquakes, floods, typhoons, etc. These natural events can damage the Internet's physical layer, negatively affecting the whole system. These negative effects over the system can end up disrupting the user's ability to access the Internet. Severe damage to the Internet's functionality due to natural catastrophes has been observed before: after the 8.8 Mw earthquake in Chile in 2010 a 1 day Internet outage was reported [30]. Hence, we are interested on measuring the effects over the Internet's robustness as the user's ability to access the Internet, after physical node removal. We are also interested in the effects that the physical environment in which the physical layer is built has over the robustness.

A question that naturally emerges for the case of a country with Chile's peculiar geography is: *does an elongated country shape such as that of Chile have an impact over the Internet's robustness?* (See Fig. 1.) We would like to know whether the shape of a country affects in some way the robustness of the Internet as a physical-logical interdependent network. This question was initially explored in a previous article [3]. In that article we proposed a physical model and an interdependencies model and found that the shape of the area where physical network is built affects the overall robustness, making the interdependent networks more fragile as the shape becomes narrower. We found that this effect can be diminished on interdependent systems by increasing the amount of inter-links. However, the robustness was tested

Fig. 1 Chile's geography. The continental portion of Chile has a proportion of approximately 1:25 width to length. Chile is 4270 km long

using only one physical model. To further understand the phenomena observed in [3], we should explore other physical network models and observe the system's behaviour under physical constraints and physical node failures.

In this work, we test the robustness of the Internet modeled as an interdependent system that comprises the BGP network (also referred in this work as logical network) and the physical Internet network. We test the robustness of six network models as physical networks of the Internet based interdependent system under random physical node removal. More specifically, we study the effect over the Internet system's robustness of the space shape (also referred as geometry) in which the physical network is built in, the maximum amount of interdependencies, and the amount of edges that the physical network has. Our findings suggest that there is a non-trivial relation between the Internet system robustness under physical node failure, the model used to generate the physical network, and the number of edges in the physical network. Our findings also suggest that systems whose robustness is affected by the space geometry (geometry-sensitive system) such as the one presented in [3] can become more robust, to a point where the system is no longer affected by the space geometry (geometry-insensitive system).

2 Related Work

Previously Willinger and Roughan have mentioned the need for a way of modeling the Internet that considers real-world AS-connection policies, multiple links, and geographic location among others, instead of considering just the AS-level Internet in isolation as a simple connected di-graph [36]. However, few studies have included more complex interactions and characteristics when modeling the Internet [3, 10].

Here, we have modeled the Internet as an interdependent network system composed of the physical Internet network, and the BGP or logical Internet network. Particularly, we aim to assess the Internet's robustness under physical damage.

Interdependent networks are systems comprising two or more interacting networks, with behaviours that emerge from such interactions. This emerging behaviour cannot be observed by studying single isolated networks. During the last decade the study of the robustness of interdependent networks has been explored. Those studies have led to the development of several frameworks to tackle the robustness assessing problem. The first work published in the area was presented by Buldyrev et al. [6] in 2010. There, they proposed the “one to one” model, which considers two interacting networks where each node depends on exactly one node in the other network with mutual dependency, meaning that if a node fails, then necessarily its interdependent neighbour will also fail.

Since then several other interdependent models have been developed. Among these, we can find variations of the “one to one” model [21, 23, 29, 33, 35, 39], new models such as those with “many to many” interdependencies where each node may be interconnected to 0 or more nodes in the other network [9, 13, 26, 28, 31], as well as models that focus on specific networks such as power grids [7, 20, 22, 25], and spatially constrained networks [1, 4, 12].

As for modeling the Internet as an interdependent network system we have not found many models where the Internet is explicitly part of an interdependent system. An example is the recent work of Chen et al. where the AS-level Internet is coupled with a power-grid [10]. However, up to the authors knowledge there haven’t been other works that attempt to model the Internet itself as an interdependent model composed of the physical and the logical layers aside from our previous work [3].

3 Internet Modeling

In this work, we used an interdependent model [3] based on a model previously proposed by Parandehgheibi et al.[27]. Our model considers a physically constrained network representing the physical Internet layer, and a network representing the logical Internet layer (with no physical constraints). Both of these networks are interdependent: on the one hand we have that each AS node in the logical network may be allocated in one or more nodes in the physical network. If all the physical nodes where a logical node is allocated fail, none of its physical systems is able to communicate, thus the logical node fails. Conversely, a physical node may route several ASes. If all the logical nodes allocated in a physical node fail, then that physical node becomes unable to function as it cannot answer.

Both layers were modeled to have a consumer-provider behaviour as stated by Parandehgheibi’s work [27], that is, consumer nodes must have a path that connects them to a provider node within their own network. This allow us to model the fact that in order for a user to have Internet access the user has to have a path to an Internet Service Provider (ISP). This consumer-provider behaviour naturally emerges on the AS-level BGP network since in order for a node to have access to the Internet service it must be connected through at least one path to an ISP and to an International gateway to have access to the worldwide network. This behaviour is replicated on the physical layer since each node in the logical layer has one or more physical counterparts.

3.1 Interdependencies Model

Based on the model presented by Parandehgheibi et al. [27] we considered an interdependence model where each node in each network can be bidirectionally interdependent with 0 or more nodes in the other network. Here, if a node loses all of its inter-connections then it fails. In the Parandehgheibi model adaptation presented in [3], we set a maximum amount of interdependencies for each logical node $I_{max} \in \mathbb{N}$, with $I_{max} \geq 1$, instead of fixed a amount of inter-connections for all the logical nodes. Here, each logical node u is inter-connected to $I(u) \in \{1, \ldots, I_{max}\}$ nodes in the physical network, where $I(u)$ is selected at random from the set $\{1, \ldots, I_{max}\}$ for each node u in the logical network. Each of these $I(u)$ inter-connections is randomly established between u and a node in the physical network. Here, node u must be inter-connected to exactly $I(u)$ nodes in the physical network.

3.2 Logical Network Model

Since the purpose of this work is to study the effect of using different physical models over the system's robustness, a simple scale-free network was used to model the BGP network. scale-free networks have been previously used to model AS-level networks [17], and it was used in the work preceding this current article [3]. However, we must note that this approach has been criticized before [16, 24, 37]. Alternative models that can be applied to represent more accurately the BGP network that we should consider for future works are HOT-Nets (highly organized/optimized tolerances/tradeoffs [8, 16]). HOT-Nets allow the incorporation of optimized aspects that engineered networks, such as the Internet's logical layer, usually present.

3.3 Physical Network Model

To further test the effects over the system's robustness that a country's shape has, we used the following existing models: Yao graphs [38], peometric preferential Attachment [18], Erdős-Rényi [15], n-nearest neighbours [14], and Gabriel Graphs [19]. All of these models were tested and compared to previous results obtained using a modified relative neighbourhood model [3].

For Yao Graphs [38] we have that each node "selects" their neighbors using the following procedure: first the node partitions the space into six parts of equal area. This partition is done taking the node as center point. Once the partition is done, the node will select the closest node for each area an will connect to that node.

The geometric preferential attachment (GPA) model was described by Flaxman et al. [18]. Here, networks are generated using an iterative algorithm. First, a random node u is selected. Within a specific radius r, taking u as center, the usual rules of preferential attachment are used. Thus, node u will be connected to another node with a probability that is proportional to the degree of such node, and that node has to be located within radius r. On each iteration five edges are added in order to resemble the original experiment. In case of ending up with more than one connected component due to the narrowness of the spaces we tested, we have opted to use an elliptical area to find u's neighbors instead of a circular one.

Erdős-Rényi models (ER) [15] allocate n nodes at random where each possible edge is added with probability p. The probability p was set to $\frac{log(n)}{(n)}$ since this value should generate a single connected component.

For the case of n-nearest neighbours models (nNN) [14] each node is connected to the n nodes closest to it. In this work we used $n = 5$ in order to obtain a single connected component, thus we have a (5NN) model.

Gabriel graphs (GG) [19] are generated as follows: given two points u and v, we create an edge between them if the circular area between them does not have other nodes. This circular area has as diameter the distance between u and v, and u and v are located on the circumference of this area.

The modified nelative neighborhood model (MRN) [3] is a model based on the interdependent network model proposed in [34]. In this model given two nodes u and v, and given $d(u, v)$ the euclidean distance between u and v, we create an edge between them if the intersection area between the two circles of radius $d(u, v)$, centered on u and v respectively, is empty. That is, there are no other nodes in that intersection area.

4 Results

Our objective was to study the country's shape effect over the Internet's robustness. For this we tested 5 new network models as physical networks built over spaces of different shapes, and contrasted these results with the ones obtained using the MRN model [3].

To do this we tested the effect of three different shapes: a square shape (with an aspect ratio of 1:1), a Chile-like shape (with an aspect ratio of 1:25), and a shape in between a Chile-like shape and a square shape (with an aspect ratio of 10:125). We considered $P_l = 6$ to be the amount of ISPs, as there are 6 ISP nodes in Chile's BGP network. We also tested the robustness of the system for $I_{max} \in \{1, 3, 5, 7, 10\}$. As for the number of nodes for each network we followed an approximation of the amount of BGP nodes that Chile has, that is, 300 nodes. The same approximation was used for the amount of physical nodes, that is, 2000 nodes. On each experiment each layer started with a single connected component. Thus, if two logical nodes u_l and v_l are connected by a path in the logical network, then their interdependent physical nodes are also connected through a path in the physical network.

We defined the Internet's robustness as the ability to keep users with Internet access in case of failure. For that we measured the fraction of functional nodes after failure. Here, a node is said to be functional if it has Internet access (i.e. a path to a provider node), and at least one of their original interdependent nodes is still functional.

To measure the robustness we used the G_L index proposed in [3]. Here, $G_L = \frac{N_l^f}{N_l}$, where N_l is the amount of logical nodes that are functional before damaging the system, and N_l^f is the amount of functional logical nodes after the system has been damaged, and cascading failures have stopped.

To test the robustness of each interdependent system under each scenario (shape, physical network model, and I_{max}) we randomly removed nodes from the physical network and any edge or inter-edge connecting the removed node with other nodes. The results were obtained from 10 interdependent systems simulated for each possible case (geometry, I_{max}, physical model). For each interdependent system 100 iterations of random physical node removal were considered. The results shown in Sects. 4.1, and 4.2 were obtained averaging those results. In particular we focus on results obtained when 15% of the physical nodes have been removed on the following subsections, since once 15% of the physical nodes have been removed, the most fragile Internet based interdependent system tested in this work reaches a state of total failure.

4.1 Country Shape Effect

As we can see in Fig. 2, although every model tested is generated using geometrical parameters such as areas and distances, most of them are not notoriously affected by the geometry of the space in which they are built in.

For Internet based interdependent systems generated using Yao and GPA physical models (see Fig. 2a and b) we observe that the space geometry does not have a significant effect over the system's robustness. On these systems we can also observe that a higher I_{max} value increases the robustness to some extent.

In the case of the Internet based interdependent systems generated using GG and ER models (see Fig. 2e and c) we observe a slightly more pronounced effect of the space shape over the system's robustness. However, these effects do not show a clear trend regarding whether a narrower space decreases (or increases) the system's robustness.

The cases of Internet based interdependent systems built using 5NN and MRN models as physical networks (see Fig. 2d and f respectively) show a clear effect of the geometry in which they where built in over the system's robustness. Specifically, a narrower space geometry leads to a less robust Internet system. For the case shown in Fig. 2d we observe that as the I_{max} value increases, the effect of the space shape over the robustness becomes less noticeable. In particular, we see that once $I_{max} \geq 5$, a narrow Chile-like space does not seem to behave so much worse than a square space. As for the case shown in Fig. 2f we observe a relatively constant effect of the space shape over the Internet system robustness.

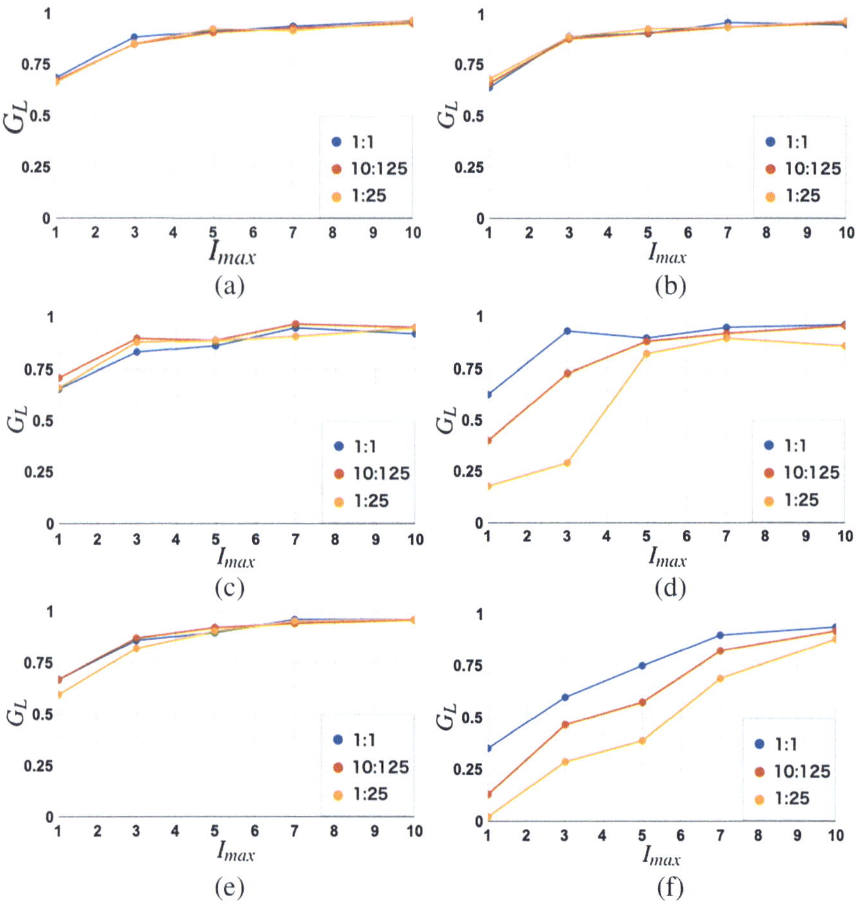

Fig. 2 System's robustness after 15% of the physical nodes have been removed. Blue indicates physical networks built in a square geometry, yellow indicates systems over a Chile-like geometry, and red represents system built in an geometry between a square space and a Chile-like space. (**a**) Yao physical network. (**b**) GPA physical network. (**c**) ER physical network. (**d**) 5NN physical network. (**e**) GG physical network. (**f**) MRN physical network

As a general observation we see in Fig. 2 that the robustness in every Internet system tested improves with a greater I_{max} value. In particular all the systems improve their robustness to a similar level once an $I_{max} \geq 7$ is reached.

It is worth asking ourselves the reason why such dissimilar robustness outcomes are observed despite the fact that all the physical networks used to test the Internet based interdependent system were generated using spatial parameters. In an attempt to understand this phenomena we decided to study the effect of the amount of edges that each system has over the Internet based interdependent system.

4.2 Effects of the Number of Edges

To further understand the phenomena observed in Sect. 4.1 we studied whether the differences observed among system's with different physical network models were due to the amount of edges, or some other network characteristics, such as the way in which the edges are allocated in the physical network.

In Fig. 2 we observe that Internet based interdependent systems generated using Yao and GPA physical models have a similar robustness under physical attacks given the different space geometries and I_{max} values tested. However, it is interesting to note that these two system have a relatively big difference on the amount of edges they have (approximately 4200 edges), this can be appreciated in Table 1 and Fig. 3.

A similar phenomenon is observed on Internet based interdependent systems generated using GG and ER physical models. Here, systems created using ER networks as physical networks have 3700–3900 edges more than those using GG networks as physical network. Just as the previously described case, systems generated using GG and ER physical models have a similar robustness under physical attacks given the different space geometries and I_{max} values tested (see Fig. 3, and Table 1).

On the other hand, we observe in Fig. 3 and Table 1 that Internet based interdependent systems generated using 5NN and GG physical models do not present a great difference on their edges amount (between 500 and 600 edges more in 5NN systems). Nonetheless, GG systems are more robust than 5NN systems under the same adverse scenarios despite GG systems having fewer edges than 5NN systems.

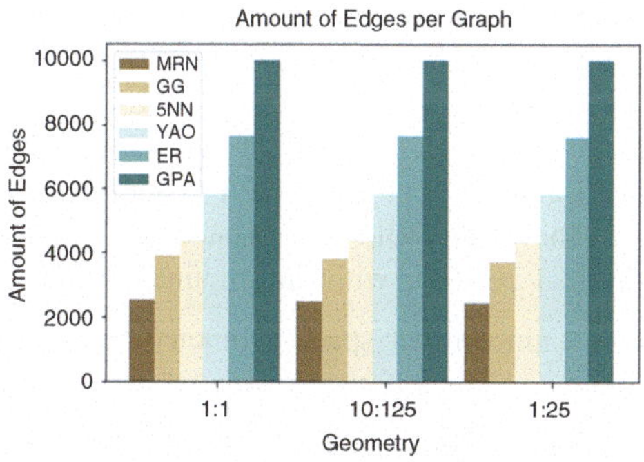

Fig. 3 Average number of edges of each physical network tested

Table 1 Average number of edges per physical network

Model\space	1:1	10:125	1:25
MRN	2531	2487	2451
GG	3893.82	3819.8	3740.96
5NN	4406.92	4369.82	4339.16
Yao	5810.32	5817.52	5815.98
ER	7634.8	7649.8	7619
GPA	9999	9999	9999

Regarding Internet based interdependent systems generated using 5NN and MRN physical models, we can see in Fig. 3 and Table 1 that 5NN systems have between 1280 and 1370 edges more than MRN systems. For this pair of systems it is interesting to note that for the case of a space with 1:25 proportions and for an $I_{max} = 3$. Both 5NN and MRN systems have a similar G_L, meaning a similar robustness level is achieved by MRN systems with approximately 1300 less edges than 5NN systems. However this is an isolated case and 5NN systems are on average much more robust than MRN systems.

Another observation is that MRN based physical networks have approximately 500 edges more than the minimum amount of edges needed to have a single connected component given that all the physical networks have 2000 nodes. This minimum amount of edges is given by the amount of edges needed to generate a tree graph which is $n - 1$, given a graph with n nodes. This contrasts with all the other networks models tested as physical networks, which have at least 1280 edges more than the minimum amount of edges needed to have a single connected component for the same 2000 nodes.

4.3 Adding Edges to the MRN-Based System

To understand the effect that the amount of edges in the physical Internet network has over the Internet's robustness under random physical node removal, we tested the effect of randomly adding edges to the most fragile system, in this case the MRN-based Internet based interdependent system.

To test this we randomly added edges to the MRN-based Internet based interdependent system without changing anything else, that is: no interdependent edges are changed or added, no logical nodes or edges were added, and no previously existing physical edges were changed.

Given $g \in \{1{:}1, 10{:}125, 1{:}25\}$ a space geometry, we define E_g^{GPA}, and E_g^{MRN} to be the amount of edges of GPA systems and MRN systems on a space with geometry g as shown in Table 1, we tested two cases: (1) Adding enough edges to obtain a physical network with an amount of edges between the highest amount and the lowest amount per space geometry (i.e. $\frac{E_g^{\mathrm{GPA}}+E_g^{\mathrm{MRN}}}{2}$). (2) Adding enough edges to obtain a physical network with an amount of edges equivalent to half of the amount of edges obtained on case (1) (i.e. $\frac{E_g^{\mathrm{GPA}}+E_g^{\mathrm{MRN}}}{4}$) as this amount is roughly between the amount of edges of GG and MRN (see Table 1).

Both cases were tested over fifteen initial Internet based interdependent systems generated using MRN physical models, one per space geometry in {1:1, 10:125, 1:25} and $I_{max} \in \{1, 3, 5, 7, 10\}$. For case (1) the final amount of edges of the physical network for geometry (1:1) was 6260, for geometry (10:125) was 6231, and for geometry (1:25) was 6238. While, for case (2) the final amount of edges of the physical network for geometry (1:1) was 3130, for geometry (10:125) was 3116, and for geometry (1:25) was 3119. Thus, on case (1) an average of 3757 edges were

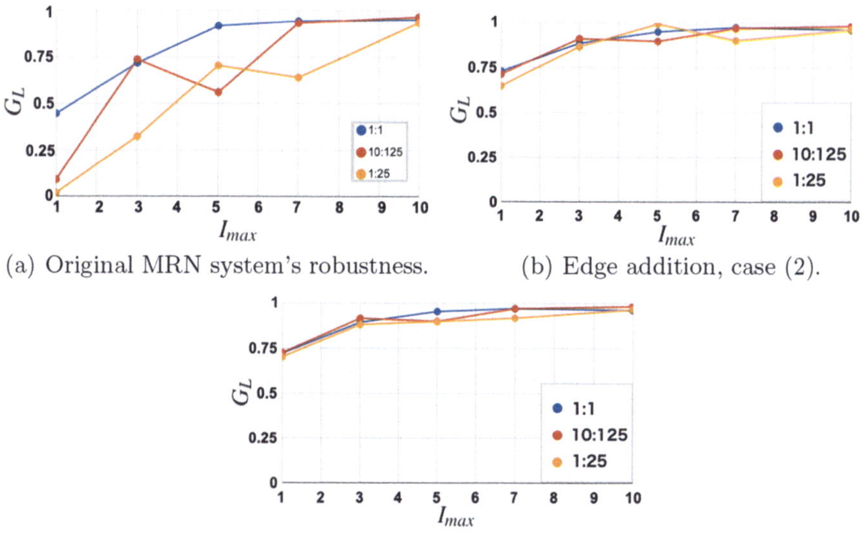

(a) Original MRN system's robustness.

(b) Edge addition, case (2).

(c) Edge addition, case (1).

Fig. 4 Test case of MRN-based Internet based interdependent systems with no added edges (**a**), an average addition of 635 edges (**b**), and an average addition of 3757 edges (**c**)

added, and on case (2) an average of 635 edges were added. In Fig. 4a we can see the robustness of the originally generated systems, i.e, systems with no added edges. In Fig. 4c and b we have the robustness of the cases (1) and (2) respectively.

In Fig. 4 we observe that adding edges at random to the Internet based interdependent systems generated using MRN as physical model with edges added according to case (1) greatly improves the system's robustness. Indeed, by adding an amount of edges according to case (1), the robustness of the Internet system becomes similar to the robustness shown by Internet based interdependent systems generated using CPA physical models, while having on average 3756 fewer edges.

As for case (2), we see in Fig. 4 that by adding an average of 635 edges at random the robustness of MRN-based Internet systems improves the robustness of the system even more so than 5NN-based systems. Thus, MRN-based systems are more robust than 5NN-based systems while having 1250 fewer edges on average. In particular the robustness shown by the system generated from case (2) edge addition presents a robustness similar to GG-based systems while having 697 fewer edges on average.

5 Discussion

From the results shown in Sect. 4.1 we observe that the relation between space shape in which the physical network is built in and the robustness of the Internet based interdependent system under random physical node failures greatly depends on the structure of the physical network, and the I_{max} value. Specifically, we observe that in all the cases tested a greater I_{max} value is positively correlated with the robustness.

In particular, we observe in Sect. 4.2 that the relation between the amount of edges of the physical network and the robustness of the system under physical node removal is non-trivial. Indeed, we observe that systems with similar robustness can have relatively large differences on their amount of edges. This suggests that the strategy used to generate the physical network, and thus the way in which we allocate the edges between nodes can have an enormous impact over the system's robustness under physical node failure. In addition, this suggests that it is possible to obtain Internet systems with a high robustness to physical attack with fewer physical edges. This is important given how costly it can be to create a new edge on the Internet physical network.

The results presented in Sect. 4.3 are especially interesting as they contrast with the results shown in Sect. 4.2. These results suggest that it is possible to greatly improve the robustness of a geometry-sensitive system such as the one generated using a MRN physical network by *randomly adding* edges. Improving to the point of becoming geometry-insensitive. Even more, this is achieved by adding relatively few edges (an average of 635 edges added). We see this on the Internet system generated by case (2) presented in Sect. 4.3 (see Fig. 4b). This system shows a robustness similar to ER and GG-based Internet systems while still having fewer edges than all the other physical models tested.

In particular, given that no specific strategy was followed when adding edges, our results suggest that with a smart edge allocation strategy we could get a similar robustness improvement, while adding even fewer edges to physical Internet network. Therefore suggesting that improving the robustness of the Internet based interdependent system against physical failures may be not that expensive.

6 Conclusion

Here, we have tested the Internet's robustness under physical node failures by measuring the amount of BGP or logical network nodes that are able to access the Internet after physical node failure. We showed that the geometry or shape in which the physical Internet network is built in may affect the Internet's robustness against physical node failure depending on factors such as the physical model, edge allocation strategy, and the amount of inter-connections between the physical and the logical layer.

We also showed that the relation between the robustness, edge allocation strategy, and edge density is non-trivial, thus this relation should be further studied. In particular we showed that geometry-sensitive Internet based interdependent systems can become geometry-insensitive, and that we can greatly improve the Internet's robustness by adding relatively few edges at random to the physical network. Thus, suggesting it is possible to improve the robustness of the Internet against physical failures with relatively few resources.

Here, we have shown that robustness improvement can be achieved by adding a small amount of edges without a particular strategy, that is, adding edges uniformly at random. The next steps would be to test the effect of non-random failures, as well as using smart edge addition strategies to maximize the robustness gain while minimizing the amount of edges added, hence minimizing the resource expenditure.

Acknowledgement This work was partially funded by CONICYT Doctorado Nacional 21170165.

References

1. Adler, C.O., Dagli, C.H.: Study of the use of a genetic algorithm to improve networked system-of-systems resilience. Proc. Comput. Sci. **36**, 49–56 (2014)
2. Autonomous system (1930). https://tools.ietf.org/html/rfc1930. Accessed 29 Sep 2019
3. Bachmann, I., Bustos-Jiménez, J.: Improving the Chilean Internet robustness: increase the interdependencies or change the shape of the country? In: International Conference on Complex Networks and Their Applications, pp. 646–657. Springer, Berlin (2017)
4. Berezin, Y., Bashan, A., Danziger, M.M., Li, D., Havlin, S.: Localized attacks on spatially embedded networks with dependencies. Sci. Rep. **5**, 8934 (2015)
5. Border gateway protocol (2006). https://tools.ietf.org/html/rfc4271. Accessed 29 Sep 2019
6. Buldyrev, S.V., Parshani, R., Paul, G., Stanley, H.E., Havlin, S.: Catastrophic cascade of failures in interdependent networks. Nature, **464**(7291), 1025–1028 (2010)
7. Cai, Y., Li, Y., Cao, Y., Li, W., Zeng, X.: Modeling and impact analysis of interdependent characteristics on cascading failures in smart grids. Int. J. Electr. Power Energy Syst. **89**, 106–114 (2017)
8. Carlson, J.M., Doyle, J.: Complexity and robustness. Proc. Natl. Acad. Sci. **99**(suppl. 1), 2538–2545 (2002)
9. Chattopadhyay, S., Dai, H.: Towards optimal link patterns for robustness of interdependent networks against cascading failures. In: 2015 IEEE Global Communications Conference (GLOBECOM), pp. 1–6. IEEE, Piscataway (2015)
10. Chen, Z., Wu, J., Xia, Y., Zhang, X.: Robustness of interdependent power grids and communication networks: a complex network perspective. IEEE Trans. Circ. Syst. Express Briefs **65**(1), 115–119 (2017)
11. Cowie, J.H., Ogielski, A.T., Premore, B., Smith, E.A., Underwood, T.: Impact of the 2003 blackouts on internet communications. Preliminary Report, Renesys Corporation (updated March 1, 2004) (2003)
12. Danziger, M.M., Bashan, A., Berezin, Y., Havlin, S. (2013). Interdependent spatially embedded networks: dynamics at percolation threshold. In: 2013 International Conference on Signal-Image Technology & Internet-Based Systems (SITIS), pp. 619–625. IEEE, Piscataway (2013)

13. Dong, G., Tian, L., Du, R., Fu, M., Stanley, H.E.: Analysis of percolation behaviors of clustered networks with partial support-dependence relations. Phys. A: Stat. Mech. Appl. **394**, 370–378 (2014)
14. Eppstein, D., Paterson, M.S., Yao, F.F.: On nearest-neighbor graphs. Discrete Comput. Geom.**17**(3), 263–282 (1997)
15. Erdős, P.,Rényi, A.: On the evolution of random graphs. Publ. Math. Inst. Hung. Acad. Sci **5**(1), 17–60 (1960)
16. Fabrikant, A., Koutsoupias, E., Papadimitriou, C.H.: Heuristically optimized trade-offs: a new paradigm for power laws in the internet. In: International Colloquium on Automata, Languages, and Programming, pp. 110–122. Springer, Berlin (2002)
17. Faloutsos, M., Faloutsos, P., Faloutsos, C.: On power-law relationships of the internet topology. In: ACM SIGCOMM Computer Communication Review, vol. 29, pp. 251–262. ACM, New York (1999)
18. Flaxman, A.D., Frieze, A.M., Vera, J.: A geometric preferential attachment model of networks. Internet Math. **3**(2), 187–205 (2006)
19. Gabriel, K.R., Sokal, R.R.: A new statistical approach to geographic variation analysis. Syst. Zool. **18**(3), 259–278 (1969)
20. Han, Y., Li, Z., Guo, C., Tang, Y.: Improved percolation theory incorporating power flow analysis to model cascading failures in cyber-physical power system. In: Power and Energy Society General Meeting (PESGM), 2016, pp. 1–5. IEEE, Piscataway (2016)
21. Huang, X., Gao, J., Buldyrev, S.V., Havlin, S., Stanley, H.E.: Robustness of interdependent networks under targeted attack. Phys. Rev. E **83**(6), 065101 (2011)
22. Huang, Z., Wang, C., Nayak, A., Stojmenovic, I.: Small cluster in cyber physical systems: Network topology, interdependence and cascading failures. IEEE Trans. Parallel Distrib. Syst. **26**(8), 2340–2351 (2014)
23. Kazawa, Y., Tsugawa, S.: On the effectiveness of link addition for improving robustness of multiplex networks against layer node-based attack. In: 2017 IEEE 41st Annual Computer Software and Applications Conference (COMPSAC), vol. 1, pp. 697–700. IEEE, Piscataway (2017)
24. Li, L., Alderson, D., Doyle, J.C., Willinger, W.: Towards a theory of scale-free graphs: definition, properties, and implications. Internet Math. **2**(4), 431–523 (2005)
25. Matsui, Y., Kojima, H., Tsuchiya, T.: Modeling the interaction of power line and scada networks. In: 2014 IEEE 15th International Symposium on High-Assurance Systems Engineering, pp. 261–262. IEEE, Piscataway (2014)
26. Nguyen, D.T., Shen, Y., Thai, M.T.: Detecting critical nodes in interdependent power networks for vulnerability assessment. IEEE Trans. Smart Grid **4**(1), 151–159 (2013)
27. Parandehgheibi, M., Modiano, E.: Robustness of interdependent networks: the case of communication networks and the power grid. In: 2013 IEEE Global Communications Conference (GLOBECOM), pp. 2164–2169. IEEE, Piscataway (2013)
28. Qiu, Y.: The effect of clustering-based and degree-based weighting on robustness in symmetrically coupled heterogeneous interdependent networks. In: 2013 IEEE International Conference on Systems, Man, and Cybernetics, pp. 3984–3988. IEEE, Piscataway (2013)
29. Radicchi, F.: Percolation in real interdependent networks. Nat. Phys. **11**(7), 597–602 (2015)
30. Ramiro, V., Piquer, J., Barros, T., Sepúlveda, P.: The Chilean Internet: did it survive the earthquake? WIT Trans. State-of-the-art in Sci. Eng. **58**, 133–151 (2012)
31. Reis, S.D., Hu, Y., Babino, A., Andrade Jr, J.S., Canals, S., Sigman, M., Makse, H.A.: Avoiding catastrophic failure in correlated networks of networks. Nat. Phys. **10**(10), 762–767 (2014)
32. Rosato, V., Issacharoff, L., Tiriticco, F., Meloni, S., Porcellinis, S., Setola, R.: Modelling interdependent infrastructures using interacting dynamical models. Int. J. Crit. Infrastruct. **4**(1–2), 63–79 (2008)
33. Schneider, C.M., Yazdani, N., Araújo, N.A., Havlin, S., Herrmann, H.J.: Towards designing robust coupled networks. Sci. Rep. **3**, 1969 (2013)
34. Wang, X., Kooij, R.E., Van Mieghem, P.: Modeling region-based interconnection for interdependent networks. Phys. Rev. E **94**(4), 042315 (2016)

35. Watanabe, S., Kabashima,Y.: Cavity-based robustness analysis of interdependent networks: influences of intranetwork and internetwork degree-degree correlations. Phys. Rev. E **89**(1), 012808 (2014)
36. Willinger, W., Roughan, M.: Internet topology research redux. ACM SIGCOMM eBook: Recent Advances in Networking. CiteSeer (2013)
37. Willinger, W., Alderson, D., Doyle, J.C.: Mathematics and the internet: a source of enormous confusion and great potential. Notices Am. Math. Soc. **56**(5), 586–599 (2009)
38. Yao, A.C.C.: On constructing minimum spanning trees in k-dimensional spaces and related problems. SIAM J. Comp. **11**(4), 721–736 (1982)
39. Zhou, D., Stanley, H.E., D'Agostino, G., Scala, A.: Assortativity decreases the robustness of interdependent networks. Phys. Rev. E **86**(6), 066103 (2012)

Part IV
Applications

The Power of Communities: A Text Classification Model with Automated Labeling Process Using Network Community Detection

Minjun Kim and Hiroki Sayama

Abstract Text classification is one of the most critical areas in machine learning and artificial intelligence research. It has been actively adopted in many business applications such as conversational intelligence systems, news articles categorizations, sentiment analysis, emotion detection systems, and many other recommendation systems in our daily life. One of the problems in supervised text classification models is that the models' performance depends heavily on the quality of data labeling that is typically done by humans. In this study, we propose a new network community detection-based approach to automatically label and classify text data into multiclass value spaces. Specifically, we build networks with sentences as the network nodes and pairwise cosine similarities between the Term Frequency-Inversed Document Frequency (TFIDF) vector representations of the sentences as the network link weights. We use the Louvain method to detect the communities in the sentence networks. We train and test the Support Vector Machine and the Random Forest models on both the human-labeled data and network community detection labeled data. Results showed that models with the data labeled by the network community detection outperformed the models with the human-labeled data by 2.68–3.75% of classification accuracy. Our method may help developments of more accurate conversational intelligence and other text classification systems.

M. Kim (✉)
Department of Systems Science and Industrial Engineering, Binghamton, NY, USA

Center for Collective Dynamics of Complex Systems, Binghamton University, State University of New York, Binghamton, NY, USA

Pypestream, Inc., New York, NY, USA
e-mail: mkim151@binghamton.edu

H. Sayama
Department of Systems Science and Industrial Engineering, Binghamton, NY, USA

Center for Collective Dynamics of Complex Systems, Binghamton University, State University of New York, Binghamton, NY, USA

Waseda Innovation Lab, Waseda University, Tokyo, Japan
e-mail: sayama@binghamton.edu

N. Masuda et al. (eds.), *Proceedings of NetSci-X 2020: Sixth International Winter School and Conference on Network Science*, Springer Proceedings in Complexity,
https://doi.org/10.1007/978-3-030-38965-9_16

1 Introduction

Text data is a great source of knowledge for building many useful recommendation systems, search engines as well as conversational intelligence systems. However, it is often found to be a difficult and time-consuming task to structure the unstructured text data especially when it comes to labeling the text data for training text classification models. Data labeling, typically done by humans, is prone to make mislabeled data entries, and hard to track whether the data is correctly labeled or not. This human labeling practice indeed impacts on the quality of the trained models in solving classification problems.

Some previous studies attempted to solve this problem by utilizing unsupervised [1–3] and semisupervised [4] machine learning models. However, those studies used a pre-defined keyword list for each category in the document, which provides the models with extra referential materials to look at when making the classification predictions, or included already labeled data as a part of the entire data set from which the models learn. In case of using unsupervised algorithms such as K−means and LDA [2, 3], it is very much possible that frequently appearing words in multiple sentences can be used as features for multiple classes. This leads the models to render more ambiguity and to result in a poor performance in classifying documents. Also, the number of distinct classes (K) to be made is not determined systematically using the data, but heuristically by trying out many different values of K which is not a reliable optimization.

Although there are many studies in text classification problems using machine learning techniques, there have been a limited number of studies conducted in text classifications utilizing network science. Network science is actively being adopted in studying biological networks, social networks, financial market prediction [5] and more in many fields of study to mine insights from the collectively inter-connected components by analyzing their relationships and structural characteristics. Only a few studies adopted network science theories to study text classifications and showed preliminary results of the text clustering performed by network analysis specially with the network community detection algorithms [6–8]. However, those studies did not clearly show the quality of community detection algorithms or other possible useful features. Network community detection [9] is graph clustering methods used in complex networks analysis from large social networks analysis [10] to RNA-sequencing analysis [11] as a tool to partition a graph data into multiple parts based on the network's structural properties such as betweenness, modularity, etc.

In this paper, we study further to show the usefulness of the network community detection on labeling unlabeled text data that will automate and improve human labeling tasks, and on training machine learning classification models for a particular text classification problem. We finally show that the machine learning models trained on the data labeled by the network community detection model outperform the models trained on the human-labeled data.

2 Method

We propose a new approach of building text classification models using a network community detection algorithm with unlabeled text data, and show that the network community detection is indeed useful in labeling text data by clustering the text data into multiple distinctive groups and also in improving the classification accuracy. This study takes the following steps (see Fig. 1), and uses Python packages such as `NLTK`, `NetworkX` and `SKlearn`.

- Gathered a set of text data that was used to develop a particular conversational intelligence system from an artificial intelligence company, Pypestream. The data contains over 2000 sentences of user expressions on that particular chatbot service such as ["is there any parking space?", "what movies are playing?", "how can I get there if I'm taking a subway?"]
- Tokenizing and cleaning the sentences by removing punctuations, special characters and English stopwords that appear frequently without holding much important meaning. For example, ["how can I get there if I'm taking a subway?"] becomes ['get', 'taking', 'subway']
- Stemmizing the words following a suffix stripping algorithm [12], and adding synonyms and bigrams of the sequence of the words left in each sentence to enable the model to learn more kinds of similar expressions and the sequences of the words. For example, ['get', 'taking', 'subway'] becomes ['get', 'take', 'subway', 'tube', 'underground', 'metro', 'take metro', 'get take', 'take subway', 'take underground', ...]
- Transforming the preprocessed text data into a vector form by computing TFIDF of each preprocessed sentence with regard to the entire data set, and computing pair-wise cosine similiarity of the TFIDF vectors to form the adjacency matrix of the sentence network
- Constructing the sentence network using the adjacency matrix with each preprocessed sentence as a network node and the cosine similarity of TFIDF representations between every node pair as the link weight.
- Applying a network community detection algorithm on the sentence network to detect the communities where each preprocessed sentence belong, and build a labeled data set with detected communities for training and testing machine learning classification models.

2.1 *Data, Preprocessing and Representation*

The data set obtained from Pypestream is permitted to be used for the research purpose only, and for a security reason, we are not allowed to share the data set. It was once originally used for creating a conversational intelligence system (chatbot) to support customer inquiries about a particular service. The data set is a two-column

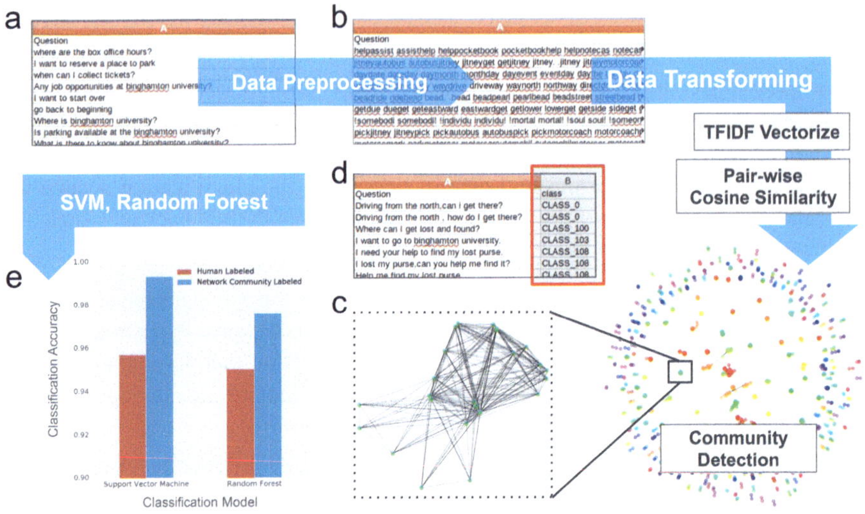

Fig. 1 Analysis process. (**a**) Preprocess the text data by removing punctuations, stopwords and special characters, and add synonyms and bigrams, (**b**) transform the prepocessed sentence into TFIDF vector, and compute pair-wise cosine similairy between every sentence pair, (**c**) construct the sentence networks, and apply the Louvain method to detect communities of every sentence, (**d**) label each sentence with the detected communities, (**e**) train and test the support vector machine and the random forest models on the labeled data

comma-separated value format data with one column of "sentence" and the other column of "class". It contains 2212 unique sentences of user expressions asking questions and answering to the questions the chatbot asked the users. The sentences are all in English without having any misspelled words, and labeled with 19 distinct classes that are identified and designed by humans. Additional data set that only contains the sentences was made for this study by taking out the "class" column from the original data set.

From each sentence, we removed punctuations, special characters and English stopwords to keep only those meaningful words that serve the main purpose of the sentence and to avoid any redundant computing. We then tokenized each sentence into words to process the data further in word level. For words in each sentence, we added synonyms of the words to handle more variations of the sentence as a typical method of increasing the resulting classification models' capability of understanding more unseen expressions with different words that describe similar meanings. The synonyms we added to the data are not context-specific synonyms, but all predefined synonyms of particular words. Although we used the predefined synonyms from the Python NLTK package, one might develop it's own synonym data to use under the context of the particular data to achieve better accuracy. We also added bigrams of the words to deal with those cases where the tokenization breaks the meaning of the word that consists of two words. For example, if we tokenized the sentence "go to Binghamton University" and process the further steps without adding bigrams of them, the model is likely to yield lower confidence on classifying unseen sentences with "Binghamton University" since the meaning of "Binghamton University" is lost in the data set [13].

With the preprocessed text data, we built vector representations of the sentences by performing weighted document representation using the TFIDF weighting scheme [14–16]. The TFIDF, as known as Term frequency inversed document frequency, is a document representation that takes account of the importance of each word by its frequency in the whole set of documents and its frequency in particular sets of documents. Specifically, let $D = \{d_1, \ldots, d_n\}$ be a set of documents and $T = \{t_1, \ldots, t_m\}$ the set of unique terms in the entire documents where n is the number of documents in the data set and m the number of unique words in the documents. In this study, the documents are the preprocessed sentences and the terms are the unique words in the preprocessed sentences. The importance of a word is captured with its frequency as $tf(d, t)$ denoting the frequency of the word $t \in T$ in the document $d \in D$. Then a document d is represented as an m-dimensional vector $\vec{t_d} = (tf(d, t_1), \ldots, tf(d, t_m))$. However, In order to compute more concise and meaningful importance of a word, TFIDF not only takes the frequency of a particular word in a particular document into account, but also considers the number of documents that the word appears in the entire data set. The underlying thought of this is that a word appeared frequently in some groups of documents but rarely in the other documents is more important and relevant to the groups of documents. Applying this concept, $tf(d, t)$ is weighted by the document frequency of a word, and $tf(d, t)$ becomes $tfidf(d, t) = tf(d, t) \times log\frac{|D|}{df(t)}$ where $df(t)$ is the number of documents the word t appears, and thus the document d is represented as $\vec{t_d} = (tfidf(d, t_1), \ldots, tfidf(d, t_m))$.

2.2 Sentence Network Construction

With the TFIDF vector representations, we formed sentence networks. In total, 10 sentence networks (see Figs. 2 and 5) were constructed with 2212 nodes representing sentences and edge weights representing the pairwise similarities between sentences with 10 different network connectivity threshold values. The networks we formed were all undirected and weighted graphs. Particularly, as for the network edge weights, the cosine similarity [16, 17] is used to compute the similarities between sentences. The cosine similarity is a similarity measure that is in a floating number between 0 and 1, and computed as the angle difference between two vectors. A cosine similarity of 0 means that the two vectors are perpendicular to each other implying no similarity, on the other hand, a cosine similarity of 1 means that the two vectors are identical. It is popularly used in text mining and information retrieval techniques. In our study, the cosine similarity between two sentences i and j is defined as below equation.

$$SIM_C(\vec{t_{d_i}}, \vec{t_{d_j}}) = \frac{\vec{t_{d_i}} \cdot \vec{t_{d_j}}}{|\vec{t_{d_i}}||\vec{t_{d_j}}|} \tag{1}$$

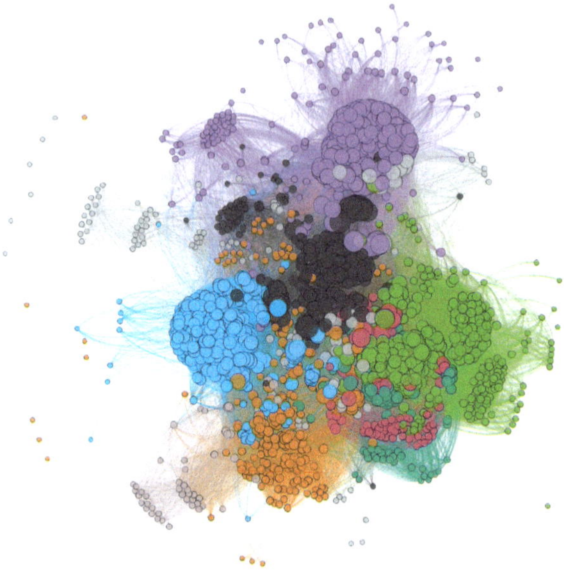

Fig. 2 A sentence network and its communities. The sentence network with no threshold on the node connectivity has 18 distinct communities including three single node communities

where:

$\overrightarrow{t_{d_i}} = (tfidf(d_i, t_1), \ldots, tfidf(d_i, t_m))$, *the TFIDF vector of i-th sentence*

$\overrightarrow{t_{d_j}} = (tfidf(d_j, t_1), \ldots, tfidf(d_j, t_m))$, *the TFIDF vector of j-th sentence*

$d =$ *a preprocessed sentence in the data set*

$t =$ *a unique word appeared in the preprocessed data set*

To build our sentence networks, we formed a network adjacency matrix for 2212 sentences, M, with the pairwise cosine similarities of TFIDF vector representations computed in the above step.

$$M = \begin{array}{c} \\ d_1 \\ d_2 \\ \vdots \\ d_{2212} \end{array} \begin{array}{c} \begin{array}{cccc} d_1 & d_2 & \cdots & d_{2212} \end{array} \\ \begin{pmatrix} SIM_C(\overrightarrow{t_{d_1}}, \overrightarrow{t_{d_1}}) & SIM_C(\overrightarrow{t_{d_1}}, \overrightarrow{t_{d_2}}) & \ldots & SIM_C(\overrightarrow{t_{d_1}}, \overrightarrow{t_{d_{2212}}}) \\ SIM_C(\overrightarrow{t_{d_2}}, \overrightarrow{t_{d_1}}) & SIM_C(\overrightarrow{t_{d_2}}, \overrightarrow{t_{d_2}}) & \ldots & SIM_C(\overrightarrow{t_{d_2}}, \overrightarrow{t_{d_{2212}}}) \\ \vdots & \vdots & \ddots & \vdots \\ SIM_C(\overrightarrow{t_{d_{2212}}}, \overrightarrow{t_{d_1}}) & SIM_C(\overrightarrow{t_{d_{2212}}}, \overrightarrow{t_{d_2}}) & \ldots & SIM_C(\overrightarrow{t_{d_{2212}}}, \overrightarrow{t_{d_{2212}}}) \end{pmatrix} \end{array}$$

2.3 Network Community Detection and Classification Models

The particular algorithm of network community detection used in this study is the Louvain method [18] which partitions a network into the number of nodes—every node is its own community, and from there, clusters the nodes in a way to maximize each cluster's modularity which indicates how strong is the connectivity between the nodes in the community. This means that, based on the cosine similarity scores—the networks edge weights, the algorithm clusters similar sentences together in the same community while the algorithm proceeds maximizing the connectivity strength amongst the nodes in each community. The network constructed with no threshold in place was detected to have 18 distinct communities with three single node communities. Based on the visualized network (see Fig. 2), it seemed that the network community detection method clustered the sentence network as good as the original data set with human-labeled classes although the communities do not look quite distinct. However, based on the fact that it had three single node communities and the number of distinct communities is less than the number of classes in the human-labeled data set, we suspected possible problems that would degrade the quality of the community detection for the use of training text classification models.

Quality of Network Community Detection Based Labeling We checked the community detection results with the original human-labeled data by comparing the sentences in each community with the sentences in each human-labeled class to confirm how well the algorithm worked. We built class maps to facilitate this process (see Fig. 3) that show mapping between communities in the sentence networks and classes in the original data set. Using the class maps, we found two notable cases where; (1) the sentences from multiple communities are consist of the sentences of one class of the human-labeled data, meaning the original class is splitted into multiple communities and (2) the sentences from one community consist of the sentences of multiple classes in human-labeled data, meaning multiple classes in the original data are merged into one community. For example, in the earlier case (see blue lines in Fig. 3) which we call Class-split, the sentences in `COMMUNITY_1`, `COMMUNITY_2`, `COMMUNITY_5`, `COMMUNITY_8`, `COMMUNITY_10`, `COMMUNITY_14` and `COMMUNITY_17` are the same as the sentences in `CHAT_AGENT` class. Also, in the later case (see red lines in Fig. 3) which we call Class-merge, the sentences in `COMMUNITY_7` are the same as the sentences in `GETINFO_PARKING`, `GETINFO_NEARBY_RESTAURANT`, `GETINFO_TOUR`, `GETINFO_EXACT_ADDRESS`, `STARTOVER`, `ORDER_EVENTS`, `GETINFO_JOB`, `GETINFO`, `GETINFO_DRESSCODE`, `GETINFO_LOST_FOUND` as well as `GETINFO_FREE_PERFORMANCE`.

The Class-split happens when a human-labeled class is divided into multiple communities as the sentence network is clustered based on the semantic similarity. This actually can help improve the text classification based systems to perform more sophisticatedly as the data set has more detailed subclasses to structure the systems with. Although it is indeed a helpful phenomenon, we would like to minimize the number of subclasses created by the community detection algorithm simply because we want to avoid having too many subclasses that would add more complexity in designing any applications using the community data. On the other hand, the Class-merge happens when multiple human-labeled classes are merged into one giant community. This Class-merge phenomenon also helps improve the original data set by detecting either mislabeled or ambiguous data entries. We will discuss more details in the following subsection. Nonetheless, we also want to minimize the number of classes merged into the one giant community, because

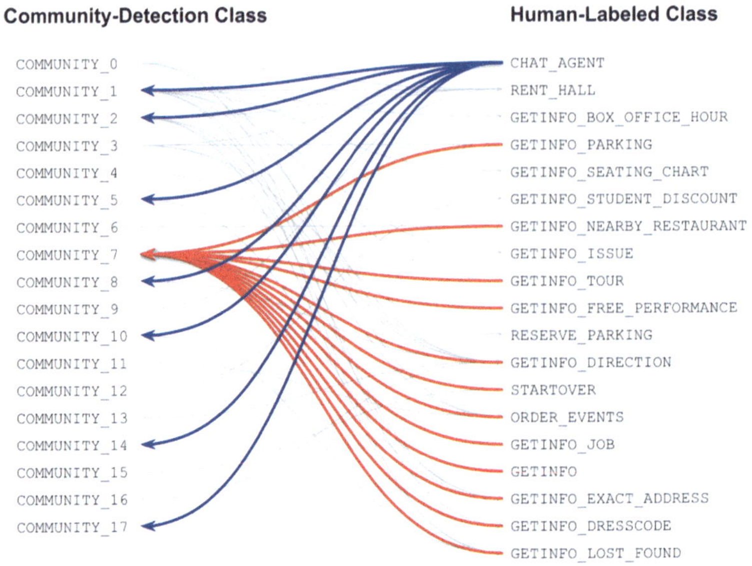

Fig. 3 A class map between detected communities and human-labeled classes. The class map shows a mapping (all lines) between communities detected by the Louvain method and their corresponding human-labeled classes of the sentence network with no threshold

when too many classes are merged into one class, it simply implies that the sentence network is not correctly clustered. For example, as shown in Fig. 3 red lines, 12 different human-labeled classes that do not share any similar intents are merged into COMMUNITY_7. If we trained a text classification model on this data, we would have lost the specifically designed purposes of the 12 different classes, expecting COMMUNITY_7 to deal with all the 12 different types of sentences. This would dramatically degrade the performance of the text classification models.

In order to quantify the degree of Class-split and Class-merge of a network, and to find out optimal connectivity threshold that would yield the sentence network with the best community detection quality, we built two metrics using the class map. We quantified the Class-split by counting the number of communities splitted out from each and every human-labeled class, and the Class-merge by counting the number of human-labeled classes that are merged into each and every community. We then averaged the Class-splits across all the human-labeled classes and Class-merges across all the communities. For example, using the class map of the sentence network with no threshold, we can easily get the number of Class-split and Class-merge as below. By averaging them, we get the Class_split and Class_merge scores of the sentence network, which is 2.7368 and 2.8333 respectively.

$$Class\_split = [2, 1, 4, 5, 1, 2, 2, 1, 1, 4, 1, 9, 1, 1, 4, 2, 2, 2, 7]$$

$$Class\_merge = [1, 1, 1, 1, 4, 1, 2, 1, 2, 4, 1, 9, 2, 1, 1, 1, 6, 12]$$

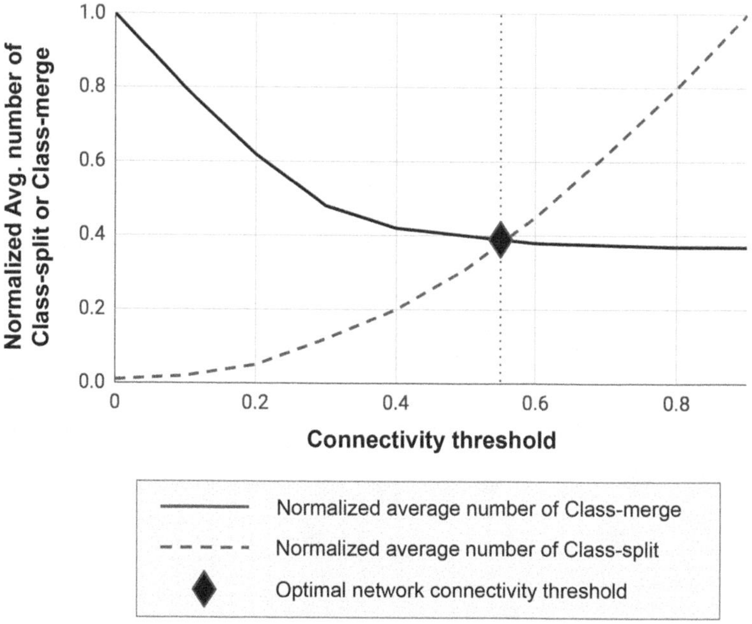

Fig. 4 Optimal connectivity threshold point based on Class-split and Class-merge metrics. The normalized Class-split score (blue line) increases as the threshold gets larger. On the other hand, normalized Class-merge (red line) decreases as the threshold gets larger. The optimal connectivity threshold is the point where both scores are minimized which is 0.5477

We computed the normalized Class_split and Class_merge scores for all 10 sentence networks (see Fig. 4). Figure 4 shows the normalized Class-split and Class-merge scores of the 10 sentence networks with different connectivity thresholds ranging from 0.0 to 0.9 (shown in Fig. 5). With these series of Class_split and Class_merge scores, we found out that at 0.5477 of connectivity threshold we can get the sentence network that would give us the best quality of community detection result particularly for our purpose of training text classification models.

Detecting Mislabeled or Ambiguous Sentences in Human-Made Data Set Using the Class_merge information we got from the class map, we were able to spot out those sentences that are either mislabeled or ambiguous between classes in the original data set. This is extreamly helpful and convenient feature in fixing and improving text data for classification problems, because data fixing is normally a tedious and time consuming task which takes a great amount of human labor. For example, by looking at the class map, in our sentence network with no threshold, `COMMUNITY_5` contains sentences appeared in `GETINFO_EXACT_ADDRESS` and `CHAT_AGENT` classes. We investigated the sentences in `COMMUNITY_5`, and were able to spot out one sentence ["I need to address a human being!"] which is very ambiguous for machines to classify between the two classes. This sentence is originally designed for `CHAT_AGENT` class, but because of its

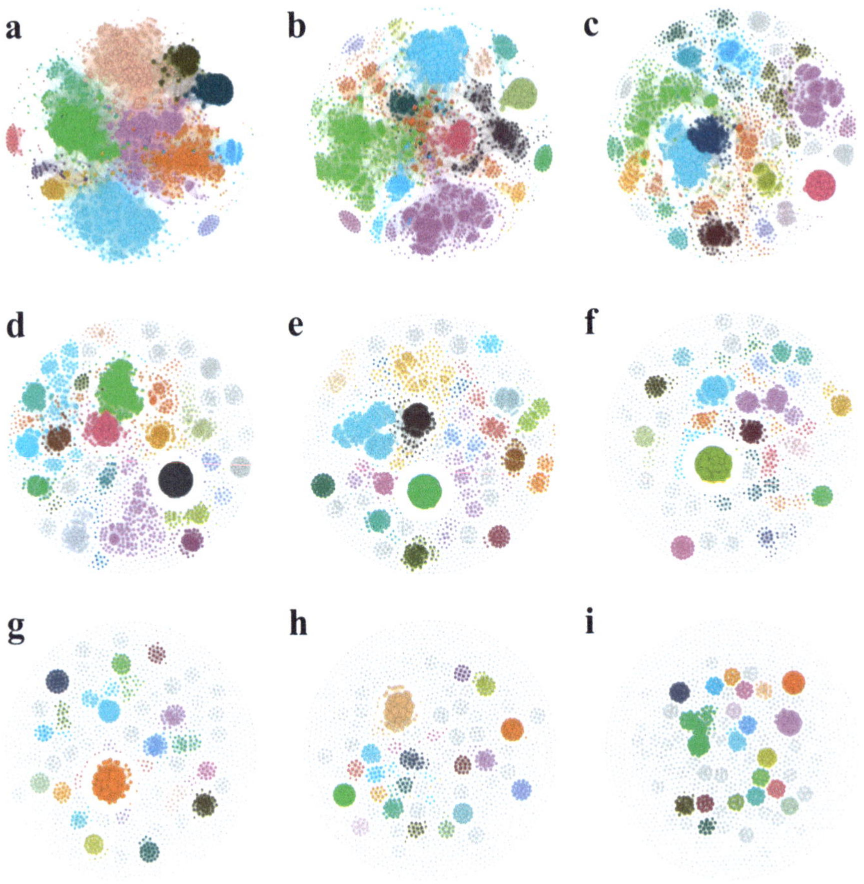

Fig. 5 Nine sentence networks with different connectivity thresholds. Each node represents a sentence and an edge weight between two nodes represents the similarity between two sentences. In this study, we removed edges whose weight is below the threshold. (**a**) Network with threshold of 0.1 has 29 distinct communities with 11 single node communities. (**b**) Network with threshold of 0.2 has 45 distinct communities with 20 single node communities, (**c**) network with threshold of 0.3 has 100 distinct communities with 58 single node communities, (**d**) network with threshold 0.4 has 187 distinct communities with 120 single node communities, (**e**) network with threshold 0.5 has 320 distinct communities with 204 single node communities, (**f**) network with threshold of 0.6 has 500 distinct communities with 335 single node communities, (**g**) network with threshold of 0.7 has 719 distinct communities with 499 single node communities, (**h**) network with threshold of 0.8 has 915 distinct communities with 658 single node communities, (**i**) network with threshold of 0.9 has 1140 distinct communities with 839 single node communities. Based on the visualized sentence networks, as the threshold gets larger it is shown that each network has more distinct communities

ambiguous expression with the word 'address', it is together with sentences in GETINFO_EXACT_ADDRESS in COMMUNITY_5. After fixing the ambiguity of that sentence by correcting it to ["I need to talk to a human being!"], we easily improved the original data set.

Classification Models Once we got the optimal connectivity threshold using the Class_split and Class_merge scores as shown in previous sections, we built the sentence network with the optimal threshold of 0.5477. We then applied the Louvain method to detect communities in the network, and to automatically label the data set. The network with the threshold of 0.5477 has 399 communities with 20,856 edges. Class_split and Class_merge scores of the network were 22.3158 and 1.0627 respectively. We finally trained and tested machine learning based text classification models on the data set labeled by the community detection outcome to see how well our approach worked. Following a general machine learning train and test practice, we split the data set into a train set (80% of the data) and a test set (20% of the data). The particular models we trained and tested were the Support Vector Machine [19] and the Random Forest [20] models that are popularly used in natural language processing such as spam e-mail and news article categorizations. More details about the two famous machine learning models are well discussed in the cited papers.

3 Result

Table 1 shows the accuracy of the four Support Vector Machine and the Random Forest models trained on the original human-labeled data and the data labeled by our method. The accuracies are hit ratios that compute the number of correctly classified sentences over the number of all sentences in the test data. For example, if a model classified 85 sentences correctly out of 100 test sentences, then the accuracy is 0.85. In order to accurately compute the ground truth hit ratio, we used the ground truth messages in the chatbot. The messages are the sentences that are to be shown to the chatbot users in response to the classification for a particular user query as below.

$$input\ sentence \rightarrow detected\ class \rightarrow output\ message$$

For example, for a question of "how do I get there by subway?", in the chatbot, there is a designed message of "You can take line M or B to 35th street" to respond to that particular query. Using these output

Table 1 Accuracies of text classification models

Data labeling	SVM	Random forest
Human labeled	0.9572	0.9504
Network community labeled	0.9931	0.9759

It is shown that the models trained on the community data resulted in higher accuracy in classifying the sentences in the test data messages in the chatbot, we were able to compute the ground truth accuracy of our classification models by comprehending the input sentences in the test sets, the detected classes from the models and linked messages. In our test, the Support Vector Machine trained on human-labeled data performed 0.9572 while the same model trained on the data labeled by our method resulted in 0.9931. Also, the Random Forest model trained on human-labeled data resulted in an accuracy value of 0.9504 while the same model trained on the data labeled by our method did 0.9759.

4 Discussions and Conclusion

In this study, we demonstrated a new approach of training text classification models using the network community detection, and showed how the network community detection can help improve the models by automatically labeling text data and detecting mislabeled or ambiguous data points. As seen in this paper, we were able to yield better results in the accuracy of the Support Vector Machine and the Random Forest models compared to the same models that were trained on the original human-labeled data for the particular text classification problem. Our approach is not only useful in producing better classification models, but also in testing the quality of human-made text data. One might be able to get even better results using this method by utilizing more sophisticatedly custom-designed synonyms and stopwords, using more advanced natural language processing methods such as word-embeddings, utilizing higher n-grams such as trigrams, and using more balanced data sets. In the future, we would like to expand this study further to use the network itself to parse out classifications of unseen sentences without training machine learning models.

References

1. Ko, Y., Seo, J.: Automatic text categorization by unsupervised learning. In: Proceedings of the 18th Conference on Computational linguistics, vol. 1, pp. 453–459 (2000)
2. Zhou, X., Hu, Y., Guo, L.: Text categorization based on clustering feature selection. Proc. Comput. Sci. **31**, 398–405 (2014)
3. Lancichinetti, A., Sirer, M.I., Wang, J.X., Acuna, D., Körding, K., Amaral, L.A.N.: High-reproducibility and high-accuracy method for automated topic classification. Phys. Rev. X **5**, 011007 (2015)
4. Dorado, R., Sylvie, R.: Semisupervised text classification using unsupervised topic information. In: Proceedings of the Twenty-Ninth International Florida Artificial Intelligence Research Society Conference (FLAIRS), pp. 210–213 (2016)
5. Kim, M., Sayama, H.: Predicting stock market movements using network science: an information theoretic approach. Appl. Netw. Sci. **2**, 35 (2017)
6. dos Santos, C.K., Evsukoff, A.G., de Lima, B.: Cluster analysis in document networks. WIT Trans. Inf. Commun. Technol. **40**, 95–104 (2008)

7. Mikhina, E.K., Trifalenkov, V.I.: Text clustering as graph community detection. Proc. Comput. Sci. **123**, 271–277 (2018)
8. Martin, G., Tiago, P., Eduardo, G.A.: A network approach to topic models. Sci. Adv. **4**, eaaq. 1360 (2018)
9. Fortunato, S., Hric, D.: Community detection in networks: a user guide. Phys. Rep. **659**, 1–44 (2016)
10. Steinhaeuser, K., Chawla, N.V.: Community detection in a large real-world social network. In: Social Computing, Behavioral Modeling, and Prediction, pp. 168–175. Springer, Boston (2008)
11. Kanter, I., Yarri, G., Kalisky, T.: Applications of community detection algorithms to large biological datasets (2019). bioRxiv 547570
12. Porter, M.: An algorithm for suffix stripping. In: Program: Electronic Library and Information Systems , vol. 40, pp. 211–218 (2006)
13. Bekkerman, R., Allan, J.: Using bigrams in text categorization.Technical Report IR-408, Center of Intelligence Information Retrieval, pp. 161–175. UMass Amberst (2004)
14. Karen, J.: A statistical interpretation of term specificity and its application in retrieval. J. Doc. **28**, 11–21 (1972)
15. Trstenjak, B., Mikac, S., Donko, D.: KNN with TF-IDF based framework for text categorization. Proc. Eng. **69**, 1356–1364 (2014)
16. Huang, A.: Similarity measures for text document clustering. In: Proceedings of the Sixth New Zealand Computer Science Research Student Conference (NZCSRSC2008), vol. 4, pp. 9–56 (2008)
17. Li, B., Han, L.: Distance weighted cosine similarity measure for text classification. In: International Conference on Intelligent Data Engineering and Automated Learning, pp. 611–618. Springer, Berlin (2013)
18. Blondel, V.D., Guillaume, J.L., Lambiotte, R., Lefebvre, E.: Fast unfolding of communities in large networks. J. Stat. Mech.: Theory Exper. **2008**, P10008 (2008)
19. Drucker, H., Wu, D., Vapnik, V.N.: Support vector machines for spam categorization. IEEE Trans. Neural Netw. **10**, 1048–1054 (1999)
20. Wu, Q., Ye, Y., Zhang, H.,Ng, M.K., Ho, S.S.: ForesTexter: an efficient random forest algorithm for imbalance text categorization. Knowledge-Based Syst. **67**, 105–116 (2014)

Effective Implementation of Energy Aware Polarization Diversity for IoT Networks Using Eigenvector Centrality

Sakil Chowdhury, Laurent Hébert-Dufresne, and Jeff Frolik

Abstract The Internet of Things (IoT) is one the most promising area of applications for complex networks since we know that both the efficiency and fidelity of information transmission rely critically on our understanding of network structure. While antenna diversity schemes improve reliability and capacity for point-to-point links of an IoT network that employs multi-polarized antennas, it is currently unclear how implementation should depend on the network structure of the IoT and what impact structure-dependent implementations will have on the energy consumption of IoT devices. We propose an antenna diversity scheme that leverages local network structure and a distributed calculation of centrality to reduce power consumption by 13% when compared to standard selection diversity technique. The proposed approach exploits distributed eigenvector centrality to identify the most influential nodes based on data flow and then limits their antenna switching frequency proportionally to their centrality. Our results also demonstrate that by taking routers' centrality metric into account, a network can reduce antenna switching frequency by 17% while ensuring approximately 99% packet delivery rate. More broadly, this study highlights how network science can contribute to the development of efficient IoT devices.

1 Introduction

The Internet of Things (IoT) interconnects heterogeneous entities like sensors, actuators, wearable items and phones to develop an integrated system where these multipurpose devices can monitor their surrounding environment, react to a certain event, collect sensory data and forward the data in multi-hop fashion to back-end systems for further processing [1]. The applications of IoT span from small scale implementation such as patient monitoring, smart homes, to large scale

S. Chowdhury (✉) · L. Hébert-Dufresne · J. Frolik
University of Vermont, Burlington, VT, USA
e-mail: sakil.chowdhury@uvm.edu

N. Masuda et al. (eds.), *Proceedings of NetSci-X 2020: Sixth International Winter School and Conference on Network Science*, Springer Proceedings in Complexity,
https://doi.org/10.1007/978-3-030-38965-9_17

implementations of industrial monitoring, smart farming, smart cities, etc. [2, 3]. In many of these potential applications, IoT devices are deployed in environments which are not ideal for wireless communication. Environments such as industrial facilities are particularly harsh where reflection, diffraction and scattering from metal structures cause distortion to the radio signal, known as multipath fading [4]. Signal attenuation, phase shifting and inter-symbol interference caused by multipath fading significantly degrade reliability and throughput of the network.

Multi-polarized antennas are an effective solution to overcome multipath effects as they allow the receiver to have multiple copies of the transmitted signal by using orthogonally polarized antenna elements [5]. However, the problem is then to choose which antenna polarization should be used given local conditions, including network structure. Selection diversity is the simplest diversity technique used in conjunction with multi-element antennas in which the antenna polarization having the highest signal strength is chosen for transmission or reception. Due to cost constraints and limited processing capabilities of IoT devices, selection diversity uses a single radio-frequency (RF) chain and switches between polarizations to determine the 'best' polarization using a RF switch. Existing works in the literature related to selection diversity schemes primarily focus on ensuring link reliability, minimizing low bit error rate (BER) and attaining high signal-to-noise (SNR) ratio. For example, the authors of Ref. [6] developed an algorithm with quartic complexity to select optimal subset of antennas that ensures maximum SNR for systems with many transmit antennas. In Ref. [7], a low-complexity generalized selection combining (GSC) scheme is introduced, which is able to match the performance of a full diversity system in terms of outage probability and symbol error rate while utilizing only a subset of the available antennas to transmit and receive. In Ref. [8], capacity maximizing suboptimal antenna selection algorithm for medium to high SNRs is proposed to determine the transmit antenna in a Rayleigh fading environment. However, all the mentioned works required either multiple RF chains to be active simultaneously or the device to solve complex optimization problems, which is not suitable for low-cost, constrained IoT devices. Moreover, the network structures of IoT systems are often complex and hierarchical, suggesting that diversity technique might be an interesting avenue of research.

In multi-hop communication based routing, router nodes that are near the base station relay the data collected by the sensor nodes that are further away from the base. Thus, in this network, the closer a router is to the base, the higher its data traffic load will be, resulting in frequent use of selection diversity to select antenna polarization. This will cause faster depletion of energy of the routers with high data traffic compared to the routers with less traffic, i.e., far away routers. Intuitively, in an IoT network operating in multipath environment, the time between consecutive data transmission by a sensor node can be large compared to the coherence time (time over which the channel changes significantly) of the channel and thus, each transmission sees independent fading. On the contrary, as routers manage packets from multiple sensor nodes, the coherence time for routers is large relative to the time between consecutive transmission/reception which implies that the fading seen by packets are correlated. For example, IoT networks aimed at wildfire detection, forest environment and agriculture monitoring require geographically dispersed sensor nodes to transmit sensed information periodically at a low data rate. The base station can provide valuable forecast, improve safety and efficiency by integrating the sensed data that is relayed through routers [9, 10]. This motivates us to consider controlling the use selection diversity according to nodes' data traffic load as approximated by their position in the network structure. Indeed, an IoT network can be effectively represented as a complex network [11], a graph object whose vertices correspond to sensor or router nodes while edges stand for data transmission between nodes. More specifically, we consider the problem of finding routers with high data traffic in an IoT network as a problem of finding the crucial nodes in a complex network. Then, we can leverage centrality metrics [12], which rank the nodes of a network based on their importance in a network, to identify highly congested routers. Our focus in this paper is to apply ideas from complex network science in order to implement a device-specific diversity scheme that considers nonuniform depletion of energy of routers in an IoT network.

By combining complex network theory and the concept of antenna diversity, we propose a network-wide diversity technique, where devices will use selection diversity in a periodic manner instead of using it before every transmission or reception and the period will be proportional to their centrality. In summary, the main contributions of this paper are as follows.

1. We employ the concept of eigenvector centrality to determine crucial nodes in an IoT network consisting of a large number of stationary nodes from the view point of data packet transmission and reception. The centrality is calculated by autonomous sensor and router nodes in a distributed manner which reduces computation complexity and ensures low-memory usage for low-resource, energy-constrained IoT devices compared to centralized computation.
2. In contrast to the conventional selection diversity technique that allows all devices to switch antenna element before every transmission or reception, our proposed energy-aware diversity scheme controls the switching of devices such that low-scoring routers are allowed to switch antenna more frequently compared to the high-scoring ones and hence, reduces excessive switching and is able to minimize antenna switching by at least 17%.
3. We demonstrate through simulation that the reduction of excessive antenna switching achieved by our Distributed Eigenvector Centrality (DEC) diversity approach decreases energy consumption of routers by at least 13% compared to simple network-wide selection diversity approach, without degrading network reliability.

The paper is organized as follows: Sect. 2 reviews related works. In Sect. 3, we give an overview of the type of target IoT networks and deployment environment considered. Section 4 introduces a distributed calculation of eigenvector centrality and proposes an implementation for IoT network in which an individual antenna switching rate is controlled based on its centrality in the network structure. Section 5 describes the comparison between our proposed centrality based diversity scheme and simple selection diversity scheme and Sect. 6 concludes the paper.

2 Related Works

A network consists of a set of nodes connected by edges which can be directed or undirected, weighted or unweighted. Centrality is often used in complex network systems to identify the relative influence of a node or edge with respect to the entire network. Various centrality measures such as betweenness, closeness and eigenvector centrality have been studied in the literature based on application context and different characteristics of a network. Betweenness centrality determines the amount of influence a node has over the information flow of a network. It first calculates the shortest path between every pair of nodes in a network and assigns a centrality to nodes based on how frequently they lie along shortest paths [13]. Closeness centrality is defined as the inverse of the average distance between a given node and all other nodes in the network [14] such that high closeness centrality indicates central nodes that have shorter distances to other nodes. However, most centrality measures are calculated based on global topology information which is prohibitive for memory-constrained, low-cost devices of an IoT network with a large number of nodes. Another popular measure is eigenvector centrality, which calculates a node's importance in a network by summing the importance of its neighbors [12]. Eigenvector centrality is defined based on the eigenvector of the network adjacency matrix such that the centrality $\mathbf{x}$ satisfies $A\mathbf{x} = \lambda\mathbf{x}$ where A is the $N \times N$ adjacency matrix, $\mathbf{x}$ is the eigenvector associated to the greatest eigenvalue λ of A and N is the number of nodes.

Although a node which is central by one centrality measure may be central by other centrality measures, this is not necessarily always true. Compared to betweenness centrality (measures the number of paths that pass through each node) and closeness centrality (based on average distances), eigenvector centrality is based on the idea that a central node is connected to other central nodes, which is a natural definiton for centrality in an IoT network. However, one of the major disadvantage of eigenvector centrality measure is that the calculation is quite complex and complexity grows as N increases which is challenging for battery-powered nodes with limited storage and processing capabilities. In this present work, we utilize the concept of eigenvector centrality and leverage the tree structure of our IoT networks for a distributed computation of centrality, where a node relies on its next hop neighbors only to compute its individual centrality. Restricting the topology means nodes do not have to obtain information about far-away nodes which reduces resource usage.

Recently, several studies have focused on exploiting eigenvector centrality in a distributed way. For example, Ref. [15] presented a reception-equal rate allocation strategy for cooperative streaming so that all nodes receive the stream with the minimal global use of resources by using a distributed version of the eigenvector centrality. Although the proposed centrality measure can be computed distributedly, every node still needs to be aware of the full network topology to calculate the centrality. In Ref. [16], the authors studied a distributed computation of the PageRank algorithm, a variant of the eigenvector centrality. In our work, we focus on a distributed version of the classic eigenvector centrality, which can be measured individually by each node of a directed loop-free wireless network consisted of resource constrained devices.

3 System Model

Due to scalability, low cost and ease of deployment, IoT networks are gaining increasing interests in the research community. Depending on the particular application, different network architecture may be of interest. We consider an IoT network, where both nodes and routers are autonomous and characterized as energy-constrained devices with limited memory and poor processing capabilities. Routers function as data aggregators and relay the received data to the base station, which has unlimited power supply and is far from the sensing area, in a directed multi-hop fashion through other routers. In addition, all the deployed devices are only aware of their next hop neighbors and have no global knowledge of network. An example of such network is a time-driven IoT network, used to collect spatio-temporal readings of various environmental parameters through densely deployed sensor nodes.

We assume that all devices are equipped with tripolar antenna consisting of three orthogonal mutual collocated antenna elements to create vertical (V) polarization, horizontal (H) polarization and a third polarization (W) which is perpendicular to the other two [16].

Figure 1 demonstrates available channel gains for such systems which can be described using a 3 × 3 complex channel matrix. During transmission, we assume that the signal gets affected by Rayleigh fading, which is independent and identically distributed on each antenna element. Both nodes and routers use selection diversity to determine the best polarization for transmission and reception.

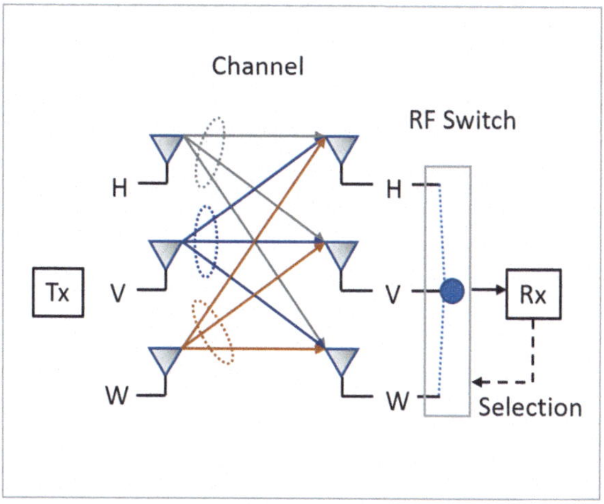

Fig. 1 Block diagram of transmission and reception using tripolar antenna

To reduce hardware complexity, a single RF chain is used by the tripolar antenna which changes antenna element using a RF switch. IoT devices receive pilot symbols using different polarization from their next hop router to estimate the channel gain of all three antenna elements by means of received signal strength. The receiver antenna then selects one of the polarizations based on its estimates. The base is assumed to be unaffected by multipath fading and uses vertical polarization only for transmission.

4 Distributed Eigenvector Centrality

Classic eigenvector centrality, which measures how well connected a node is to other well-connected nodes in the network, is computed globally. To facilitate faster computation and reduce memory usage of resource-constrained IoT devices, we use distributed eigenvector centrality (DEC), where each device (sensor or router) will calculate their own centrality. To model the IoT network, we let $G(V, E)$ be a directed graph with N sensor nodes and R router nodes, where V is a set of vertices representing all devices of the network and E is a set of edges representing links between the devices. To calculate the centrality of node k with neighbor set $\{1, 2, \ldots, j\}$, we define an edge-weight matrix $\mathbf{W}$, which is a $j \times 1$ column matrix, and neighbor-centrality matrix $\bar{\mathbf{C}}_{v_k}$, which is a $1 \times j$ row matrix, as,

$$\mathbf{W} = \begin{bmatrix} w_{1,k} \\ w_{2,k} \\ \vdots \\ w_{j,k} \end{bmatrix}_{j \times 1} \quad \text{and} \quad \bar{\mathbf{C}}_{v_k} = [c_{v_1} \quad c_{v_2} \quad \ldots \quad c_{v_j}]_{1 \times j} \tag{1}$$

here weight of each edge $w_{i,k}$ is either 1 or 0 and i is one hop neighbor of node k. In the context of our network, a directed edge from node i to node k indicates data packet flow direction from i to k. If there is an edge from node i to node k, then $w_{i,k} = 1$, otherwise $w_{i,k} = 0$. Also, c_{v_i} denotes the centrality of the node i. The proposed centrality scheme in initialized by awarding one centrality point to each vertices. After that each node calculates its own centrality by summing the centrality of its neighbor nodes that have edges directed towards them. Thus, DEC for node k is defined as the weighted sum of the centralities of all its neighbor sensor nodes and routers and can be written as

$$c_{v_k} = 1 + \mathbf{W}\bar{\mathbf{C}}_{v_k} = 1 + \sum_{i=1, i \neq k}^{j} w_{i,k} c_{v_i} \tag{2}$$

Figure 2 illustrates an example of centrality calculation using two routers and five sensors. Sensor nodes n_1, n_2 and n_3 do not have any directed edge towards them and hence each sensor has centrality 1. On the other hand, R_1 has a centrality of 4 since

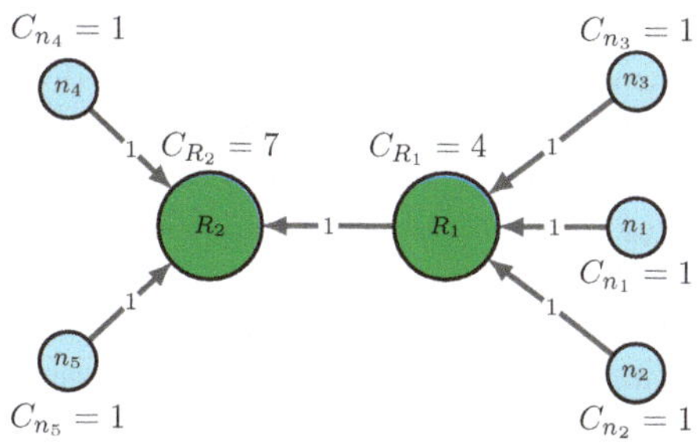

Fig. 2 Sample network of two routers and five sensor nodes with routers depicted in green and sensors depicted in light blue color

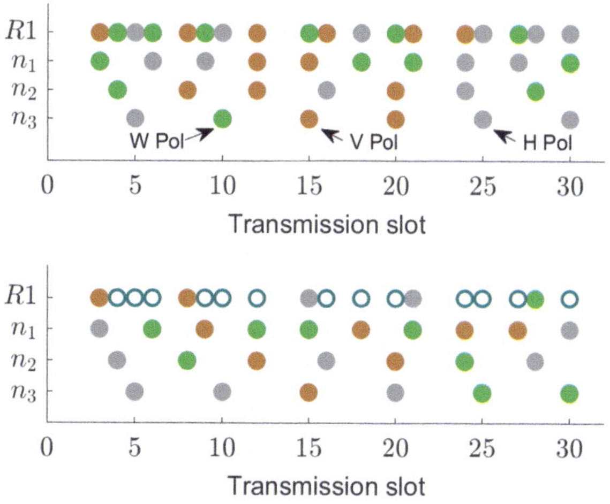

Fig. 3 An illustration of data transmission by sensors and router. Solid circles indicate usage of selection diversity before transmission while empty circles indicate no antenna switching occurred and colors represent different polarizations. *Top:* Router $R1$ uses conventional selection diversity *Bottom:* $R1$ uses centrality based selection diversity

there are three directed links from three neighbor nodes each having a centrality of 1. Although R_2 is a neighbor of R_1, it does not contribute to the centrality of R_1 as there is no directed edge from R_2 to R_1. Similarly R_2 has a centrality of 7 since it has directed edges from neighbors with centrality 1, 1 and 4.

Under the assumption that each device knows their type and total number of devices present in the network, it can compute their centrality by only using local interactions with its neighbor. Our goal is to allow nodes to limit their antenna switching based on their centrality. We can then define the interval slot for node k as

$$\lfloor s_k \rfloor = \frac{1}{(N+R)} \alpha c_{v_k} \tag{3}$$

where, N and R are the total number of sensors and routers, respectively. Also, s_k is the number of transmissions during which a node will not use selection diversity unless the signal strength of the currently used antenna branch falls below the threshold and α is an integer that denotes the interval parameter. We note that the interval slot, i.e., the waiting period between two consecutive antenna switching is proportional to a node's centrality and it increases for large values of α.

Figure 3 presents an illustration of transmission rates between nodes and a router for the example network presented in Fig. 2, where sensor nodes (denoted as n_1, n_2, and n_3) are transmitting data packets to the router R_1 at different rates. We note that, when R_1 uses conventional selection diversity (see Fig. 3 *Top*), it requires antenna switching before every transmission. On the other hand, when R_1 employs centrality based switching (see Fig. 3 *Bottom*), antenna checks for best polarization among the three elements only after some fixed (3 in this example) transmission slots. For high centrality routers, the interval between consecutive receptions and transmissions will be smaller and hence it's highly likely that the channel conditions will not change between consecutive transmissions. Thus, restricting the use of selection diversity for such routers before every transmission will reduce excessive switching and minimize energy consumption at the same time. With a time complexity scaling linearly with the number of vertices in the network, DEC offers fast computation and requires little memory usage. Moreover, with DEC, any changes in network topology can be dealt locally as only a part of nodes need to recalculate their centrality.

4.1 Centrality Based Diversity Scheme

We now describe the infrastructure of the IoT network that is used for simulation and also how experimental data is incorporated to assess the performance of the proposed scheme in a Rayleigh-fading environment. The network is initialized with random sensor node deployment and the base is located at one corner of the monitoring area. The routers are equidistant form one another and when a router joins the network, it sends a multicast packet to discover its adjacent sensors and routers and creates a routing table based on the received response. The time difference between two consecutive data packet transmission by sensor nodes is varied randomly between 1–10 s. Centrality is calculated in a bottom-up approach, where each sensor and router use their own routing table to calculate their centrality and share the score to their next level router only. Once calculated, devices will keep using the centrality unless there are changes in their neighborhood. If a new sensor or router joins, then their neighboring devices update centrality. After computing centrality, devices determine their individual switching rate, which defines how often a device will use selection diversity to select the best antenna element. Once a device selects a polarization for transmission/reception, it may need to wait for a couple of transmission slots to use selection diversity again and, importantly, this waiting period is chosen proportionally to its centrality. During the interval, the antenna will keep monitoring the signal strength of the currently used antenna branch and if the branch falls below a predetermined threshold, it will use selection diversity to select the best branch among the three branches. To asses the performance under a setting similar to real world environment, we exploit the signal strength and energy consumption data obtained experimentally, as described in Ref. [16], using embedded devices equipped with tripolar antennas in a high multipath environment.

5 Performance Evaluation

In this section, we describe the simulation parameters used to evaluate the performance of the proposed diversity scheme. Furthermore, we also compare the results with existing selection diversity technique.

5.1 *Simulation Model*

We present the results for a case with 50 sensors and 10 routers as depicted in Fig. 4a, where devices are using different antenna polarization at a certain time. We note that, routers that are closer to the base station see substantially more data traffic compared to the routers that are far away from the base or on the edge of the sensing area. Figure 4b demonstrates the use of DEC, where high centrality is assigned to the routers that are closer to the base and tend to aggregate more data packets compared to routers that are far from the base.

In order to evaluate the performance of the proposed centrality based diversity scheme, we consider an IoT network that performs periodic data collection through sensor nodes based on IEEE 802.15.4 protocol. Sensor nodes are static and unable to relay data from other nodes. Routers receive data from other nodes and forward the data to the next hop routers in a tree-based routing fashion. We built a discrete event simulator based on Matlab where a rectangle region is used to deploy the nodes. The default parameters used in our simulation are presented in Table 1.

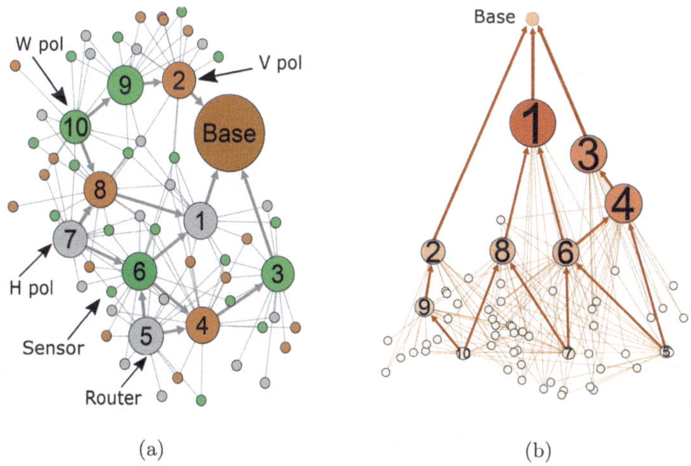

Fig. 4 (**a**) Basic architecture of an IoT network consisting of 50 sensors and 10 routers. Colors represents different polarizations, sizes represents different type of IoT devices. (**b**) Representation of the network presented in (**a**) using DEC. Color coding and size indicates centrality of sensor and router nodes. Less central nodes have smaller size and lighter color compared to more central nodes which have larger size and darker colors

Table 1 Simulation parameters

Parameter	Value
Area of deployment	$300 \times 300\,\mathrm{m}^2$
Number of sensors	50
Number of routers	10
Energy: transmission	0.01 J
Energy: reception	0.008 J
Energy: switching	0.001 J
Energy: pilot packets (transmission/reception)	0.002 J
Data packet size	32 bytes
Data rate	250 kbit/s
Pilot packet size	16 bytes
Battery capacity	18.7 kJ
Frequency	2.4 GHz
MAC protocol	802.15.4
Number of repetitions	10

We then run a comparative analysis between our proposed scheme and selection diversity technique. Three performance metrics are used: switching frequency, packet delivery ratio and energy consumption. In the baseline scenario, we consider a network, where each device uses selection diversity to determine the best antenna element for transmission and reception. To analyse the performance of our proposed model, we experiment with different network sizes in terms of the number of sensors and routers.

5.2 Simulation Results

Figure 5 presents our results on the impact of the centrality metric in decreasing antenna switching rate. We focus on the routers only since sensor nodes are assumed to be unable to perform data forwarding. Figure 5a illustrates the centrality of routers calculated based on Eq. (2) and normalized by the total number of devices 60. We note that few routers stand out amongst other routers due to high centrality and thus, serve as central points of data aggregations. We also see that the high scoring routers receive and forward more data traffic, which deplete their energy rapidly, compared to other routers with low centrality values. The results also show the heterogeneity among routers in terms of data traffic through them and hence, reinforces the requirement of node-specific diversity scheme. Figure 5b shows the effect of using interval parameter α, where the antenna switching of routers with high centrality are restricted compared to other routers. Even though the number of switching varies for routers for different simulation runs, we observe that when α is set to 2, our proposed diversity scheme decreases antenna switching approximately by 17% compared to the conventional selection diversity.

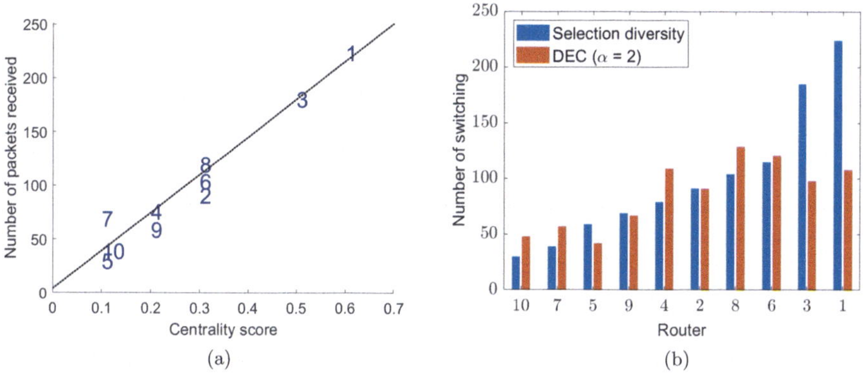

Fig. 5 (**a**) The number of packets received by routers, plotted against their normalized centrality. It can be seen that routers which receive more data packets have higher centrality. (**b**) Comparison between selection diversity and the proposed technique in terms of switching frequency. Routers are plotted in ascending order based on the number of switching. Note that the number of switching is decreased for high scoring routers

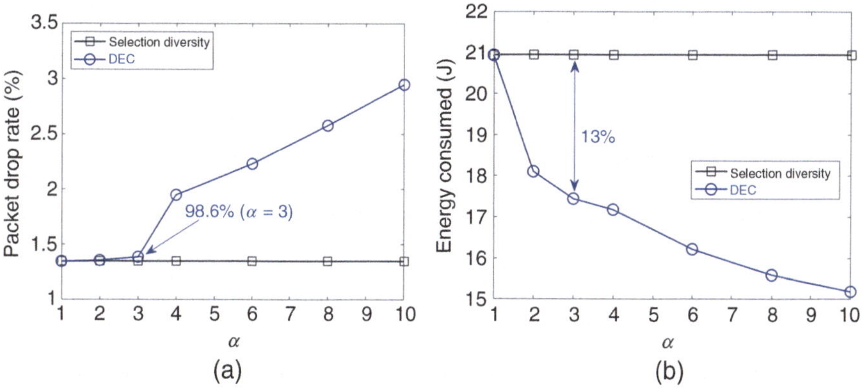

Fig. 6 Comparison between selection diversity and the proposed technique for different values of α in terms of (**a**) packet drop rate and (**b**) energy consumption rate, for a network consisting of 50 sensor nodes and 10 routers. As can be seen in the figure, for $\alpha = 3$, our proposed scheme has approximately 99% successful packet delivery rate and reduces energy consumption by 13% compared to the selection diversity technique

Figure 6 demonstrates the use of interval parameter by comparing the centrality based diversity scheme with selection diversity technique in terms of packet delivery and energy consumption for different values of α. From Fig. 6a, we note that when $2 \leq \alpha \leq 3$, the proposed centrality based diversity scheme is on par with selection diversity technique in terms of packet delivery rate. However, as α increases, packet drop rate increases for our proposed scheme compared to the selection diversity. Since a large value of α increases the waiting time between consecutive antenna polarization selection, network reliability decreases. Figure 6b demonstrates the influence of α on the energy consumption of routers, where energy consumption includes power consumed due to antenna switching, transmission and reception of both pilot packets and data packets. Since a large value of α implies that more routers have reduced switching rate, the energy consumption decreases considerably. However, restriction in updating antenna polarization for longer period results in greater packet loss compared to the selection diversity. Therefore, selecting an appropriate value of α is crucial for achieving satisfactory performance in terms of reliability and energy efficiency.

6 Conclusion

In this work, we present an energy-aware polarization diversity scheme based on node centrality metric for IoT networks. We consider a typical IoT network composed of sensor devices that periodically sense data and utilizes tripolar antenna to forward it to the base station through routers in a multi-hop fashion. The proposed diversity scheme leverages distributed eigenvector centrality metric, calculated by all IoT devices individually without requiring global information about the network topology, to measure a router's importance based on the importance of its connected neighbors. The identification of most influential router nodes allows us to employ a node-specific diversity scheme that lets low scoring routers to switch polarization more frequently compared to high scoring routers and hence decreases excessive switching over the whole network.

Our results suggest that methods to rank the influence of different nodes in complex networks can be applied in IoT networks to save energy consumption without compromising fidelity. Indeed, our simulation results demonstrate that the proposed centrality based approach reduces switching by at least 17% compared to the approach of utilizing selection diversity for all sensor and router nodes irrespective of their roles. This shows that the proposed scheme is able to lessen energy consumption by at least 13% compared to the conventional selection diversity while offering similar network reliability. In future work, we plan to implement the proposed scheme in real devices using various topologies and routing strategies.

References

1. Hussain, M.I.: Internet of Things: challenges and research opportunities. CSI Trans. ICT **5**(1), 87–95 (2017). https://doi.org/10.1007/s40012-016-0136-6
2. Hammoudi, S., Aliouat, Z., Harous, S.: Challenges and research directions for Internet of Things. Telecommun. Syst. **67**(2), 367–385 (2018). https://doi.org/10.1007/s11235-017-0343-y
3. Wortmann, F., Flchter, K.: Internet of Things. Bus. Inf. Syst. Eng. **57**(3), 221–224 (2015). https://doi.org/10.1007/s12599-015-0383-3

4. Qin, W., Chen, S., Peng, M.: Recent advances in industrial internet: insights and challenges. Digital Commun. Netw. (2019). https://doi.org/10.1016/j.dcan.2019.07.001
5. Kwon, S.C., Stuber, G.L.: Geometrical theory of channel depolarization. IEEE Trans. Veh. Technol. **60**(8), 3542–3556 (2011). https://doi.org/10.1109/tvt.2011.2163094
6. Gkizeli, M., Karystinos, G.N.: Maximum-SNR antenna selection among a large number of transmit antennas. IEEE J. Sel. Top. Sign. Proces. **8**(5), 891–901 (2014). https://doi.org/10.1109/jstsp.2014.2328329
7. Wang, X., Yang, N., Zhang, H., Hoang, T.M., Gulliver, T.A.: Generalised selection at multi-antenna sources in two-way relay networks. IET Commun. **10**(7), 824–831 (2016). https://doi.org/10.1049/iet-com.2015.0110
8. Bemanali, S., Eslami, M.: Low complexity capacity maximizing transmit antenna selection schemes for massive MIMO wireless communications. Wirel. Pers. Commun. **96**(3), 3873–3887. https://doi.org/10.1007/s11277-017-4355-4
9. Molina-Pico, A., Cuesta-Frau, D., Araujo, A., Alejandre, J., Rozas, A.: Forest monitoring and wildland early fire detection by a hierarchical wireless sensor network. J. Sens. **2016**, 8325845 (2016). https://doi.org/10.1155/2016/8325845
10. Othman, M.F., Shazali, K.: Wireless sensor network applications: a study in environment monitoring system. Proc. Eng. **41**, 1204–1210 (2012). https://doi.org/10.1016/j.proeng.2012.07.302
11. Latora, V., Nicosia, V., Russo, G.: Complex Networks: Principles, Methods and Applications. Cambridge University Press, Cambridge (2017). https://doi.org/10.1017/9781316216002
12. Newman, M.: Networks. Oxford University Press, Oxford (2018). https://doi.org/10.1093/oso/9780198805090.001.0001
13. Bloodgood, J.M., Hornsby, J.S., Rutherford, M., McFarland, R.G.: The role of network density and betweenness centrality in diffusing new venture legitimacy: an epidemiological approach. Int. Entrep. Manag. J. **13**(2), 525–552 (2017). https://doi.org/10.1007/s11365-016-0412-9
14. Aytaç, V., Turac, T.: Closeness centrality in some splitting networks. Comput. Sci. J. Moldova, **26**(3), 251–269 (2018).
15. Baldesi, L., Maccari, L., Cigno, R.L.: On the use of eigenvector centrality for cooperative streaming. IEEE Commun. Lett. **21**(9), 1953–1956 (2017). https://doi.org/10.1109/lcomm.2017.2713361
16. Ramirez, R.A., Lugo, D., Weller, T.M., Golmohamadi, M., Frolik, J.: Additive manufactured tripolar antenna system for link improvement in high multipath environments. In: 2017 IEEE International Symposium on Antennas and Propagation & USNC/URSI National Radio Science Meeting, pp. 2539–2540. IEEE, Piscataway (2017) https://doi.org/10.1109/apusncursinrsm.2017.8073312

Using Network Science to Quantify Economic Disruptions in Regional Input-Output Networks

Emily P. Harvey and Dion R. J. O'Neale

Abstract Input-output (IO) tables provide a standardised way of looking at interconnections between all industries in an economy, and are often used to estimate the impact of disruptions or shocks on economies. IO tables can be thought of as networks—with the nodes being different industries and the edges being the flows between them. We develop a network-based analysis to consider a multi-regional IO network at regional and sub-regional level within a country. We calculate both linear matrix-based IO measures ('multipliers') and new network theory-based measures, and contrast these measures with the results of a disruption model applied to the same IO network. We find that path-based measures (betweenness and closeness) identify the same priority industries as the simulated disruption modelling, while eigenvector-type centrality measures give results comparable to traditional IO multipliers, which are dominated by overall industry strength.

1 Introduction

Economic disruptions such as those due to natural hazards have a large impact on local and global economies. There is evidence that the flow-on impacts of disruptions will have an increasing impact as the world becomes more globalised and inter-connected [1, 4]. In order to build resilience and prioritise investment to mitigate impacts, it is crucial to identify key industry sectors and regions that play

E. P. Harvey (✉)
M.E Research, Takapuna, Auckland, New Zealand

Te Pūnaha Matatini, Centre of Research Excellence, Auckland, New Zealand
e-mail: emily@me.co.nz

D. R. J. O'Neale
Physics Department, University of Auckland, Auckland, New Zealand

Te Pūnaha Matatini, Centre of Research Excellence, Auckland, New Zealand
e-mail: d.oneale@auckland.ac.nz

N. Masuda et al. (eds.), *Proceedings of NetSci-X 2020: Sixth International Winter School and Conference on Network Science*, Springer Proceedings in Complexity,
https://doi.org/10.1007/978-3-030-38965-9_18

a role in amplifying (or dampening) the flow-on impacts of disruptions or shocks. When considering disruption impacts, economists and disruption planners are faced with the need to evaluate a number of different measures of economic impacts that are not necessarily comparable and which almost certainly can not all be minimised simultaneously.

Internationally, studies of flow-on impacts of disruptions on economic systems have most commonly been based on the data from input-output (IO) tables, which are readily available, at least at a national, and often also at regional levels. Many years of research has gone into using IO tables in economic impact analysis [14] and a common approach is to use 'multipliers', based on linear algebra matrix formulations, to estimate the indirect impacts of a change in demand (or supply) for an industry. In response to natural hazard events, the most popular approach has been inoperability input-output models, which are based on standard 'IO multiplier analysis' with minor modifications [6].

Recently IO tables have been thought of as networks, with the nodes being the different industries and the edges being the flows between the industries. This has enabled network science techniques to be used to attempt to identify crucial nodes (industry sectors) within an economy and other industry structures. Existing work in this field began with calculating the properties of IO networks [2, 3, 13], and is now beginning to investigate the propagation of shocks on the networks [4, 11, 13, 16].

In this work we seek to consider the connections between industries as a network, to determine whether the structure of the network can provide useful extra information for quantifying the importance of industries or regions in propagating disruptions. We consider a multi-regional IO network at local (Territorial Authority) level within the Waikato Region in New Zealand, and calculate both linear matrix-based IO measures (e.g. 'multipliers') and network theory-based measures at this higher spatial resolution. We compare these network-based measure with results from a disruption model applied to the same IO data, which gives us further information about disruption impacts.

Research to date has considered global IO networks, looking at flows within and between each country, or looking at a single country of interest. However, when considering more fine-grained economic data regions are often heterogeneous and impacts can be highly local. By comparing and contrasting the analysis at both Regional and Local spatial resolutions, we are also able to investigate the impact of spatial resolution on the results obtained.

2 Background and Data

2.1 Level of Spatial and Industry Aggregation

In this work, our starting data is a Multi-regional Input-Output table (MRIO) which partitions the 10 Territorial Authorities (TAs) in the Waikato Region into separate subregions and breaks the rest of New Zealand into 'North of the Waikato Region'

(Auckland and Northland) and 'South of the Waikato Region' (all other Regions). This gives us 12 different spatial regions which span a large range of sizes, both geographically and economically. The number of industry sectors is aggregated to 106, which is the maximum allowed from the reference data used to construct the IO tables.

2.2 Economic Network Setup

IO networks are weighted, directed networks, where the weighting indicates the size ($) of inter-industry-location flows, and the direction depends on which industry-location the flow is from and to. In this work we have 1272 industry-location nodes (12 regions (locations) and 106 (industry) sectors) in the network. In addition to the flows between industry sectors, the IO tables we use include inputs from four Value Added categories (labour and capital inputs, taxes/subsidies on products and production) and Imports, and outputs to three Final Demand categories (household and government consumption, capital formation) and Exports. This adds another three nodes to each of the 12 'regions' for Final Demand, four nodes for Value Added at whole country level, and a node each for Imports and Exports.

It is a requirement of IO tables that flows in and out of an industry-location must balance, so this places restrictions on the network, specifically the row and column sums of the weights in the adjacency matrix must much once value added and final demand components are included. Additionally, because the nodes are the grouping of all industries of the same 'sector' in the TA, self-links are possible and will typically account for a significant fraction of monetary flows. There are some other features of IO networks that it is worth noting; one of which is that they are very dense (nearly-complete) with most industries having connections with most other industries, though not all of these flows (link weights) will be significant. These features mean that a lot of the standard approximations and simplifications for weighted, directed network analysis are often not able to be applied [12].

3 Method

3.1 Network Analysis

In network science *centrality* broadly refers to the 'importance' or 'influence' of a node in the network. Centrality measures can range from local node properties (e.g. node strength); to more extensive properties that consider the properties of those nodes connected to a node of interest (e.g. PageRank centrality); to measures that account for the structure or topology of the network (e.g. betweenness centrality). In this work we analyse the network using a range of different centrality measures, including those that have been identified as potentially important in economic networks. Where possible we consider the network as a weighted, directed network, with self-loops, but not all algorithms allow for this.

The centrality measures used here were chosen to cover a range different types of measures while keeping the range of measures manageable. Specifically, we consider the following (from the igraph package [5]):

Node Strength An equivalent of node degree that accounts for the differing edge weights in a network. The strength of a node is simply the sum of the weights of edges connected to that node.

PageRank A popular variant of eigenvector centrality. Eigenvector-based measures consider not only the strength of a node but also how well connected a node is to other nodes with high node strength.

Kleinburg Centrality A generalisation of eigenvector centrality. Nodes are imbued with two attributes: *Authority*—how much information/influence is held by the node; and *Hubness*—how well a node connects to nodes with high authority. If A is the adjacency matrix of a network, the hub score of nodes is calculated as the principle eigenvector of AA^T.

Authority Related to hub centrality, the authority centrality of vertices is calculated as the principle eigenvector of $A^T A$.

Diversity The scaled Shannon entropy of the edge weights of a node. Given the context of IO networks it is worth noting that the diversity measure used here is a version of the species diversity measure commonly used in ecology to quantify the diversity of a habitat [9], not the measure of diversity sometimes used in economic geography and popularised by Hausmann and Hidalgo in [8] which is simply the node degree of a binary matrix that measures whether a region is strongly associated with particular products or exports.

Betweenness The number of weighted shortest paths that pass through a node, given all possible paths between pairs of nodes in the network.

Closeness The number of steps required to reach every other node in a network from a given node following edge-weighted paths.

Having calculated this selection of centrality measures, we use Kendall's τ [10] to calculate the correlation between the importance rankings of the industries between any two centrality measures.

3.2 *Multiplier Analysis*

IO Multipliers In calculating the economic effects of changes in an economy (positive or negative shocks), 'IO multipliers' are the most commonly used approach. When the change is considered on the demand side the 'Leontief inverse' is used, which propagated the change to all downstream industries. When the change is considered on the supply side the 'Ghosh inverse' is used which effectively propagates the change to all upstream industries. See [14] for a full description.

The main technique used for quantifying economic impacts of a disaster (for example a natural hazard such as an earthquake or volcanic eruption) is known as Inoperability Input-Output Model (IIM) or the Dynamic IIM as a time varying extension [6]. In this model, the inoperability of industries is assumed to follow a smooth logistic curve from the disaster induced loss of productive capacity back to full capacity over a specified recovery period. The direct loss of production due to industry-location inoperability is calculated and used to modify the final demand by the same amount; that is, if production halved from $20,000 to $10,000 then the final demand vector for that industry-location in that region would be reduced by $10,000. Then the flow on impacts from this reduced demand would be calculated using the Leontief inverse [14]. This continues through time, until full operability is restored.

In this work we calculate both Type I (industry to industry spending only) multipliers and Type II (including household spending and labour income) multipliers, for the whole multi-regional IO table, following [14].

Disruption Multipliers There are many issues with this IIM approach [15], including that it can lead to double counting and not only inaccurate quantitative results, but more importantly it can lead to inaccurate rankings for prioritisation of industries. Harvey et al. [7] have instead developed a dynamic model that propagates short-term (days to weeks) disruptions through the multi-regional IO network. Using this model 'disruption multipliers' can be calculated by disrupting one industry-location at a time in each region and working out the ratio of direct effects to flow-on (indirect) effects throughout the whole of New Zealand.

3.3 Comparing Network Centralities with IO and Disruption Multipliers

We use Kendall's τ [10] to calculate the correlation between industry-location rankings based on the multiplier measures compared to the centrality measures.

3.4 Comparing Spatial Aggregation

In parallel with this, we also construct a network at the level of the Waikato Region (not separated at TA level) and the same 'North of the Waikato Region' and 'South of the Waikato Region' regions (3 network regions) to investigate the impact of spatial aggregation of IO tables on the economic multipliers and the network properties.

4 Results

4.1 Network Centrality Measures and Their Correlations

Figure 1 shows a heatmap of the Kendall correlation coefficients [10] between all the centrality measures considered here. Before comparing the centrality measure rankings, we first remove the industries from the rest of New Zealand (North and South of the Waikato Region), as these have much larger inputs/outputs than those broken down by TA within the Waikato Region and risk dominating the results. Furthermore, we are focused here on identifying important industry-location pairs within the Waikato Region.

We find that overall the different eigenvector-based centrality measures are highly correlated, in particular the Kleinburg Authority, Kleinburg Hub, and PageRank, and that these are strongly driven by the node strength (total inputs/outputs) of the industry-location pairs. The different path-based measures are also highly correlated, for example, the closeness and betweenness measures. These path-based measures highlight different industry-location pairs to those identified by the eigenvector-based methods, as shown by the high level of anti-correlation (red) between these types of measures. More importantly, the path-based measures identify industry-location pairs that would not be immediately revealed by eigenvector-based methods, or by linear economic multiplier measures that are strongly linked to their size (strength) in the local economy. We elaborate on this in the next section.

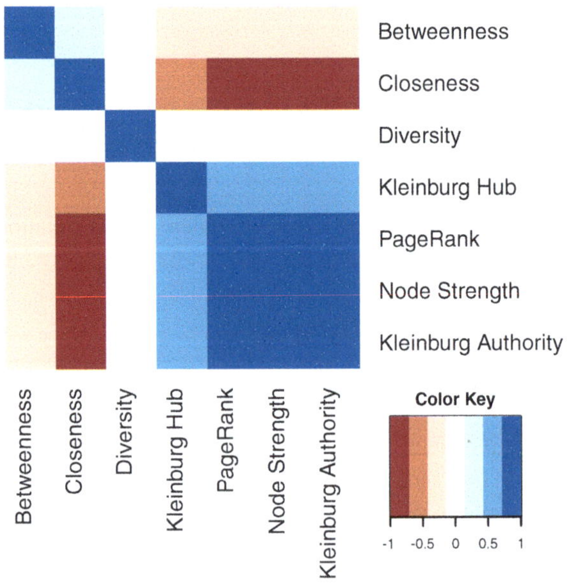

Fig. 1 Centrality measure correlations (τ values) for industries in the ten TAs in the Waikato Region

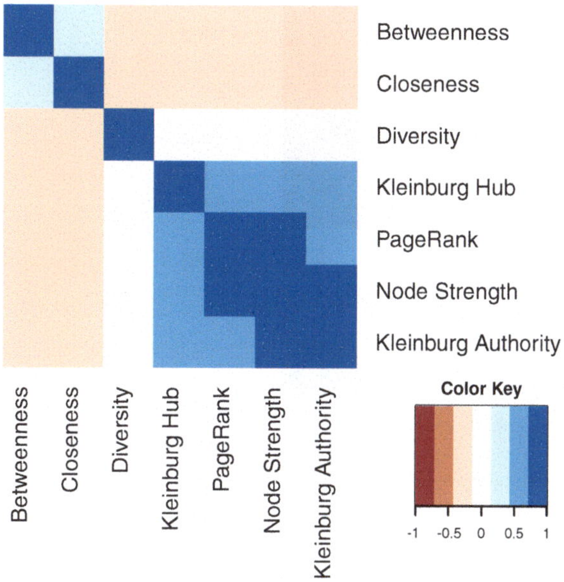

Fig. 2 Centrality measure correlations (τ values) for industries in the Waikato Region as a single network region

If we aggregate the ten TAs up to a single Waikato Region, we can compare centrality measure correlations over a much smaller set of industry-location pairs (106 nodes instead of 1060). This produces slightly weaker correlations, as shown in Fig. 2, but overall the pattern remains.

4.2 Comparing Multiplier and Centrality Measures

Comparing rankings for the different multipliers calculated, we find that the correlation between the Industry (Type I) and Household (Type II) disruption multipliers is $\tau = 0.21$. This matches the literature which shows that the inclusion of the household sector has a large impact on the results [14]. When we look at the Disruption multiplier, we find that this has a correlation of $\tau = 0.38$ with the Industry multiplier and $\tau = -0.01$ with the Household multiplier. This shows that the Disruption model, which simulates a disruption propagating through the IO network, is identifying different key industries to the existing IO multiplier analysis. This has implications for regional disruption planning.

Comparing the three multipliers with the network centrality measures (Fig. 3a), we find that the overall correlations are lower (-0.33 to 0.36) but that overall the eigenvector-based centralities and the overall industry-location strengths tend to match up with the traditional IO multipliers. This is not unexpected as they are both based on linear algebra matrix calculations that are mathematically similar, and the numerics agree with this. More interestingly the path-based measures (betweenness and closeness) are much more strongly correlated with the Disruption multipliers.

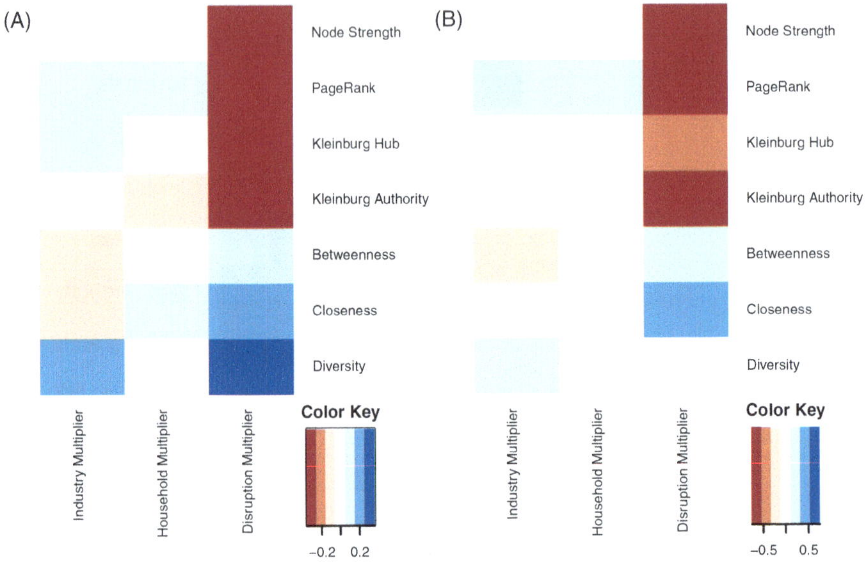

Fig. 3 Comparing the network centrality measures with: the two IO multipliers, the mean of the IO multipliers, and the disruption multiplier. This shows correlations between eigenvector-based centrality measures and IO multipliers, whereas path-based centrality measures correlate well with Disruption multipliers. There is a negative correlation between the two. For (**a**) industries in the ten TAs in the Waikato Region, and (**b**) industries in the Waikato Region as a whole

This makes intuitive sense as they are both concerned with flows and bottlenecks, that is with quantifying how disruptions to specific nodes flow on to impact the activity of dependent nodes. Another point to note is that we find the having a high diversity score is connected to having both a high Industry multiplier and a high Disruption multiplier. This highlights the potential importance of rarer industries within economic networks.

Aggregating to a single Waikato Region, we find much the same results (Fig. 3b), but with slightly weaker correlations (and anti-correlations). This is due to the disruption modelling becoming more homogenous in terms of industry-location distribution and activity when looking at the aggregated Region. A feature of the Disruption model is that it was designed to consider lower levels of aggregation, with the aim to be able to provide detailed results at single industry-location level resolution.

4.3 *Impact of Spatial Aggregation*

For all the analyses performed we considered the IO tables and economic networks with the Waikato Region broken down into 10 subregions (TAs) as well as with the whole Waikato Region considered together. This allowed us to look at the impact of the spatial aggregation on the industries identified as as important from the disruption analysis.

We find that the eigenvector-based (strength-based) network measures identify the same key industries at both Region and TA levels of resolution. We find the same pattern for the Industry and Household multipliers. The value in analysing the system at the TA level disaggregation then becomes simply the ability to identify which TA the identified industry is most important to—it does not change which industries are identified. The exception is for industries that are disproportionately (or uniquely) represented in one or two TAs; for example, Coal Mining in the Waikato District, and Hospitals in Hamilton City. In these cases looking at TA level allows these to be ranked higher in importance than they would be if aggregating up to Regional level. Selected examples are given in Table 1.

However, for any path-based measures and for the Disruption multipliers, the level of spatial aggregation has a large impact on which industries are identified as important. Examples are given in Table 2. This can be explained as follows: aggregating the network up changes its structure—for example, at TA level there are fewer individual businesses within each industry categorisation, so the self-loops are smaller. Furthermore, the proportion of inter-industry flows that are within the TA itself is quite low (14–32%), with the majority of flows into (or out) of each industry coming from (or going to) other TAs within the Waikato Region and the rest of NZ. When considering the whole region, the proportion of inter-industry flows that stay within the region increases to around 60%. This is still far below the equivalent proportions that are typically observed in the literature when looking at IO networks at a whole country level [12]. It is therefore worth noting that metrics that are applicable for national level analysis may not behave as expected when working with disaggregated regional data, such as that considered here.

5 Discussion

In this work we have considered the question of how to identify industries that have a large impact on an economic system when they are disrupted. A goal of this paper was to show that network science measures can provide new useful tools for targeting interventions to reduce the impacts of disruptions on regional economies. In order to approach this, we have considered traditional IO multipliers, a new disruption model multiplier, and a range of network centrality measures. We have found that although traditional IO measures and eigenvector-based centrality measures are good at picking out the largest industries in terms of gross output or value-added, they do not match up with the industries that disruption modelling shows to have a large amplifying effect. We find that path-based measures, such as betweenness and closeness centrality, are far better at identifying industries that would have large flow-on impacts. These path-based methods explicitly consider the flow of money through the economy, and we find that the industries identified on these measures depend strongly on the level of spatial resolution.

Table 1 Selection of industries identified as important that are the same at Regional and TA level

Measure	Regional level	TA level
Total industry value added	Dairy cattle farming	Owner-occupied property operation (Hamilton City)
	Owner-occupied property operation	Hospitals (Hamilton City)
		Dairy cattle farming (Matamata-Piako)
Node strength	Dairy product manufacturing	Dairy product manufacturing (3 TAs)
	Dairy cattle farming	
	Electricity generation and on-selling	Electricity generation and on-selling (1 TA)
		Dairy cattle farming (1 TA)
		Coal mining (Waikato District)
PageRank	Dairy product manufacturing	Dairy product manufacturing (3 TAs)
	Dairy cattle farming	
	Electricity generation and on-selling	Meat and meat product manufacturing (1 TA)
		Dairy cattle farming (1 TA)
		Coal mining (Waikato District)
Industry Multiplier	Electricity generation and on-selling	Electricity generation and on-selling (6 TAs)
	Primary metal and metal product manufacturing	Primary metal and metal product manufacturing (7 TAs)
	Dairy product manufacturing	Dairy product manufacturing (7 TAs)
	Meat and meat product manufacturing	Meat and meat product manufacturing (7 TAs)
Household Multiplier	Preschool education	Preschool education (10 TAs)
	Postal and courier pick up and delivery services	Postal and courier pick up and delivery services (10 TAs)
	Specialised food retailing	Specialised food retailing (10 TAs)

In considering a natural hazard disruption, both the total size of the industry and the proportion its impact gets amplified by will play a role in determining the resulting impacts. By taking a network science approach, we are able to get a fuller picture of the potential targets for mitigation investment (e.g. stockpiling goods, having back-up generators in case of electricity outages).

In most disruption events, the impact will not be homogeneous through space. In most cases we would like to be able to consider the impact of a disruption on the well-being of communities, instead of just at national or even regional level. This is especially true for smaller localised events, which will not have a large impact at a national or regional level, but that could devastate a community. We have found that by considering smaller spatial units (in this case TA level) it is possible to get a better estimate of where the impacts will fall, as well as where susceptibilities are.

Table 2 Selection of industries identified as important on measures that differ Regional and TA level

Measure	Regional level	TA level
Betweenness	Petroleum and coal product manufacturing	Motor vehicle and motor vehicle parts wholesaling (Hauraki)
	Defence	Sewerage and drainage services (Thames-Coromandel)
	Sewerage and drainage services	Waste collection, treatment and disposal services (2 TAs)
	Air and space transport	Health and general insurance (Taupō)
		Petroleum and coal product manufacturing (Matamata-Piako)
		Warehousing and storage services (Waitomo)
Closeness	Air and space transport	Other transport (Otorohanga)
	Petroleum and coal product manufacturing	Motor vehicle and motor vehicle parts wholesaling (Hauraki)
	Beverage and tobacco product manufacturing	Warehousing and storage services (Waitomo)
	Warehousing and storage services	Health and general insurance (Taupō)
		Waste collection, treatment and disposal services
		Polymer product and rubber product manufacturing (2 TAs)
		Machinery manufacturing
		Electricity transmission and distribution
Disruption Multiplier	Defence	Other transport (Otorohanga)
	Life insurance	Motor vehicle and motor vehicle parts wholesaling (Hauraki)
	Petroleum and coal product manufacturing	Warehousing and storage services (Waitomo)
	Warehousing and storage services	Waste collection, treatment and disposal services (Waitomo)
		Polymer product and rubber product manufacturing (3 TAs)
		Health and general insurance (Taupō)

Even for the measures that do not change much between Region and sub-regional (TA) level (Table 1), looking at a higher granularity allows one to identify the unique (spatially specific) industries e.g. Hospitals and Coal Mining, that would be missed at a Regional level.

In future, increased data collection will make it possible to create networks at individual firm level. Making sure that we understand how different measures scale from National to Regional all the way to individual firm level will be an important focus of future research.

Acknowledgements Resilience to Nature's Challenges - National Science Challenge contestable funding 2017–2019. Ministry of Business, Innovation & Employment, NZ. http://resiliencechallenge.nz
Te Pūnaha Matatini, Centre of Research Excellence, Tertiary Education Commission, NZ. https://www.tepunahamatatini.ac.nz

References

1. Acemoglu, D., Carvalho, V.M., Ozdaglar, A., TahbazSalehi, A.: The network origins of aggregate fluctuations. Econometrica **80**, 1977–2016 (2012). https://doi.org/10.3982/ECTA9623
2. Blöchl, F., Theis, F.J., Vega-Redondo, F., Fisher, E.O.N.: Vertex centralities in input-output networks reveal the structure of modern economies. Phys. Rev. E **83**(4), 046127 (2011)
3. Cerina, F., Zhu, Z., Chessa, A., Riccaboni, M.: World input-output network. PLoS One **10**(7), e0134025 (2015)
4. Contreras, M.G.A., Fagiolo, G.: Propagation of economic shocks in input-output networks: a cross-country analysis. Phys. Rev. E **90**(6), 062812 (2014)
5. Csardi G., Nepusz T.: The igraph software package for complex network research. InterJ. Complex Syst. **1695** (2006). http://igraph.org
6. Dietzenbacher, E., Miller, R.E.: Reflections on the inoperability input-output model. Econ. Syst. Res. **27**(4), 478–486 (2015). https://doi.org/10.1080/09535314.2015.1052375
7. Harvey, E.P., Mead, S., Ayers, M.A., Smith, N.S., McDonald, G.M.: Quantifying the flow-on impacts from short-term natural hazard disruptions in local economies within New Zealand. (in preparation)
8. Hausmann, R., Hidalgo, C.: The network structure of economic output. J. Econ. Growth **16**(4), 309–342 (2011)
9. Hill, M.O.: Diversity and evenness: a unifying notation and its consequences. Ecology **54**, 427–432 (1973)
10. Kendall, M.: A new measure of rank correlation. Biometrika **30**(1–2), 8189 (1938). https://doi.org/10.1093/biomet/30.1-2.81
11. Liang, S., Qu, S., Xu, M.: Betweenness-based method to identify critical transmission sectors for supply chain environmental pressure mitigation. Environ. Sci. Technol. **50**, 1330–1337 (2016)
12. McNerney, J.: Network properties of economic input-output networks. IASA Interim Report. IIASA, Laxenburg (2009)
13. McNerney, J., Fath, B.D., Silverberg, G.: Network structure of inter-industry flows. Phys. A Stat. Mech. Appl. **392**(24), 6427–6441 (2013)
14. Miller, R.E., Blair, P.D.: Input-Output Analysis Foundations and Extensions, 2nd edn. Cambridge University Press, Cambridge (2009)
15. Oosterhaven, J.: On the limited usability of the inoperability IO model. Econ. Syst. Res. **29**(3), 452–461 (2017). https://doi.org/10.1080/09535314.2017.1301395
16. Xu, M., Allenby, B.R., Crittenden, J.C.: Interconnectedness and resilience of the US economy. Adv. Complex Syst. 14(5), 649–672 (2011)

Correction to: An Allometric Scaling for the Number of Representative Nodes in Social Networks

Liang Zhao and Tianyi Peng

Correction to:
Chapter 4 in: N. Masuda et al. (eds.), *Proceedings of NetSci-X 2020: Sixth International Winter School and Conference on Network Science*, Springer Proceedings in Complexity, https://doi.org/10.1007/978-3-030-38965-9_4

The original version of this chapter was inadvertently published with middle initial of the second author "Tianyi Y. Peng". The middle initial has been removed as per author's request and updated as "Tianyi Peng".

The updated online version of this chapter can be found at
https://doi.org/10.1007/978-3-030-38965-9_4

N. Masuda et al. (eds.), *Proceedings of NetSci-X 2020: Sixth International Winter School and Conference on Network Science*, Springer Proceedings in Complexity,
https://doi.org/10.1007/978-3-030-38965-9_19

The manufacturer's authorised representative in the EU is Springer Nature Customer Service Centre GmbH, Europaplatz 3, 69115 Heidelberg, Germany. If you have any concerns regarding our products, please contact ProductSafety@springernature.com

Printed and bound by CPI Group (UK) Ltd, Croydon, CR0 4YY

30/09/2026

02225458-0012